T0310003

LEAD-FREE SOLDER PROCESS DEVELOPMENT

IEEE Press
445 Hoes Lane
Piscataway, NJ 08854

LEAD-FREE SOLDER PROCESS DEVELOPMENT

Edited by

Gregory Henshall
Jasbir Bath
Carol A. Handwerker

A JOHN WILEY & SONS, INC., PUBLICATION

Published by John Wiley & Sons, Inc., Hoboken, New Jersey.
Published simultaneously in Canada.

Library of Congress Cataloging-in-Publication Data:

Henshall, Gregory.
Lead-free solder process development / Gregory Henshall, Jasbir Bath, Carol Handwerker.
p. cm.
Includes bibliographical references and index.
ISBN 978-0-470-41074-5 (cloth)
1. Lead-free electronics manufacturing processes. 2. Solder and soldering. I. Henshall, Gregory Arthur. II. Handwerker, Carol A. III. Title.
TK7836.B38 2010
621.9'77—dc22

2010028407

Printed in Singapore.

10 9 8 7 6 5 4 3 2 1

CONTENTS

Technical Reviewers xi

Preface xiii

Introduction xv

Contributors xix

1 REGULATORY AND VOLUNTARY DRIVERS FOR ENVIRONMENTAL IMPROVEMENT: HAZARDOUS SUBSTANCES, LIFE-CYCLE DESIGN, AND END OF LIFE 1

John Hawley

1.1 Introduction 1
1.2 Substances of Environmental Concern 2
1.2.1 Hazardous Substances Legislation: EU-RoHS 2
1.2.2 Chemical Substances Legislation: EU-REACH 4
1.2.3 Hazardous Substances Legislation: China-RoHS 6
1.2.4 Hazardous Substances Legislation: Korea-RoHS 7
1.2.5 Hazardous Substances Legislation: United States 7
1.2.6 Hazardous Substances: Material Declarations 8
1.2.7 Hazardous Substances: Voluntary Initiatives 9
1.3 Design for Environment/Energy Efficiency 9
1.3.1 Energy Efficiency: United States Legislation 10
1.3.2 Energy Efficiency: United States Voluntary Program 10
1.3.3 Energy Efficiency: Australia Legislation 10
1.4 Recycling and Take-Back 10
1.4.1 Recycling Legislation: EU-WEEE 11
1.4.2 Recycling Legislation: United States 11
1.4.3 Recycling Legislation: China 12
1.4.4 Other Legislation: EU Packaging and Batteries 12
1.5 Summary 13
References 13

2 LEAD-FREE SURFACE MOUNT TECHNOLOGY **15**
Jasbir Bath, Jennifer Nguyen and Sundar Sethuraman
2.1 Introduction 15
2.2 No-Clean and Water-Soluble Lead-Free Pastes 15
2.3 Solder Paste Handling 17
2.4 Board and Stencil Design 18
2.5 Screen Printing and Printability of Lead-Free Solder Pastes 21
2.6 Paste Inspection 22
2.7 Component Placement (Paste Tackiness) 23
2.8 Reflow Soldering and the Reflow Profile 25
2.9 Effect of Nitrogen versus Air Atmosphere during Lead-Free Reflow 29
2.10 Head-in-Pillow Component Soldering Defect 31
2.11 Visual Inspection of Solder Joint 35
2.12 Automated Optical Inspection (AOI) 37
2.13 X-ray Inspection 38
2.14 ICT/Functional Testing 38
2.15 Conclusions 39
2.16 Future Work 40
Acknowledgments 41
References 41

3 LEAD-FREE WAVE SOLDERING **45**
Denis Barbini and Jasbir Bath
3.1 Wave-Soldering Process Boundaries 45
3.1.1 Fluxer 46
3.1.2 Function of the Flux in Wave Soldering 47
3.1.3 Flux Chemistries 48
3.1.4 Keeping the Flux on the Board during Soldering 49
3.1.5 Correct Flux Amount 49
3.1.6 Flux Application Methods 50
3.1.7 Preheating 50
3.1.8 Wave Height 51
3.1.9 Wave Contact Length 52
3.1.10 Wave Machine Conveyor Speed 53

3.2 Soldering Temperatures on the Chip and Main Soldering Waves 53
3.2.1 Wetting during Wave Soldering and Wave Capillary Action 56
3.3 Alloys for Lead-Free Wave Soldering 57
3.4 Function of Nitrogen in Wave Soldering 58
3.5 Effect of PCB Design on Wave Solder Joint Formation 61
3.5.1 Solder Bridging 61
3.5.2 Large Fillet Formation 62
3.5.3 Lead-Free Wave Hole-Fill 62
3.5.4 Open/Skipped Solder Joints 66
3.5.5 Solder Balls 66
3.6 Standards Related to Wave Soldering 67
3.7 Conclusions 67
3.8 Future Work 68
Acknowledgments 68
References 68

4 LEAD-FREE REWORK **71**
Alan Donaldson
4.1 Introduction 71
4.2 Surface Mount Technology (SMT) Hand Soldering/Touch-Up 72
4.2.1 Passive Component Rework 72
4.2.2 QFP/Gull Wing Rework 72
4.2.3 ECO Wiring 75
4.2.4 Lead-Free Solder Pad Repair 76
4.3 BGA/CSP Rework 76
4.4 BGA Socket Rework 84
4.5 X-raying 86
4.6 Through-Hole Hand-Soldering Rework 86
4.7 Through-Hole Mini-Pot/Solder Fountain Rework 87
4.7.1 Board Preparation 87
4.7.2 Component Removal 88
4.7.3 Component Replacement 89
4.8 Best Practices and Rework Equipment Calibrations 91
4.9 Conclusions 91
4.10 Future Work 92
References 92

5 LEAD-FREE ALLOYS FOR BGA/CSP COMPONENTS **95**
Gregory Henshall
5.1 Introduction 95
5.1.1 Issues with Near-Eutectic SnAgCu Alloys 95
5.2 Overview of New Lead-Free Alloys 97
5.2.1 Range of New Lead-Free Alloys 98
5.2.2 Alloys in Current Use for Area Array Packaging 99
5.3 Benefits of New Alloys for BGAs and CSPs 101
5.3.1 Mechanical Shock Resistance 101
5.3.2 Resistance to Bend/Flex Loading 106
5.4 Technical Concerns 107
5.4.1 Surface Mount Assembly 107
5.4.2 Thermal Fatigue Resistance 110
5.5 Management of New Alloys 112
5.5.1 Overview of Concerns 112
5.5.2 Assessment of New Alloys for Acceptability 113
5.5.3 Updating Industry Standards 114
5.5.4 Industry Activities 116
5.6 Future Work 117
5.7 Summary and Conclusions 119
Acknowledgments 120
References 121

6 GROWTH MECHANISMS AND MITIGATION STRATEGIES OF TIN WHISKER GROWTH **125**
Peng Su
6.1 Introduction 125
6.2 Role of Stress in Whisker Growth 127
6.3 Understanding Standard Acceleration Tests 134
6.3.1 Air-to-Air Thermal Cycling (AATC) 135
6.3.2 Low-Temperature/Low-Humidity Storage 135
6.3.3 High-Temperature/High-Humidity Storage 137
6.4 Plating Process Optimization and Other Mitigation Strategies 139
6.4.1 Electrolyte and Plating Process 139
6.4.2 Underplate and Heat Treatment 141
6.5 Whisker Growth on Board-Mounted Components 142
6.6 Summary 148
References 149

7 TESTABILITY OF LEAD-FREE PRINTED CIRCUIT ASSEMBLIES **151**
Rosa D. Reinosa and Aileen M. Allen

7.1 Introduction 151
7.2 Contact Repeatability of Lead-Free Boards 151
7.2.1 Probeability Methodology 156
7.2.2 SMT Target Results 159
7.2.3 Through-Hole Test Target Results 161
7.2.4 Trends and Solutions 165
7.3 Probe Wear and Contamination 166
7.4 Board Flexure 167
7.5 Conclusions 171
Acknowledgments 171
References 172

8 BOARD-LEVEL SOLDER JOINT RELIABILITY OF HIGH-PERFORMANCE COMPUTERS UNDER MECHANICAL LOADING **173**
Keith Newman

8.1 Introduction 173
8.2 Establishing PWB Strain Limits for Manufacturing 174
8.3 SMT Component Fracture Strength Characterization 177
8.3.1 Bend 177
8.3.2 Drop 178
8.3.3 High-Speed Solder Ball Shear/Pull 180
8.3.4 Other Test Method Examples 185
8.4 PWB Fracture Strength Characterization 186
8.5 PWB Strain Characterization 187
8.5.1 Board Assembly and Test 187
8.5.2 Optical Methods 189
8.5.3 Unpackaged and Packaged Drop 189
8.6 Solder Joint Fracture Prediction—Modeling 191
8.7 Fracture Strength Optimization 194
8.7.1 Surface Finish 194
8.7.2 Solder Alloy 196
8.7.3 PWB Laminate 198
8.7.4 Thermal Exposure 198
8.7.5 Epoxy Reinforcement 200
8.8 Conclusions 200
Acknowledgments 202

Note 202
References 202

9 LEAD-FREE RELIABILITY IN AEROSPACE/MILITARY ENVIRONMENTS 205
Thomas A. Woodrow and Jasbir Bath
9.1 Introduction 205
9.2 Aerospace/Military Consortia 206
9.2.1 Executive Lead-Free Integrated Process Team (ELF IPT) 206
9.2.2 Pb-Free Electronics Risk Management (PERM) Council 206
9.2.3 Tin Whisker Alert Group 208
9.2.4 Center for Advanced Life-Cycle Engineering (CALCE) 208
9.2.5 Joint Council on Aging Aircraft/Joint Group on Pollution Prevention (JCAA/JG-PP) Lead-Free Solder Project 209
9.2.6 NASA/DoD Lead-Free Electronics Project 210
9.2.7 Crane/SAIC Repair Project 211
9.3 Lead-Free Control Plans for Aerospace/Military Electronics 211
9.4 Aerospace/Military Lead-Free Reliability Concerns 212
9.4.1 Thermal Fatigue Resistance 212
9.4.2 Vibration Fatigue Resistance 215
9.4.3 Mechanical Shock Resistance 235
9.4.4 Tin Whiskers and Tin Pest 236
9.5 Summary and Conclusions 239
References 240

10 LEAD-FREE RELIABILITY IN AUTOMOTIVE ENVIRONMENTS 243
Richard D. Parker
10.1 Introduction to Electronics in Automotive Environments 243
10.2 Performance Risks and Issues 245
10.3 Legislation Driving Lead-Free Automotive Electronics 246
10.4 Reliability Requirements for Automotive Environments 247
10.5 Failure Modes of Lead-Free Joints 248
10.6 Impact to Lead-Free Component Procurement and Management 250
10.7 Change versus Risks 251
10.8 Summary and Conclusions 253
References 254

Index 255

TECHNICAL REVIEWERS

The editors would like to acknowledge the following persons who helped to review chapters in this book.

Jennifer Shepherd, Canyon Snow Consulting LLC

Quyen Chu, Jabil

Doug Watson, Flextronics International

Craig Hamilton, Heather McCormick, Joel Trudell, Zenaida Valianu, Celestica

Richard Coyle, Alcatel-Lucent

Linda Woody, Lockheed Martin

Alan McAllister, Intel Corporation

Andy Ganster, Crane Division, Naval Surface Warfare Center (NSWC)

Mark Fulcher, Continental Automotive

TECHNICAL REVIEWERS

The authors would like to thank the following persons who helped to review [illegible]

[illegible]

[illegible]

[illegible]

[illegible]

PREFACE

The transition to lead-free soldering in electronics started its first main push in the late 1990s in Japan, with accelerated efforts to convert lead-free consumer products that were subject to RoHS legislation of the European Union in the two to three years before 2006. This lead to major evaluations of component temperature requirements, lead-free solder process development, and reliability evaluations, and thus required a significant amount of company resources.

The change to lead-free soldering for consumer products also affected other segments of the industry, such as IT infrastructure, telecom, military and automotive that were not directly subject to the EU RoHS legislation because most of the electronics supply chain (consisting of component, soldering material, and board suppliers) had started to move to lead-free technology.

This supply-chain transition, along with anticipated changes in future regulations, is forcing the telecom, enterprise server, and military/ defense segments of the electronics industry to conduct evaluations on lead-free technology. For example, the anticipated shortage of tin-lead component and board materials has led to intensive investigation of the processes and reliability impacts of "mixed" lead-free/tin-lead interconnect solutions.

Based on the evolution of lead-free technology over the past few years, there is a need to provide an update on the concerns and developments in these high-reliability areas as well as a review of more recent developments in lead-free process assembly and reliability. Many of the authors contributing to this book are involved in these high-reliability segments and have concentrated their efforts on addressing the continuing challenges associated with lead-free electronics. Their contributions make this book a useful resource to those transitioning to lead-free soldering as well as to a wider audience.

The book covers a list of key topics, including legislation, SMT, wave, rework, alloy development, component and solder joint reliability. Each subject area is discussed by those who have conducted work in the field and can provide insight into what are the most important areas to consider. The book gives updates in areas for which research is ongoing, and addresses new topics that are relevant to lead-free soldering. A practicing engineer will find the book of use because it goes into these (and other) topics in sufficient detail to provide a practical guide to address issues of concern in lead-free technology.

As already mentioned, the book is particularly timely as companies that include the IT infrastructure, telecom, defense, and aerospace industries are moving to lead-free implementation. Some of the authors are seeing lead-free issues that have not been reported widely before, so the book provides up-to-date assessments of these issues.

INTRODUCTION

In the past few years major regulatory changes have spurred the development of lead-free soldering. The book covers the evolution of lead-free soldering technology in key areas written by researchers specializing in these areas. The chapters also address issues of concern in lead-free solder technology. The topics listed below give an overview of the issues related to the areas of interest.

Environmental regulatory and voluntary efforts in electronics products
Challenges of lead-free surface mount technology, wave-soldering and rework
Research results on various lead-free alloys for BGA/CSP components
Tin whisker growth and mitigation strategies
Testability of lead-free soldered printed circuit assemblies
Board-level solder joint reliability under mechanical loading
Reliability of lead-free electronics in aerospace and military environments
Reliability of lead-free electronics in automotive environments

The legislation chapter (Chapter 1) looks at some of the regulatory trends and voluntary efforts that are affecting the electronics industry worldwide in the move to make product designs more environmentally friendly. The challenge facing the industry is to ensure that product design, manufacture and end-of-life activities are environmentally sound. The chapter covers substances of environmental concern with a review of EU-RoHS, EU REACH, and other areas that include China-RoHS and Korea-RoHS. It also covers the areas of Design for Environment, Energy Efficiency, Recycling and Take-Back.

Lead-free surface mount technology (SMT) development has been fairly rapid over the last few years with improvements being made in soldering materials, processes and inspection techniques. Chapter 2 reviews the development and testing of halogen-free lead-free solder pastes, solder paste handling techniques, and board and stencil design. It discusses solder paste printing parameters as well as paste inspection techniques. Solder paste reflow profiles are assessed by comparing tin-lead with lead-free with a discussion on using air versus a nitrogen reflow atmosphere. The issues discussed related to the reflow process include the head-in-pillow component soldering defect. Solder joint inspection techniques are also reviewed along with test techniques for the assembled product board.

Various developments in lead-free wave-soldering optimization (Chapter 3) have proved to achieve good wave-soldering results. Areas discussed include wave solder machine conveyor speed, solder wave contact length, flux chemistry and flux application, preheating, solder temperature, solder alloy, and the use of nitrogen. The effect of PCB design on wave soldering is also reviewed with a discussion of optimization that can help reduce solder bridging, solder opens, and solder balls. A review is provided of evaluations used to improve lead-free wave hole-fill especially on thicker product boards. Standards related to lead-free wave soldering are discussed as well.

Various developments in lead-free rework (Chapter 4) from passive and lead-frame component rework to BGA/BGA socket rework are discussed, including through-hole hand and solder fountain rework. Some other areas covered are ECO wiring, solder pad repairs, and temperatures and times used during the lead-free rework processes.

In Chapter 5, the impact of increasing lead-free alloy choice for BGA and CSP components is explored. The drivers to replace the first-generation lead-free alloys (near-eutectic SnAgCu alloys like Sn4Ag0.5Cu and Sn3Ag0.5Cu) are first presented. Chief among these is the improvement in mechanical shock resistance. Next the concerns and risks related to these new alloys, most of which are low-silver SnAgCu, are discussed. These include risks in the reflow process in 100% lead-free and "mixed" lead-free/tin-lead solders, potentially poor thermal fatigue resistance, and a lack of established properties. The chapter concludes with a description of recent industry activity that addresses these concerns, namely by establishing new alloy test standards, and of iNEMI efforts to determine the thermal fatigue properties of the new alloys.

The current understanding of the driving forces and the kinetics of tin whisker growth is described in Chapter 6 in the context of industry standard acceleration tests. The stress reversals and the specific morphologies associated with air-to-air thermal cycling are covered and related to tin whisker formation at ambient and higher humidity, higher temperature isothermal anneals. Whisker growth on board-mounted components after reflow is analyzed in detail along with some of the mitigation techniques to reduce the risk of whisker growth for electronic systems. This chapter shows how the driving force changes with testing conditions and the processing of the surface finish or the solder joint contribute to the tin whisker growth mechanism as well as the relationship between changing temperature and humidity conditions, seen in the field and the industry standard acceleration tests.

The challenges in manufacturing test of lead-free PCAs are considered in Chapter 7. Three areas of concern are reviewed: contact repeatability, wear and contamination of ICT probes, and the impact of transient bend flexure on lead-free PCAs. The chapter shows that because of the increased levels of flux residues remaining on test targets after the lead-free manufacturing process, test probes have a difficult time penetrating and making electrical contact. Improved probe tip geometries, among other solutions, can enable better performance in ICT. Unfortunately, the life of test probes is significantly reduced for lead-free PCAs compared to tin-lead due to the higher levels of contaminants, as well as the high yield strength and stiffness of the SnAgCu solders. Overflexure of the PCA is also considered, since high probe forces can cause partial cracking of the second-level interconnect. Manufacturers and component suppliers will be able to use standards such as IPC/JEDEC 9707 (when published) to determine a maximum strain level for components or PCAs that can be used as guidelines in manufacturing, test, and assembly.

The integrity of solder joints under mechanical loading, as can occur during PCA manufacture or in service, is critical to high-performance computer manufacturers (and to manufacturers of other electronic products). In Chapter 8, the author reviews the issues and solutions relevant to the various mechanical reliability concerns. The setting of appropriate strain limits during manufacturing, and how to measure them, are described. The characterization of fracture strength through a variety of tests, modeling, and fracture strength optimization are described in sufficient detail to provide a good guide to practitioners in this field.

Although aerospace and military electronics are currently exempt from legislation requiring lead-free assemblies, their components, boards, and materials are obtained through the supply chain for commercial electronics. Chapter 9 presents the current understanding of the performance of lead-free components, boards, materials, and assemblies in harsh military and aerospace environments. The military/aerospace industry and university research consortia are described along with their efforts to understand the risks associated with using lead-free solders in these environments and the standards that have been developed or are being developed to assess the performance of lead-free rather than tin-lead in military/aerospace systems. The specific concerns for using lead-free assemblies in military/aerospace systems are presented in the context of the results from the JCAA/J-GPP Lead-Free Solder Project. Based on these concerns, the chapter offers mitigation measures that can minimize the effects of shock and vibration on solder joints and control whisker growth on tin-plated components.

The challenges of manufacturing reliable, lead-free electronics for automobiles are described in Chapter 10. Automotive electronics have high-reliability requirements, long design lives, and operate under harsh conditions of temperature and vibration compared to most consumer and IT electronics. At the same time automotive manufacturers are now driven by legislation that is pushing to set the timing for products being lead-free. The chapter discusses risk management as the overriding factor in the transitioning of automotive electronics to a lead-free design. Managing the transition effectively will require discipline in the procurement portion of the business, with component database management, tracking, and error proofing. Further the design of new electronics will need to be based on new data for lead-free components and circuit board layout requirements. Finally, testing all aspects of an electronic assembly with new lead-free components will be needed to ensure a successful transition.

The authors hope that this book will provide readers with useful information from their experiences in investigating and implementing lead-free soldering.

CONTRIBUTORS

Aileen M. Allen, Hewlett-Packard Company, Palo Alto, California
Denis Barbini, Vitronics-Soltec, Stratham, New Hampshire
Jasbir Bath, Bath Technical Consultancy LLC, Fremont, California
Alan Donaldson, Intel Corporation, Hillsboro, Oregon
Carol A. Handwerker, Purdue University, West Lafayette, Indiana
John Hawley, Palm, Sunnyvale, California
Gregory Henshall, Hewlett Packard Company, Palo Alto, California
Keith Newman, Sun Microsystems, Sunnyvale, California
Jennifer Nguyen, Flextronics International, Milpitas, California
Richard D. Parker, Delphi Electronics and Safety, Kokomo, Indiana
Rosa D. Reinosa, Hewlett-Packard Company, Palo Alto, California
Sundar Sethuraman, Jabil, San Jose, California
Peng Su, Cisco Systems, San Jose, California
Thomas A. Woodrow, The Boeing Company, Seattle, Washington

1

REGULATORY AND VOLUNTARY DRIVERS FOR ENVIRONMENTAL IMPROVEMENT: HAZARDOUS SUBSTANCES, LIFE-CYCLE DESIGN, AND END OF LIFE

John Hawley (Palm)

1.1 INTRODUCTION

The market for environmentally friendly electronic products is growing rapidly. Growing just as rapidly is the responsibility companies are assuming or are being compelled to assume for products through their entire life cycle and aftermath, including end of life, product recycling, and product take-back. Energy efficiency of electronic products has also become increasingly important because of their profusion and the associated load they impose on national electrical grids.

In the past the electronics industry did not consider the environmental effects of its products through their life cycles. Its primary concerns centered on how manufacturing processes or facilities infrastructure might impact the immediate environment. It also considered hazardous substances used in manufacturing processes that could have detrimental effects in the event of human contact or exposure. One example of the former consideration was the transition to volatile organic compound–free (VOC-free) processes. Freon, a type of chlorofluorocarbon (CFC), for instance, was used to clean electronics, even though it was known to be a highly ozone-depleting chemical. The electronics industry responded by developing "water-clean" or "no-clean" processes that eliminated the need for CFCs. The challenge now facing the industry is to ensure that product design, manufacture, and end-of-life activities are equally environmentally sound.

Lead-Free Solder Process Development, Edited by Gregory Henshall, Jasbir Bath, and Carol A. Handwerker

This chapter examines some of the regulatory trends and voluntary efforts that are transforming the electronics industry worldwide in the drive to design more environmentally friendly products.

1.2 SUBSTANCES OF ENVIRONMENTAL CONCERN

For the last ten years there has been a concentrated effort to address the problem of potentially hazardous substances found in electronics products. Many parties have been involved, including governments, electronics producers, universities, and nongovernmental organizations. Providing impetus to the effort are environmental and public health issues. There are concerns, for instance, about improper disposal of electronic waste containing potentially hazardous substances. In some third world countries low-temperature burning of electronic parts in open pits for metal recovery has had seriously deleterious effects on both the health of the workers and their environment [1]. In developed, as well as developing, countries, leaching of heavy metals such as cadmium, hexavalent chromium, lead, and mercury from landfills containing electronics into the groundwater has created public health concerns. One well-known example is a case involving hexavalent chromium. An electric utility in California had used hexavalent chromium to mitigate corrosion in a cooling tower in the town of Hinkley between 1952 and 1966. The wastewater slowly dissolved the hexavalent chromium and discharged it into unlined ponds. Some of this material leached into the groundwater and eventually entered the town's drinking water. Over time the contamination resulted in serious health problems [2]. Another well-known case is that of Minamata, Japan, where a chemical company dumped mercury compounds directly into the bay between 1932 and 1968. Three thousand people developed very serious health issues and many died [3]. A third example is in Silicon Valley, where the US EPA (Environmental Protection Agency) Superfund sites were required to clean up groundwater contamination from chemicals linked to birth defects, such as trichloroethane and Freon, from certain semiconductor processing facilities [4].

One outcome of toxic substance release into the environment has been the modification of criteria used to judge risk. In the European Union, a leader in the effort to eliminate substances of environmental concern, the "precautionary principle" has been relied on for assessing risk. The precautionary principle advises caution in advance of effect: if an action or policy seems like it may cause severe or irreversible harm to the public or to the environment, and if there is no scientific consensus that harm would not ensue, the burden of proof falls on those who would adopt the action or policy. The principle implies that there is a responsibility to intervene, to protect the public from exposure to harm where scientific investigation has discovered a plausible risk.

1.2.1 Hazardous Substances Legislation: EU-RoHS

The European Union's Restriction of Hazardous Substances (RoHS) Directive (2002/95/EC)[5] became effective on July 1, 2006. This legislation has served as a model for the rest of the world where restrictions are placed on certain hazardous substances in

TABLE 1.1 Categories of electrical and electronic equipment currently within RoHS

Category 1. Large household appliances	Category 5. Lighting equipment
Category 2. Small household appliances	Category 6. Electrical and electronic tools
Category 3. IT/telecommunications	Category 7. Toys, leisure, and sports
Category 4. Consumer equipment	Category 10. Automatic dispensers

TABLE 1.2 Restricted substances and maximum concentration values

Restricted Substance	Maximum Concentration Value
Cadmium and its compounds	0.01% by weight (100 ppm)
Hexavalent chromium and its compounds	0.1% by weight (1000 ppm)
Lead and its compounds	0.1% by weight (1000 ppm)
Mercury and its compounds	0.1% by weight (1000 ppm)
Polybrominated biphenyls (PBB)	0.1% by weight (1000 ppm)
Polybrominated diphenyl ethers (PBDE)	0.1% by weight (1000 ppm)

electronic products. The scope of the RoHS Directive currently consists of eight of the ten categories found in the Waste Electrical and Electronic Equipment (WEEE) Directive as shown in Table 1.1. The two categories that are currently excluded from the EU RoHS Directive are Category 8 (medical devices) and Category 9 (monitoring and control equipment).

The legislation restricts the use of four heavy metals and two classes of brominated flame retardants and sets the maximum allowed concentration values (MCV) for each, as shown in Table 1.2. Maximum concentration values are defined to be at the homogeneous material level. A homogeneous material is defined as a material that cannot be mechanically disjointed into separate materials.

The Directive includes an exemption regime governed by a Technical Adaptation Committee (TAC) for certain specific applications where no viable technical alternatives exist. There are currently in excess of thirty technical exemptions that have been approved or are in the process of being approved. One exemption for Deca BDE was recently removed by the European Court of Justice on procedural grounds. The TAC is continuing to evaluate exemptions.

The RoHS Directive is evolving and is currently under general review. It appears likely that currently excluded categories 8 (medical devices) and 9 (monitoring and control equipment) will be put into scope by January 1, 2014; that in vitro medical devices will be phased in by January 1, 2016; and that industrial monitoring and control equipment will be phased in from January 1, 2017.

As part of a comprehensive review of exemptions, an external consultant to the EU Commission, the Öko Institute, has recommended that seven current exemptions be phased out [6]. The Institute further recommends limiting the scope of many current exemptions and setting expiration dates. Some of the proposed changes affecting the industry in the near term include the following:

TABLE 1.3 Possible hazardous substances to be added to RoHS

Name	Application
Hexabromocyclododecane (HBCDD)	Flame retardant
Bis(2-ethylhexyl) phthalate (DEHP)	Plasticizer typically used in PVC
Butylbenzylphthalate (BBP)	Plasticizer typically used in PVC
Dibutylphthalate (DBP)	Plasticizer typically used in PVC

- Reduction of mercury levels in fluorescent lamps, effective upon publication of the new Annex (Exemptions 1–4).
- Elimination of the Exemption for cadmium in plating, effective upon publication of the new Annex (Exemption 8). Other uses of cadmium will be exempt at later dates.
- Elimination of the exemption for lead oxide in glass used for bonding front and rear substrates of flat fluorescent lamps in liquid crystal displays (LCDs), effective upon publication of the new Annex (Exemption 20).
- Elimination of the exemption for lead as an impurity in rare earth iron garnet (RIG) Faraday rotators used for fiber optic communication systems, effective upon publication of the new Annex (Exemption 22). Expiration of this exemption on December 31, 2009, has recently been published in the *Official Journal of the European Union*.
- Elimination of the exemption for lead used in C-press compliant pin connector systems (Exemption 11). The recommended date for elimination of this exemption is June 30, 2010. The use of lead in other compliant pin connector systems remains exempt until a later date.

The European Commission and other European authorities are considering the recommendations of the Öko Institute, and the newly published Annex is anticipated shortly.

The expected changes in exemptions multiple times over the next several years will present challenges. The definition of being RoHS compliant for a product will be somewhat in a state of flux. There will be supply chain management challenges for the industry and enforcement challenges for EU member states. Large OEMs are currently working through methods of tracking and enforcing continued compliance by their suppliers, as suppliers work to address relevant changes to materials restrictions for their products.

Under the updated directive, RoHS becomes part of the European conformity mark, CE mark; new rules will be developed for exemption assessments; and at least four other chemicals will probably be added to the hazardous substances index, as shown in Table 1.3. The maximum concentration values will likely be 1000 ppm or 0.1% by weight.

1.2.2 Chemical Substances Legislation: EU-REACH

The terms of REACH (Registration, Evaluation, Authorization, and Restriction of Chemical Substances), Regulation EC 1907/2006 [7], a revised chemical substance regulation for the European Union, began to be implemented on June 1, 2007. It has had, and will continue to have, an enormous effect on the chemical substance and preparation industry and therefore on the electronics industry. It supersedes earlier chemical directives such as 67/548/EEC [8] and 76/769/EEC [9].

Most chemical substances will eventually be subject to registration (usage > 1 ton/year/importer/producer) under REACH, and certain substances of very high concern (SVHC) will be subject to authorization or restriction. A simplified pre-registration period for over 100,000 substances started on June 1, 2008, and concluded on December 1, 2008. A new agency, the European Chemicals Agency (ECHA), which is based in Helsinki, is charged with managing REACH. Registration is required for all substances or preparations not pre-registered with ECHA, before they can enter the market.

ECHA published an initial list of fifteen candidates for SVHC status in the summer of 2008, and it is expected to make regular biannual additions to the list starting in late 2009. A second SVHC candidate list was published in August 2009. SVHCs are defined as having at least one of the following characteristics:

1. CMR: carcinogenic, mutagenic, or toxic for reproduction
2. PBT: persistent, bioaccumulative, toxic
3. vPvB: very persistent and very bioaccumulative
4. Identified on a case-by-case basis from scientific evidence as causing probable serious effects to human health or the environment, such as endocrine disruptors

A "first candidate" list is shown in Table 1.4. Only selected candidates from this list will become SVHCs.

For the electronics industry the three phthalates (BBP, DBP, and DEHP), HBCDD, SCCP, and diarsenic trioxide are probably the most significant. Under REACH the extent of obligations for article producers (e.g., for OEM companies in electronics) is less onerous than that for chemical producers, since substance and preparations manufacturers are assigned most of the responsibility for registration. Article producers ordinarily do not have to register substances present in articles; however, article producers will need to take notice of the following:

1. Assessments must be conducted to determine whether products containing SVHCs are in quantities and concentrations requiring action. If both conditions (in excess of one ton per year per importer and present in concentrations greater than 0.1% weight by weight) are met, ECHA is to be notified by December 1, 2011.
2. The supply chain is to be queried to ensure that all components of it understand REACH and are making appropriate preparations for pre-registering. All substances and preparations in use today must be pre-registered. This simplifies the registration process, since the information required for pre-registration is fairly minimal and allows producers of substances a longer timeline to complete all required documentation. The complete documentation requires a complete dossier with accompanying extensive test data.

TABLE 1.4 First candidate list for SVHC

Substance	Reason for Selection	Possible Uses
Anthracene	PBT	Used in production of anthraquinone dyes, pigments, insecticides, wood preservatives, and coating materials
4,4′-Diaminodiphenylmethane (MDA)	CMR	Rubber preservative; also used in curing epoxy resins and neoprene
Dibutyl phthalate (DBP)	CMR	Plasticizer in adhesives, pigment
Cobalt dichloride	CMR	Desiccant indicator
Diarsenic pentaoxide	CMR	Hardener for lead, copper, or gold
Diarsenic trioxide	CMR	Preparation of elemental arsenic, arsenic alloys, and GaAs semiconductors
Sodium dichromate, dihydrate	CMR	Chromate passivation
5-Tert-Butyl-2,4.6-trinitro-m-xylene (musk xylene)	vPvB	Fragrance compositions; e.g., cleaning fluids
Bis (2-ethylhexyl) phthalate (DEHP)	CMR	Plasticizer in flexible PVC and polymers
Hexabromocyclododecane (HBCDD) and all major diasteroisomers: alpha-and beta-and gamma-hexabromocyclododecane	PBT	Brominated flame retardant, common in ABS
Alkanes, C10–13, chloro (short-chained chlorinated paraffins, SCCP)	PBT	PVC plasticizer, flame retardant
Bis(tributyltin) oxide (TBTO)	PBT	Biocide, wood, paper preservative
Lead hydrogen arsenate	CMR	Wood preservative
Triethyl arsenate	CMR	Biocide, wood, paper preservative
Benzyl butyl phthalate (BBP)	CMR	Plasticizer in adhesives, paints, sealants, inks

3. The REACH Regulation requires article producers to notify suppliers and consumers of SVHC in their articles/product if the concentration weight by weight exceeds 0.1% of the entire article. Thus the basis for measuring SVHCs in REACH is quite different from RoHS, which is based at the homogeneous level. Upon consumer request, the article producer will need to supply information concerning any SVHC above 0.1% within forty-five days of receipt.

1.2.3 Hazardous Substances Legislation: China-RoHS

The People's Republic of China adopted a phased approach to implementation in China-RoHS. The first phase went into effect on March 1, 2007. The product scope is wider than in EU-RoHS and is referred to as Electronic Information Products (EIP).

The restricted substances and maximum concentration values are the same as those governed under EU-RoHS.

The first phase focused on product labeling and disclosure of hazardous substance information, but not on hazardous substance compliance. The requirements included labels including environmental protection use period (number of years prior to a hazardous substance leaching into the environment) and a table indicating the restricted substances if present.

The second phase is focused on developing specific product catalogs and premarket entry testing. If a product has been put on a key catalog, that product will have to be pretested prior to entry into the Chinese market, and it will have to meet the maximum concentration values for the six hazardous substances. The second phase was delayed several times, but the first proposed product catalog was released in October 2009. The first products to be included are mobile phones, telephones, and printers connected to computers. Exemptions based on the EU technical exemptions were also proposed. The details of the testing requirement have not been fully defined. The initial implementation of the second phase is not expected until early 2011.

1.2.4 Hazardous Substances Legislation: Korea-RoHS

On April 2, 2007, the National Assembly of the Republic of Korea (South Korea) adopted the Act for Resource Recycling of Electrical and Electronic Equipment [10], an EU-type RoHS with additional elements, similar to the EU End-of-Life Vehicle (ELV) and Waste Electrical and Electronic Equipment (WEEE). The Act came into force on January 1, 2008.

The restricted substances and maximum concentration values for electronic products are the same as those stipulated in EU-RoHS and, for vehicles, the same as those stipulated in ELV (four heavy metals). Like EU-RoHS but unlike China-RoHS, Korea-RoHS does not include labeling requirements. The WEEE portion emphasizes improvements in products to facilitate recycling, as well as establishing collection and recycling programs that are the responsibility of manufacturers using an extended producer responsibility (EPR) model. In addition the reporting requirement for manufacturers is quite complex and must be submitted in Korean.

The proliferation of materials restriction legislation continues at an unabated pace. Some recent legislation has passed in Turkey and Ukraine. Proposals for RoHS-like materials restriction legislation are ongoing in other major countries such as Brazil, Canada, and India.

1.2.5 Hazardous Substances Legislation: United States

The United States currently has no product-level environmental legislation similar to RoHS or WEEE (Section 1.4.1) at the federal level. One recent piece of federal legislation, the Consumer Product Safety Improvement Act (CPSIA) of 2008, will regulate lead and phthalates in children's products, including electronic products designed for children under twelve years of age. The program will be managed by the US Consumer Product Safety Commission (CPSC) and is comprehensive: within the scope of the legislation are maximum concentration values, certificates of conformity, and labeling and testing requirements. Recently consideration of Resolution 2420, Environmental Design of Electrical Equipment Act (EDEE), began in the US House of Representatives. This legislation mirrors EU-RoHS but its enactment is still in doubt.

California currently has the only RoHS-like legislation in the United States. It has a narrower scope than EU-RoHS and restricts the four heavy metals (cadmium, hexavalent chromium, lead, and mercury) for electronics having a display screen greater than four inches on the diagonal. In 2008 California passed framework legislation collectively referred to as the California Green Chemistry Initiative. This legislation aims to develop a strong foundation for a comprehensive chemicals policy in the state. The legislation will identify chemicals of concern to human health or the environment, and the California Department of Toxic Substances Control (DTSC) has been granted the authority to regulate hazardous chemicals in consumer products. This is the most comprehensive state-level initiative on eliminating hazardous substances from products.

Other states have passed restricted substances legislation that has typically been very narrow in scope. Mercury legislation, for example, has passed in a number of states, including Arizona, Louisiana, Michigan, Texas, and Vermont. Certain types of brominated flame retardants, such as pentaBDE and octaBDE, have been restricted in Hawaii, Maine, Missouri, and Oregon, and an entire class (PBDE) has been restricted in Washington State.

1.2.6 Hazardous Substances: Material Declarations

As both regulatory and voluntary drivers continue to increase the number of restricted substances, managing material composition will increase in importance. The Electronic Industries Alliance (EIA), the JEDEC Solid State Technology Association, and the Japan Green Procurement Survey Standardization Initiative (JGPSSI) jointly developed the Joint Industry Guide (JIG) Material Composition Declaration. The guide is divided into JIG levels A and B. Level A materials and substances are subject to current legislation that:

1. prohibits their use;
2. restricts their use; or
3. requires reporting or results in other regulatory effects.

Level B materials and substances are those that the industry has determined relevant for disclosure because of one or more of the following criteria:

1. of significant environmental, health, or safety interest;
2. will trigger hazardous waste management requirements;
3. could have a negative impact on end-of-life management.

The Joint Industry Guide has been extensively revised and a new version recently released [11].

Other material declarations include the IPC 1752 standard, which covers RoHS, JIG, and other restricted substances [12].

1.2.7 Hazardous Substances: Voluntary Initiatives

The electronics industry, in particular the consumer electronics sector, has recently focused on the voluntary elimination of certain halogens from their products, with priority given to two common halogens, bromine (Br) and chlorine (Cl). The reasons are various and equally weighted: pressure from nongovernmental organizations (NGOs), anticipated future legislation, and the ability to control standards.

The typical definition of halogen-free is less than 0.09wt% (weight percent) Cl (900 ppm) and less than 0.09wt% Br (900 ppm) [13,14,15]. In addition the IEC and IPC standards refer to a combined maximum Cl and Br content of 0.15wt% (1500 ppm) [14,15].

Some NGOs play an important role in driving voluntary environmental initiatives, particularly in the consumer electronics space. One example of an environmentally based product rating program promoted by NGOs is the Electronic Product Environmental Assessment Tool (EPEAT), which has incorporated criteria including hazardous substance elimination, recycling, and energy efficiency for environmentally preferable purchasing [16]. Another example is Greenpeace, which has developed an environmental scorecard that includes energy efficiency criteria for rating consumer electronics companies [17]. A current focus area among some companies is the voluntary effort to eliminate antimony and beryllium and their compounds.

1.3 DESIGN FOR ENVIRONMENT/ENERGY EFFICIENCY

More and more electronics companies are employing design-for-environment methodologies to address product environmental attributes in the design phase. This approach can be significantly more proactive than simply meeting compliance requirements. Elements of design for environment can include materials selection (removal of hazardous substances), recyclability, and energy efficiency. Recyclability, for example, should consider factors such as material identification to facilitate re-use, maximizing the use of recycled material in place of virgin material, and designing to ensure that materials are easily separated. Detailed analysis of the entire product life cycle is also performed using a life-cycle assessment (LCA).

The Energy using Products (EuP) Directive (2005/32/EC) [18] of the European Union is framework legislation focused on eco-design. The objective of this Directive is to influence design in order to reduce energy use. The specific requirements for particular products are determined by separate implementing measures.

Energy consumption for the purposes of electricity production and transportation is a major cause of air pollution. Coal-, gas-, and oil-fired power plants emit greenhouse gases as well as gases that contribute to smog and acid rain. Energy efficiency addresses many objectives, including environmental benefits, cost savings, and energy security. Both voluntary and mandatory programs are in place in several jurisdictions, including the United States, the European Union, and Australia.

1.3.1 Energy Efficiency: United States Legislation

The Energy Independence and Security Act (EISA) of 2007 is comprehensive and broad in scope. A section of this legislation deals with energy efficiency for electronic products. It sets efficiency standards for external power supplies (EPS) and preempts the California Energy Commission's Appliance Efficiency Regulations. The values it sets for EPS efficiency are identical to California's Tier II EPS Standard. It also authorizes the US Federal Trade Commission (FTC) to promulgate energy use labeling requirements for certain products such as televisions, computers, and set-top TV boxes.

1.3.2 Energy Efficiency: United States Voluntary Program

The US Environmental Protection Agency's (EPA) Energy Star is a voluntary energy efficiency program that has been successful with manufacturers and consumers alike. The program is run in coordination with the US Department of Energy (DOE) and private stakeholders. It covers more than thirty-five product categories, including office equipment, consumer electronics, home appliances, lighting, and heating and cooling equipment. The objective of the program is to improve energy efficiency in a product, minimizing any deleterious environmental impact, without sacrificing features or performance. Products must meet strict energy performance criteria set by the US EPA or US DOE to be eligible for Energy Star certification and associated labeling. The program evolves constantly, with more rigorous specifications released on a timely basis. The Energy Star program is widely respected in the European Union, where some of its standards have been, or will be, adopted. The energy efficiency proposals under the Energy using Products Directive have been strongly influenced by the Energy Star Program.

1.3.3 Energy Efficiency: Australia Legislation

Australia has established Minimum Energy Performance Standards (MEPS). Products that are covered by MEPS include refrigerators, clothes washers and dryers, and dishwashers. New MEPS programs starting in November of 2008 now cover products such as external power supplies, set-top TV boxes, and home audio and video equipment. Australia's MEPS have also been adopted and implemented in New Zealand.

1.4 RECYCLING AND TAKE-BACK

Proper recycling of electronic products has become increasingly important, for a number of reasons: concern about illegal transfer of waste electronic goods to third world countries for improper disposal and metal recovery, the huge burden on municipalities charged with waste management, and a growing trend to have manufacturers assigned responsibility for end-of-life management of their products.

TABLE 1.5 EU-WEEE Directive categories: Annex 1A

Category 1. Large household appliances	Category 6. Electrical and electronic tools
Category 2. Small household appliances	Category 7. Toys, leisure, and sports
Category 3. IT/telecommunications	Category 8. Medical devices
Category 4. Consumer equipment	Category 9. Monitoring and control equipment
Category 5. Lighting equipment	Category 10. Automatic dispensers

1.4.1 Recycling Legislation: EU-WEEE

The Waste Electrical and Electronic Equipment Directive (2002/96/EC) [19] of the European Union was one of the first types of legislation to prohibit dumping electronic products in municipal landfill and mandate separate recycling of these products. Annex 1A of the legislation outlines the general categories as shown in Table 1.5, and Annex 1B provides the specific products covered within the scope of the WEEE Directive.

The Directive's aim has been to ensure the proper treatment of e-waste, and to encourage recycling and reuse. It also directs the cost of recycling to producers in member states by assessing fees based on weight of product shipped into an EU member state. The producer is thereby held accountable for end-of-life recycling. This concept is known as extended producer responsibility (EPR). In addition the Directive's Annex II enumerates materials or components that must be removed prior to recycling to ensure against any contamination to the bulk of the material to be recycled. Four years after implementation, however, only about one-third of WEEE is being properly treated, while two-thirds are being diverted to other countries or landfill. The WEEE Directive is being revised to set higher, binding targets for the collection of electrical and electronic equipment. The proposed revisions include:

1. New binding targets for the collection of electrical and electronic equipment. Member states in which the consumption of electrical and electronic equipment is widespread will be assigned more ambitious targets; member states with smaller markets will be given less ambitious targets. The targets will equal 65% of the average weight of electrical and electronic equipment placed on the market over the two previous years in each member state.
2. Recycling and recovery targets that include the reuse of whole appliances.
3. Weight-based targets to increase by 5%.
4. Targets for the recovery of medical devices.
5. Producers' registration and reporting requirements will be harmonized and national registers will be made inter-operable.

1.4.2 Recycling Legislation: United States

Recycling and take-back legislation in the United States currently is at the state rather than the federal level. The scope of covered electrical and electronic products varies somewhat from state to state. Typically covered products include video display products (televisions, monitors of dimensions four inches (10 cm) or more diagonally), computers (both desktop and laptop), general consumer electronics, and batteries.

A brief comparison of Washington State's and California's financial models for recycling and take-back programs provides a good example. California's financing scheme is called advance recovery fee (ARF). In this model consumers pay a point-of-sale fee for covered products. In the Washington model, which is the EPR (extended producer responsibility) model, the entire financial burden of recycling covered goods is borne by the producers. Under this model, producers have a tangible incentive to make recycling as efficient as possible.

More than twenty states and one city (New York) have already enacted recycling and take-back legislation. More states have legislation pending. The trend in this legislation is to place responsibility for product recycling on the producer rather than the consumer or the government (EPR model). Coordination of the directives across the states would benefit both producers and consumers.

1.4.3 Recycling Legislation: China

China-WEEE is officially known as the Regulation for the Management of the Disposal of Waste Electrical and Electronic Products. It has been developing slowly and has suffered a number of delays. When released, the regulation will be a national baseline laying out the responsibilities of various agencies. The initial catalog of products to be covered by the legislation will consist of large household goods:

1. Televisions
2. Refrigerators
3. Washing machines (including dryers) with capacities up to 13 kg
4. Air conditioners
5. Computers (desktop and laptop)

Manufacturers will be expected to pay a fee to the WEEE Management Fund, achieve recycling ratios, and maintain records.

1.4.4 Other Legislation: EU Packaging and Batteries

The EU has updated separate directives on batteries and packaging. The new Battery Directive (2006/66/EC) [20] came into force in September 2008. Both primary and secondary (rechargeable) batteries are covered. The Directive provides for establishing a WEEE-like recycling scheme, a hazardous substance section providing maximum concentration values for certain heavy metals, and ensuring ease of access to batteries in devices. The new Packaging Directive (2004/12/EC) [21] amends an earlier Packaging Directive (94/62/EC) [22]. Some of the particulars in the directives include setting up increasingly aggressive targets for energy recovery of packaging, recycling of packaging, and the reduction of heavy metals (cadmium, hexavalent chromium, lead, and mercury) in packaging. The defined maximum concentration value is 100 ppm for all four metals combined.

1.5 SUMMARY

Product-based environmental initiatives will likely accelerate and increase in scope. Due diligence must be performed by the appropriate departments within companies to ensure compliance. Proactive efforts using design-for-environment and life-cycle assessment techniques will generate eco-friendly products that will succeed in the market place, allowing environmental concerns to be mitigated by increasingly green features that are inherent and integral to the products' purpose and manufacture.

REFERENCES

1. Basel Action Network and Silicon Valley Toxics Coalition, "Exporting Harm: The High Tech Trashing of Asia," Feb. 25, 2002, www.ban.org/E-Waste/technotrashfinalcomp.pdf
2. D. Blowes, "Tracking Hexavalent Chromium in Groundwater," *Science*, vol. 295, p. 2024–2025, March 15, 2002.
3. M. Cross, "Minamata and the Search for Justice," *New Scientist*, Feb. 16, 1991.
4. B. Pimentel, "The Valley's Toxic History," *San Francisco Chronicle*, Jan. 30, 2004.
5. Directive 2002/95/EC of the European Parliament and of the Council on the Restriction of the Use of Certain Hazardous Substances in Electrical and Electronic Equipment, O.J. (L19), Jan. 27, 2003.
6. Oko Institute Final Report, "Adaptation to Scientific and Technical Progress under Directive 2002/95/EC," Feb. 2009.
7. Regulation (EC) No. 1907/2006 of the European Parliament and of the Council, concerning the Registration, Evaluation, Authorization, and Restriction of Chemicals (REACH), establishing a European Chemicals Agency, amending Directive 1999/45/EC, and repealing Council Regulation (EEC) No. 793/93 and Commission Regulation (EC) No. 1488/94 as well as Council Directive 76/769/EEC and Commission Directive 91/155/EC, 93/67/EEC, 93/105/EC AND 2000/21/EC, Dec. 18, 2006.
8. Council Directive 67/548/EEC on the approximation of laws, regulations, and administrative provisions relating to the classification, packaging, and labeling of dangerous substances, June 27, 1967.
9. Council Directive 76/769/EEC on the approximation of the laws, regulations, and administrative provisions of the member states relating to the restrictions on the marketing and use of certain dangerous substances and preparations, July 27, 1976.
10. Act for Resource Recycling of Electrical and Electronic Equipment and Vehicles, Act No. 6319, National Assembly of the Republic of Korea, Apr. 2, 2007.
11. Joint Industry Guide, "Material Composition Declaration for Electrotechnical Products," JIG 101, Ed 2.0, March 25, 2009, by CEA, JGPSSI, and Digital Europe.
12. IPC-1752 Standard, *Materials Declaration Management*, Version 1.1, Feb. 2007.
13. JPCA –ES-01-2003 Standard, *Halogen-Free Copper Clad Laminate Test Method*, 2003.
14. IEC 61249-2-21 ED. 1.0 B:2003 Standard, "Materials for Printed Boards and Other Interconnecting Structures Part 2–21: Reinforced Base Materials, Clad and Unclad—Non-halogenated Epoxide Woven E-Glass Reinforced Laminated Sheets of Defined Flammability (Vertical Burning Test), Copper Clad," 2003.

15. IPC-4101B Standard, *Specifications for Base Materials for Rigid and Multilayer Printed Boards.*
16. EPEAT criteria, http://www.epeat.net/.
17. Greenpeace Guide to Greener Electronics, http://www.greenpeace.org/international/campaigns/toxics/electronics/how-the-companies-line-up.
18. Directive 2005/32/EC of the European Parliament and of the Council establishing a Framework for the setting of Eco-design Requirements for Energy-Using Products and amending Council Directive 92/57/EC and 200/55/EC of the European Parliament and of the Council, O.J.(L191), July 6, 2005.
19. Directive 2002/96/EC of the European Parliament and of the Council on Waste Electrical and Electronic Equipment (WEEE), O.J. (L37), Jan. 27, 2003.
20. Directive 2006/66/EC of the European Parliament and of the Council on batteries and accumulators and waste batteries and accumulators and repealing Directive 91/157/EEC, O.J. (L266), Sept. 6, 2006.
21. Directive 2004/12/EC of the European Parliament and of the Council amending Directive 94/62/EC on packaging and packaging waste, O.J. (L047), Feb. 11, 2004.
22. Directive 94/62/EC of the European Parliament and of the Council on packaging and packaging waste, O.J. (L365), Dec. 20, 1994.

2

LEAD-FREE SURFACE MOUNT TECHNOLOGY

Jasbir Bath (Bath Technical Consultancy LLC), Jennifer Nguyen (Flextronics International), and Sundar Sethuraman (Jabil)

2.1 INTRODUCTION

Lead-free surface mount technology (SMT) soldering development has been fairly rapid over the last few years with improvements being made in soldering materials, processes, and inspection techniques. There are still improvements to be made, and this chapter aims to provide an overview of some of the developments as well as some of the issues that are being faced.

The chapter will review the development of halogen-free lead-free solder pastes, solder paste handling techniques and board and stencil design. It will then discuss solder paste printing parameters as well as inspection techniques. Solder paste reflow profile will be assessed with a discussion of the use of air versus nitrogen reflow atmosphere. Issues related to the reflow process will be discussed in the head-in-pillow soldering defect section. Finally, solder joint inspection techniques will be reviewed along with test techniques for the product board.

2.2 NO-CLEAN AND WATER-SOLUBLE LEAD-FREE PASTES

There are two types of lead-free solder pastes used in the industry: no-clean and water-soluble. The majority of applications use no-clean solder paste formulations where the solder flux residue remains on the board after assembly.

Lead-Free Solder Process Development, Edited by Gregory Henshall, Jasbir Bath, and Carol A. Handwerker

No-clean solder pastes can be further divided into two categories: halogen-containing and halogen-free. Traditionally the majority of no-clean solder pastes have been halogen-containing with the halogen in the flux helping to activate the flux to promote wetting of the solder and remove oxide films on the surface of the solder particles during reflow [1]. The halide content in the halogen containing flux typically varies from 0 to 0.5 wt%(5000 ppm) up to 1 to 2 wt% (10,000–20,000 ppm). Environmental pressure against the use of bromine and chlorine containing materials has started to have an impact on the use of halogens in soldering fluxes.

The typical definition of halogen-free is <0.09 wt% Cl (900 ppm) and <0.09 wt% Br (900 ppm) [2,3,4]. In addition the IEC and IPC standards refer to a combined maximum Cl and Br content of 0.15 wt% (1500 ppm) [3,4]. It should be noted that these industry definitions of halogen-free do not include fluorine, iodine, and astatine.

The main drawback of using halogen-free, lead-free solder pastes over halogen-containing solder pastes is the lack of halogens used in the activators. Development of non–halogen-containing activators is still in progress, so wetting and solderability can suffer. However, there are increasing developments for halogen-free pastes, with some formulations approaching the wetting and solderability levels for halogen-containing pastes. The likely differences in performance will be reduced spreading on more difficult to solder board finishes such as OSP (organic solderability preservative) over copper.

Testing methods for determining halogen content in solder pastes have included the following [5]:

1. *Titration method.* This method assesses the halides present in the flux as a Cl equivalent according to J-STD-004 [6] and IPC-TM-650 2.3.35 [7]. This measurement method is only affected by ionic halides, not covalently bonded halides, and so does not give a total halides content.
2. *Ion chromatography.* This test method quantifies how much halide is present and identifies the particular halides. One test method used is J-STD-004[6] and another is IPC-TM-650 2.3.28.1 [8]. This method typically only detects ionic halides, not covalently bonded halides, even during testing of reflowed flux residue, as not all the covalent halide bonds are broken down. The IPC J-STD-004 standard refers to the halide content in the flux. A classification of L0 indicates low flux activity with between 0 to 0.05 wt% (500 ppm) halide and covers Cl , Br, F, and I. However, as the ion chromatography method used to detect halide content typically only detects ionic halides, the L0 classification can be misleading. The current J-STD-004 standard [6] mentions in Appendix B-10 that the IPC-TM-650 2.3.28 test method [8] is intended for the detection of ionic halides only and is not be confused with total halogen content [ionic halide plus non-ionic (covalent) halide]. Total halogen content should be tested by oxygen bomb combustion analysis followed by ion chromatography.
3. *Oxygen bomb combustion followed by ion chromatography.* This method is growing in popularity. A sample of the flux has the organic materials burned off at high temperature, leaving the halogen remaining in the ash. This ash is run through the ion chromatography equipment to determine total halogen

content. The covalently bonded halides are detected because the covalent bonds are broken down through the oxygen bomb process. Various test methods can be used, including EPA SW-846 5050/9056 [9], EN 14582 [10] and JPCA ES-01-2003 [11] standards.

For oxygen bomb testing of solder paste, there are various samples that can be used, from the raw flux of the solder paste to the final flux residue on the board. Flux residue from a board is typically more difficult to collect, so typically the raw flux from the solder paste is used.

Lead-free water-soluble solder pastes do not have the concern for halogen-free content because the flux residue is removed during washing after assembly. The IPC J-STD-004 standard [6] refers to the halide content in the flux with water soluble solder pastes having flux classifications typically from M1 to H1. M1 fluxes have a halide content ranging from 0.5 wt% (5000 ppm) to less than 2 wt% (20,000 ppm), whereas H1 fluxes contain over 2 wt% (20,000 ppm) total halide. This compares with a typical L0 or L1 classification for no-clean solder pastes from 0 to <0.05% (500 ppm) for L0 and from 0.05 (500 ppm) to less than 0.5 wt% (5000 ppm) halide content for L1.

There is still a small demand for lead-free water-soluble pastes moving forward, especially where flux residue could not remain on the board after assembly, for example, high-frequency product applications. Unfortunately, the development of lead-free water-soluble pastes has been fairly slow because demand for lead-free no-clean paste has been far greater. Consequently printability and especially reflow behavior in terms of solder voiding for lead-free water-soluble paste has suffered in comparison with lead-free no-clean pastes. Some remaining challenges include slumping behavior of lead-free water-soluble pastes, as has been the case for tin-lead water soluble pastes. Removal of water-soluble paste flux residue from under low-profile component parts placed on the boards and the potential reliability issues is another challenge.

2.3 SOLDER PASTE HANDLING

There are many procedures for solder paste handling. If not addressed, SMT defects can occur due to the various ways the solder paste is transported, received, stored, and applied. A solder paste is a combination of solder alloy powder and flux, with the two ingredients having different densities. The paste is sensitive to heat and moisture, which affects its life and performance. Some separation of the flux from the paste is expected, but excessive heat causes a large amount of flux separation that can affect the rheological (flow) properties of the paste, affecting printing. Moisture can cause an increased solder powder oxidation, thus using an excessive amount of the activator in the flux and causing poor wetting to the component or board. Moisture can also cause paste slumping, solder balling when a paste is reflowed, solder/flux splatter, and reduce tack time [12].

Based on the above, transportation, storage, and use of solder paste should be well controlled. Transportation times are normally kept as short as possible, with paste exposed to controlled environments during transport typically using ice packs, dry ice, or other insulation material to reduce exposure to high temperatures. The actual temperature for solder paste transportation varies on the recommendation of the solder paste supplier. Typically solder paste is shipped at a temperature of 5°C to 10°C [13]. During transit it may reach up to 25°C. Cartridges and syringes may be packed and shipped tip down to prevent or reduce flux separation. Once received at the incoming site, paste should be refrigerated as soon as possible at a typical temperature of 5°C to 10°C [13,14], which prolongs the shelf life of the paste. Recommended paste storage temperatures can vary by solder paste supplier based on their specific material usage. The refrigeration typically extends the shelf life of a paste from 1 month at 25°C to 6 months at 5°C to 10°C. Paste is used on a FIRST-IN–FIRST-OUT (FIFO) basis. This applies to both lead-free and tin-lead paste. Previously, lead-free pastes had shorter shelf lives as paste suppliers were more conservative in their paste life recommendations for the new formulation lead-free pastes, but with more data from the field, shelf life for lead-free pastes is generally the same as tin-lead pastes.

Once a paste is ready to be used, it should be removed from the refrigerator and allowed to reach room temperature. This usually takes about 4 hours. Once a solder paste in the jar has reached room temperature, it is stirred for a short amount of time before applying to the stencil. Overstirring can cause excessive shear thinning of the paste, which may result in paste slumping. The typical amount of paste applied to the stencil is a solder paste bead along the full length of the printable pattern of the stencil with a bead diameter of around 0.5 to 0.7 inch (13 mm to 17 mm). Excessive paste can cause the solder paste to stick to the housing of the stencil blade and poor paste release through the stencil apertures onto the board pads. Typically fresh paste is added to the stencil when needed rather than applying a large amount at one time. Once printing has finished, the used paste on the stencil is typically kept in a jar separate from the fresh paste and used as quickly as possible or disposed of. Typically opened jars of paste should not be put back in a refrigerator as the refrigeration can cause condensation inside the opened containers that can give moisture-related problems to the paste [12,15].

2.4 BOARD AND STENCIL DESIGN

One of the areas to consider during the assembly process is the optimized use of board and stencil designs.

There are many factors affecting board pad design. Optimization of pad designs helps reduce solder defects. For example, a study on 0201 [0603 metric] pads showed that half rectangular/half oval pads gave better assembly yields versus other pad shapes considered [16]. In addition optimized pad designs help achieve adequate or good reliability.

Typically, when board pads are designed, there is some history of the board pad design used or reliability testing of the assembled component solder joint to validate the board pad design. This process is especially the case when using tin-lead paste, which has a longer history of use than lead-free pastes. For lead-free reliability testing of a 1 mm pitch PBGA680 component with pad diameters of 0.36 mm and 0.45 mm, there was no difference noticed in accelerated thermal cycle (ATC) test results [17]. A 0.65-mm pitch lead-free BGA476 component with board pad diameters varying from 0.25 mm to 0.3 mm did not affect ATC reliability results. As the pitch gets smaller, with a reduced volume of solder deposited, more attention is focused on whether the reliability of the solder joints will be adequate for the specific application. As the development period for tin-lead pastes has been longer than that of lead-free pastes, printability data are more complete for tin-lead. However, as lead-free paste development has improved, printability of lead-free pastes is now similar to tin-lead pastes.

Based on the board pad design, the stencil apertures are designed for effective paste deposition on the pads. There are many different stencil designs employed in manufacturing, and these have varied by manufacturing engineer and stencil manufacturer preference. Alignment to industry standards for board [18] and especially stencil design guidelines [19] has started to occur with the goal of minimizing assembly defects.

In general, issues during manufacturing based on stencil design typically occur for fine pitch components. For these parts, optimized stencil design guidelines help improve paste release and reduce solder paste defects, improving yield. Initially considerations were given to have stencil aperture designs for lead-free pastes opened up slightly versus those for tin-lead paste when using boards with surface finishes such as OSP. Opening of stencil apertures was done because lead-free paste did not spread as well as tin-lead pastes during reflow. Typically stencil apertures could be 1 : 1 with the board pad for lead-free compared to a 10% stencil aperture reduction with tin-lead paste. This would need to be balanced by the fact that too large a stencil aperture opening on fine pitch components can give rise to solder bridging. The solder paste release rate is typically related to the size of the stencil aperture, the thickness and type of stencil, and the area of the stencil aperture walls.

There are two common stencil aperture ratios used when discussing the likelihood of good solder paste release onto the board pad through the stencil aperture. These ratios apply to both tin-lead and lead-free pastes. They are stencil aspect ratio and stencil area ratio. The aspect ratio is the ratio of the width or diameter of the stencil opening to thickness of the stencil. If the aspect ratio is too small, the solder paste will not release properly through the solder stencil with insufficient solder on the pad. The typical aspect ratio should be >1.5 to have good release of solder paste through the stencil onto the pad [19, 20]. The area ratio is the ratio of the aperture area opening divided by the area of the stencil sidewalls. The typical area ratio should be >0.66 to have good solder paste release through the stencil onto the board pad [19].

These two factors help determine how well the solder paste will print through the stencil apertures onto the board; the higher the aspect ratio and area ratio are, the higher is the paste transfer efficiency (calculated as paste volume divided by the aperture volume). For coarse pitch components, such as 1-mm pitch and above, there are usually no issues in terms of solder paste transfer efficiency, and more care is taken not to overprint the solder paste, which would cause solder bridging.

For fine pitch components, such as 0.3 mm, 0.4 mm, and 0.5 mm pitch parts, the aspect ratios and area ratios, should be calculated based on stencil apertures and thickness so that the preferred ratios are used. The type of stencil fabrication method selected can widen or reduce the solder paste printing window, as well. The three main types of stencil fabrication method used are chemical etch, laser cut, and electroform stencils [21]. The cost typically increases from chemical etch to laser cut to electroform stencils, but the choice of method depends on the application. For general applications, laser cut stencils are used. Electroform stencils are used for very fine component pitches and small stencil aperture sizes. For laser cut stencils, as apertures are cut one at a time, the larger the number of apertures in a stencil, the higher are the manufacturing costs.

The amount of tapering and electropolishing of the stencil aperture walls also helps improve solder paste release. A medium to high degree of taper helps increase paste transfer, especially for small aperture openings. Electropolishing smooths the surface finish of the aperture walls, further enhancing paste release.

The shape of the stencil aperture also affects paste release rate. A square aperture tends to have a better paste release rate than a circular stencil aperture. Besides the stencils aperture shape, size, and method of fabrication, consideration should be given to the positional accuracy of the stencil apertures [22]. Some types of variations in the stencils aperture as well as positional accuracy of the PCB board pads and the alignment of the printing machine can affect the solder paste placement on the board pad.

The actual stencil aperture dimensions and stencil thickness choices depend on the types of component placed on the board. Typical categories of component have concentrated on CSP/BGA, QFP, and chip components. For CSP/BGA components, stencil aperture designs have been based on the board pad designs as well as consideration for the use of square stencil apertures for better paste release, especially for fine pitch CSP components from 0.3 mm to 0.5 mm pitch. The stencil area ratio has an important role in determining the aperture dimensions as well as the type of stencil fabrication method. With a fine pitch CSP component, the stencil area ratio would be close to the 0.66 area ratio guideline where the solder paste's release through the stencil apertures would be a concern. The stencil fabrication method such as the use of electroformed stencils would open the process window for printing when using these types of area ratios. For QFP components with pitches of 0.4 mm and 0.5 mm, the stencil aspect ratio also has an important role in determining the aperture dimensions. On one hand, stencil aperture widths should be narrow to avoid any solder bridging. On the other hand, too small a stencil aperture width reduces the stencil aspect ratio so that the paste transfer efficiency is reduced. This can result in exposed board pads after soldering, especially for lead-free paste where the spreading of the lead-free paste during reflow is less than for tin-lead pastes, particularly on board surface finishes such as OSP.

For chip components, excessive paste deposition can lead to solder balls created under and to the side of the chip component. Various stencil aperture shapes, such as home plate aperture designs, are being used to reduce the amount of solder paste printed on the board pad. Thus during placement of the component, paste does not accumulate under the part that can create solder balls during reflow. Consideration also should be given to the fact that if the part is not placed properly onto the board pads, one side of the chip termination may contact more solder paste than the other side; this situation can lead to tombstoning during reflow. A home plate aperture design applies less paste under the part to reduce solder balling, so the tombstoning may be more of an issue in this case. Clearly, stencil aperture design needs to balance various factors to achieve the best compromise.

For very small chip components, such as 01005 (0402 metric), thin stencils of 3-mil (75-μm) thickness and Type 4 paste can be used [23] to give better paste transfer efficiency. Thin stencils (<5 mils[<125 μm]) and Type 4 paste are also recommended for 0.3 mm and 0.4 mm pitch CSP components. Use of such stencils needs to be offset with the need for thicker stencils used for any coarse pitch components on the same board, in certain cases requiring step-up/step-down stencils.

There has also been some tendency to consider using a stencil for paste printing both surface mount and through-hole component locations. For stencil paste printing of through-hole components, the stencil design is critical to ensure sufficient solder paste is printed. The goal is to create a solder joint that ensures good hole-fill. Approximately 50% of the printed paste volume is evaporated after reflow since 50% of the paste volume is volatile flux [24]. For thick boards (>93 mil [2.3 mm]), the amount of solder paste that must be printed becomes more problematic to achieve. Excessive solder paste could lead to solder balling on the board for the through-hole part after reflow. Also the pin cross-sectional area to the barrel hole cross-sectional area should be considered. Pin-to-hole area ratios of less than 0.2 typically require too much paste and large paste overprinting requirements, whereas pin-to-hole area ratios of greater than 0.5 typically cause extensive loss of solder paste during component pin insertion [25].

2.5 SCREEN PRINTING AND PRINTABILITY OF LEAD-FREE SOLDER PASTES

Screen printing and printability concerns for lead-free pastes are similar to those for tin-lead pastes. Squeegee pressure should typically be as low as possible to scrape the solder paste clean from the surface of the stencil during paste printing. During printing, the solder paste should be soft and fluid-like but be stable after it is printed on the PCB pads. Excess pressure can cause stretching/mispositioning of the stencil and deformation of the squeegee blade, resulting in issues such as solder paste flux bleeding and paste smearing. A guideline squeegee pressure is toughly 1 lb for each linear inch of blade length, which is equivalent to 0.45 kg for each 2.5 cm of blade length.

The printing speed should be set so that the solder paste rolls well on the stencil at the highest speed without decreasing paste printing quality. Excessive squeegee printing speed due to large mechanical stress to the paste can result in lower paste viscosity and thixotropy, giving rise to solder paste slump. Certain pastes are designed for faster print speeds where high-volume production printing is needed for consumer products such as cellphones.

Stencil separation speed is another important factor to consider. Incorrect stencil separation speed after paste printing will result in clogging of the stencil apertures, tailing, or high edges around the solder paste deposits. Different pastes have different optimum settings in terms of stencil separation speed. Machine settings, such as squeegee down stop, should also be adjusted so that the squeegee just touches the stencil surface [26].

The solder paste suppliers have printer setting recommendations for their pastes but it is typically good practice to conduct a design of experiment on the specific tin-lead or lead-free paste being used to understand the process window for the paste printing parameters of squeegee pressure, squeegee speed, and stencil separation speed. For example, Shah et al. [27] did tests on print pressure, print speed, and separation speed on 01005 (0402 metric) stencil paste printing. Stencil printing parameters were found to be more sensitive to the Type 3 size paste versus the Type 4 size solder paste tested.

Although the printing parameters are important, there are some studies suggesting that they are not the only critical factors affecting the paste deposited on the board. Wright [28] showed that paste volume was not affected by print pressure, print speed, and separation speed as much as by stencil design. This finding highlights the importance of ensuring all factors are considered in developing good printing results. The paste too is important, since certain solder pastes have a wider printing process window than others for tin-lead or lead-free pastes.

During continual paste printing, there is a potential for flux bleeding to the bottom surface of the stencil. To prevent flux bleeding, the bottom side of the stencil is cleaned periodically, which is especially useful with fine/small stencil apertures where solder paste clogging is common. In a printing machine, either a dry or wet stencil wipe/cleaner is used. It is usual to conduct a dry machine wipe on the bottom side of the stencil after a few solder paste prints. Wet wipe with a solvent can be done after an increased number of prints, but care needs to be taken to avoid excess solvent through the use of a dry wipe after a wet wipe to remove the solvent. Good alignment of the stencil with the board pads is also important to prevent solder paste smearing under the stencil.

In addition to the print parameters affecting the printability of the solder paste, investigation should be done of the stencil idle time and the stencil life of the paste being printed. The stencil idle time is the ability of the solder paste to be left idle on the stencil and then recover, without paste kneading, to give acceptable printing. Idle time is influenced by the rheological properties of the solder paste. The stencil life is the length of time a solder paste can be used on a stencil while still maintaining its printing properties. Pastes with long stencil life show a small variation in viscosity during continual printing. The thixotropic agents in the solder paste flux help make the solder paste resistant to shear stress during printing and help recover the viscosity of the paste after it is deposited on the pad [29].

Typical printing problems that can occur during production include bridging and open or insufficient paste printing. Some of the possible causes of bridging are slumping of paste, an unclean bottom side of the stencil, incorrect stencil separation speed, and high squeegee print pressure. Some of the causes of open/insufficient printing are clogged stencil apertures, high squeegee print pressure, incorrect stencil separation speed, and incorrect squeegee speed.

2.6 PASTE INSPECTION

SMTA studies on SMT defects have shown that between 60% and 70% of SMT defects can be related back to the SMT solder paste printing and reflow process. Paste inspection

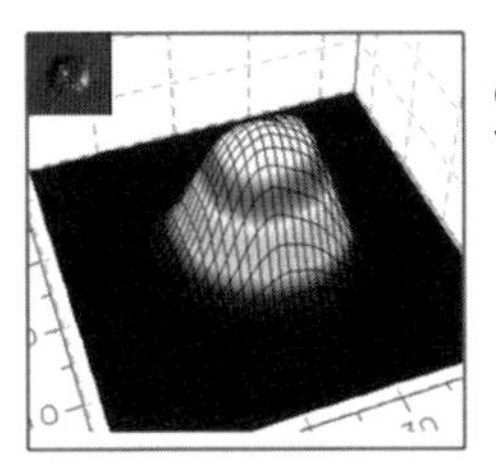

Good
Volume:101%

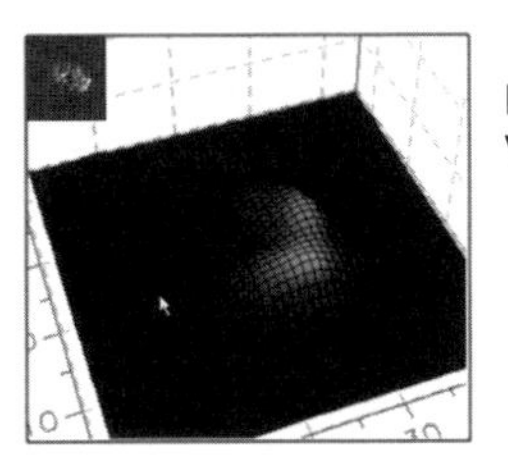

Insufficient
Volume:7%

Figure 2.1 Solder paste inspection images in 3D and 2D showing good and insufficient solder paste print deposits (Courtesy: Koh Young, Christopher Associates)

equipment to verify that the solder paste printing process is in control helps improve yields as well as to ensure repetitive defects are minimized.

Recently there has been a movement toward 3D rather than 2D paste inspection because of the improved ability to measure solder paste volume (Figure 2.1). Paste inspection is necessary to conduct during prototype builds to ensure the initial process is under control. During full-scale manufacturing, inline paste inspection has been increasing in usage, especially for complex boards where there are large I/O components and small components on the same board with a high number of paste pads deposits.

Feng et al. [30] found solder paste inspection to be effective at detecting likely solder joint defects before they occurred in reflow, which resulted in cost savings as well as having a good process monitor. There needs to be more emphasis on paste inspection, not only to filter out some badly printed boards but also as a manufacturing gate to improve process yield. Development work should be done to improve the paste volume and reduce the variation in paste volume deposited to optimize the process and improve yield.

2.7 COMPONENT PLACEMENT (PASTE TACKINESS)

For a component to be kept in position before it is reflowed, as well as to allow accurate placement by the placement machine, there needs to be a certain amount of tackiness of the printed lead-free solder paste to the component. The solder paste needs to have a certain amount of tackiness (tack force) to hold the component in place, and it should also maintain its tackiness over time (tack time) because there may be extended delays between printing and component placement, and reflow. Solder paste suppliers usually provide tack force and tack time data for their pastes. Manufacturers can also assess the tack force and tack time of the paste by evaluating the paste in production with various delays between printing, component placement, and reflow. An example of tack force results on two different solder pastes obtained by a method used to evaluate tack force is shown in Figure 2.2. There are no real differences in tackiness for tin-lead and lead-free pastes. The initial lead-free solder pastes could have had some issues with tackiness because of the development nature of the fluxes used, but with complete development any potential issues were resolved.

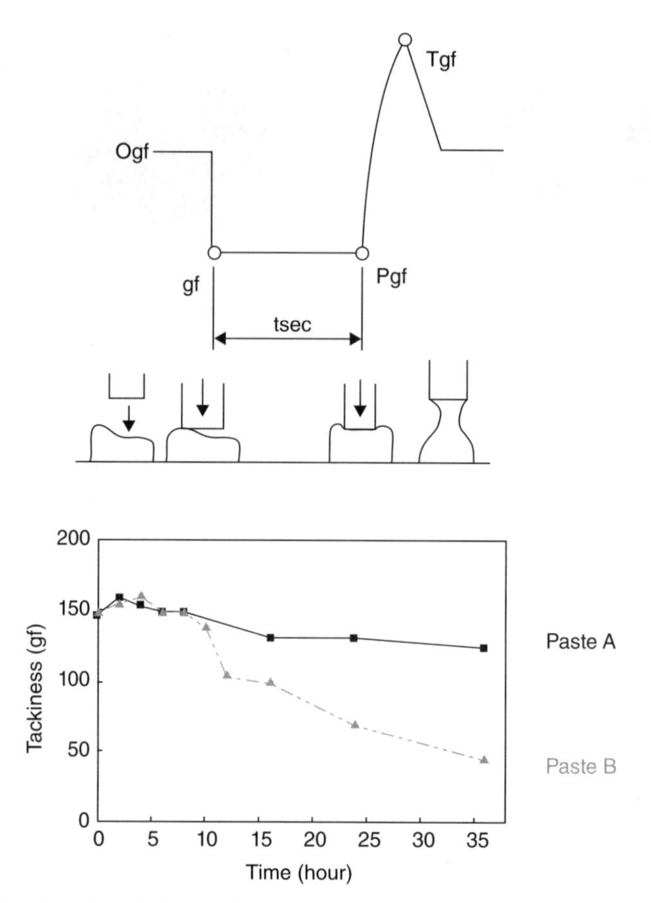

Figure 2.2 Testing of tack force of solder paste (top). An example of tack force testing results on two lead-free solder pastes (bottom). The solder paste marked A has good tack force, and it maintains a good tack force up to 35 hours. The solder paste marked B has good tack force up to 10 hours but shows a decline in tack force after that time, which means there could be issues in production. Typically the tack force is greater than 100gf for good tack of the lead-free or tin-lead solder paste. (Courtesy: Malcom/Koki Solder)

Gilbert [31] found that the solvent systems used in the solder paste flux could be developed to have slow solvent evaporation rates at room temperature, thereby increasing the length of time after printing during which they maintain good tackiness without drying out. If excessive squeegee pressure is applied during printing, the flux separation from the solder powder that can result will reduce tackiness. If in a chip component there is reduced tackiness, the component may be misplaced, and cause tombstoning.

It is also important that the solder paste not slump excessively after printing as this can affect component placement and cause bridging during reflow.

2.8 REFLOW SOLDERING AND THE REFLOW PROFILE

The purpose of reflow soldering is to melt the solder paste and create a good mechanical, thermal, and electrical bond between the component and the board pad. The solder paste, which consists of the solder particles and flux, is developed to allow for this bonding to occur. The flux in the solder paste during reflow helps remove the oxide film on the surface of the components, board, and solder powder. The flux also help cover the surface of the components, board, and solder during reflow to prevent re-oxidation during the heating stages of the reflow oven. The flux helps reduce the surface tension of the molten solder during reflow and increases solder wetting (contact area) on the board and component.

The typical reflow profile for tin-lead or lead-free solder paste consists of 4 stages: preheat, soak, reflow, and cooling. A typical lead-free reflow profile is shown in Figure 2.3, and a typical reflow profile for tin-lead paste is shown in Figure 2.4.

For lead-free SnAgCu (SAC) solders, preheating of the solder paste typically occurs from room temperature to 150°C with a ramp up or heating rate in the reflow oven of 1 to 3°C/s. For tin-lead pastes, preheat of the paste typically occurs from room temperature to 120°C with a ramp up or heating rate in the reflow oven of 1 to 3°C/s. During the preheat stage the solvent starts to evaporate. If the ramp rate is too high during preheating, the solids in the flux, which includes rosins, thixotropic agents, and activators, soften while a large portion of solvent is still present leading to slump, solder balls and bridging.

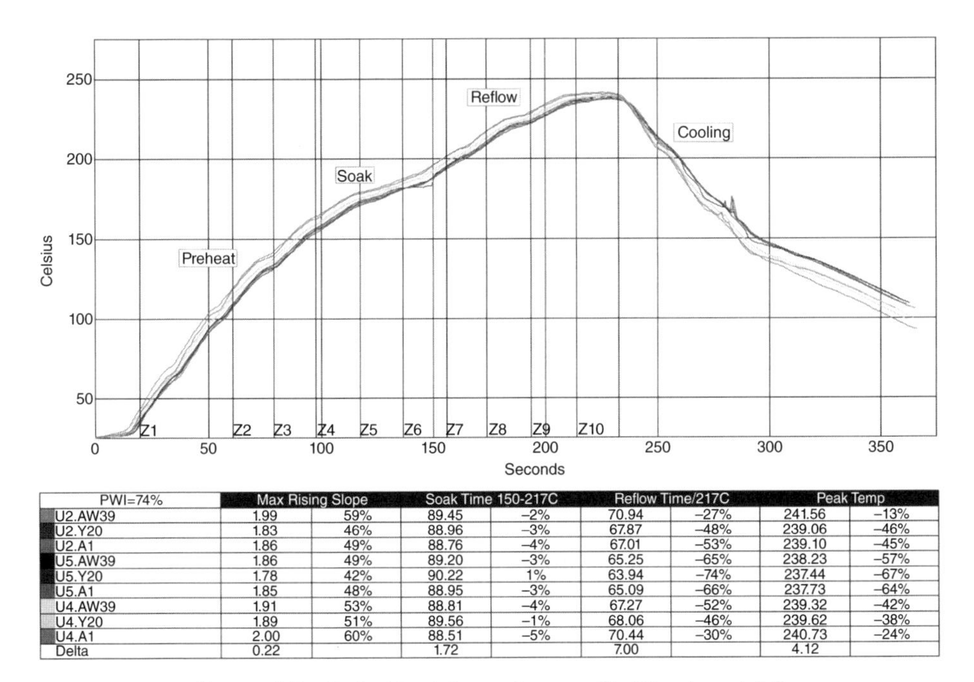

PWI=74%	Max Rising Slope		Soak Time 150-217C		Reflow Time/217C		Peak Temp	
U2.AW39	1.99	59%	89.45	−2%	70.94	−27%	241.56	−13%
U2.Y20	1.83	46%	88.96	−3%	67.87	−48%	239.06	−46%
U2.A1	1.86	49%	88.76	−4%	67.01	−53%	239.10	−45%
U5.AW39	1.86	49%	89.20	−3%	65.25	−65%	238.23	−57%
U5.Y20	1.78	42%	90.22	1%	63.94	−74%	237.44	−67%
U5.A1	1.85	48%	88.95	−3%	65.09	−66%	237.73	−64%
U4.AW39	1.91	53%	88.81	−4%	67.27	−52%	239.32	−42%
U4.Y20	1.89	51%	89.56	−1%	68.06	−46%	239.62	−38%
U4.A1	2.00	60%	88.51	−5%	70.44	−30%	240.73	−24%
Delta	0.22		1.72		7.00		4.12	

Figure 2.3 Typical lead-free reflow profile (Courtesy: Jabil)

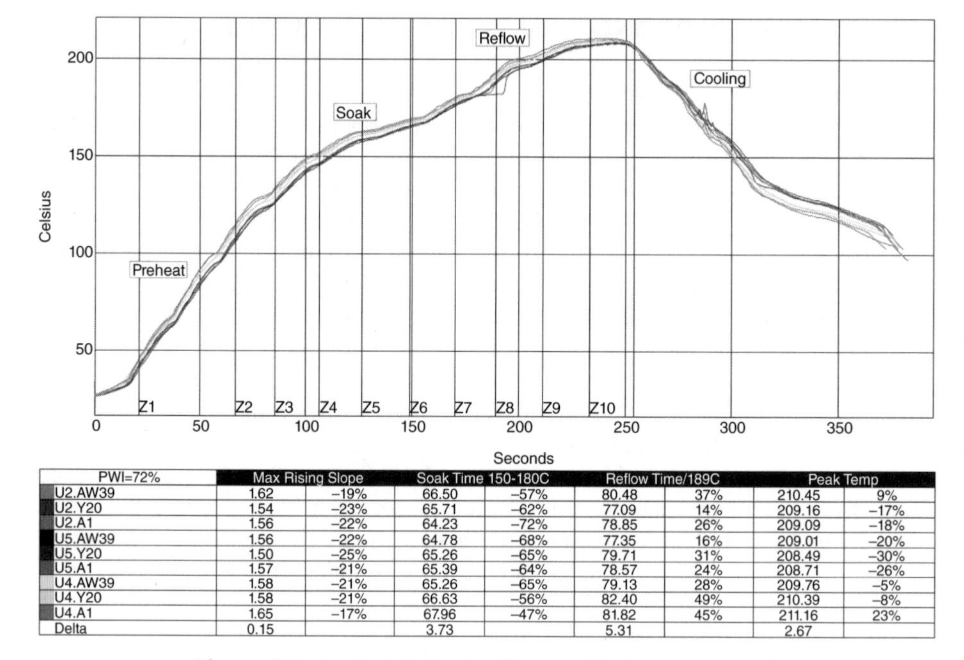

PWI=72%	Max Rising Slope		Soak Time 150-180C		Reflow Time/189C		Peak Temp	
U2.AW39	1.62	−19%	66.50	−57%	80.48	37%	210.45	9%
U2.Y20	1.54	−23%	65.71	−62%	77.09	14%	209.16	−17%
U2.A1	1.56	−22%	64.23	−72%	78.85	26%	209.09	−18%
U5.AW39	1.56	−22%	64.78	−68%	77.35	16%	209.01	−20%
U5.Y20	1.50	−25%	65.26	−65%	79.71	31%	208.49	−30%
U5.A1	1.57	−21%	65.39	−64%	78.57	24%	208.71	−26%
U4.AW39	1.58	−21%	65.26	−65%	79.13	28%	209.76	−5%
U4.Y20	1.58	−21%	66.63	−56%	82.40	49%	210.39	−8%
U4.A1	1.65	−17%	67.96	−47%	81.82	45%	211.16	23%
Delta	0.15		3.73		5.31		2.67	

Figure 2.4 Typical tin-lead reflow profile (Courtesy: Jabil)

The soak stage typically occurs from 150°C to 217°C for Pb-free solders with a time between these temperatures of 60 to 180 seconds and a heating rate in the oven of 1 to 3°C/s. Sn37Pb solder pastes use similar time values except the soak stage occurs from 120°C to 183°C with a heating rate in the oven of 1 to 3°C/s. This is the "flux activation" stage, when the activator material in the flux allows it to remove the oxides from the solder powder, and clean the PCB and component soldering surfaces.

The reflow stage typically occurs when the solder powder starts to melt, which for Sn3-4Ag0.5Cu occurs at and above 217°C. For lead-free Sn3-4Ag0.5Cu, the time over 217°C is typically 45 to 90 seconds with peak temperature between 235°C and 250°C as measured at the solder joint of the component. For Sn37Pb paste, the reflow occurs at and above 183°C. The time over 183°C is typically 45 to 90 seconds with peak temperature between 205°C and 220°C as measured at the solder joint of the component. For both tin-lead and lead-free solders, if the peak temperatures and time above liquidous are excessive, this may give rise to excessive component and board temperatures and high intermetallic compound growth rates at the solder/board or solder/component interfaces, potentially affecting reliability.

Following reflow, the cooling stage occurs, and the solder joint cools from the peak temperature to solidification of the solder joint. The rate of temperature decrease from peak temperature to solidification of the solder joint is termed as the cooling rate, which for lead-free is typically from the peak temperature down to 217°C at a rate of 1 to 4°C/s. Like the heating rate, this rate is similar to that for SnPb solders, apart from the

TABLE 2.1 Typical lead-free and tin-lead reflow profile data

	Preheat and Heating Rate	Soak and Heating Rate	Reflow Time and Peak Temperature	Cooling Rate
Tin-lead Sn37Pb profile	Room temperature to 120°C at 1–3°C/s	120–183°C, 60 to 180 seconds at 1–3°C/s	Time over 183°C, 45 to 90 seconds, Peak 205°C–215°C	Peak temperature down to 183°C at 1–4°C/s
Lead-free SnAgCu profile	Room temperature to 150°C at 1–3°C/s	150–217°C, 60 to 180 seconds, at 1–3°C/s	Time over 217°C, 45 to 90 seconds, Peak 235°C–250°C	Peak temperature down to 217°C at 1–4°C/s

fact that for SnPb solders the cooling rate is from the peak temperature down to 183°C. The cooling rates both for tin-lead and lead-free solders are dictated by considerations of component thermal shock. A comparison of reflow profile data for tin-lead and lead-free solder paste shown in Table 2.1.

There are various methods that can be used to understand the reflow profile limits of a solder paste. One technique is thermo-gravimetric analysis (TGA) to understand the thermal behavior of the flux in the solder paste. Barbini et al. [32] conducted TGA work on different lead-free pastes to understand the importance of certain parameters in reflow soldering, including the temperature gradient in the preheat zone, soak temperature and time, temperature gradient from the soak to the maximum temperature, TAL (time above liquidous), maximum peak temperature, cooling gradient in the cooling zones, and total heating time. Their results show that for the lead-free pastes tested by TGA, typically 1% of the flux evaporation occurs in the preheating stage, 20–30% in the soak stage, 45–60% in the peak stage, and 10–20% in the cooling stage. The critical stages for flux evaporation appear to be at the end of the soak zone and in the peak reflow zones, where temperatures are high and the potential for oxidation at the solder joint is high also. In general, for a longer reflow profile the amount of flux residue left on the board is less than for a shorter reflow profile, with typically between 30–60% of the flux residue remaining on the board after soldering.

For lead-free pastes the need is for the activators in the flux to be more stable at the higher preheat and higher peak reflow temperatures used [33]. The higher soak temperatures and times require thermally stable activators so that the flux does not evaporate too quickly and cause solder balls [34].

When solderability issues arise, some of the considerations include whether the reflow profile was adequately defined and optimized in terms of soak time and temperature, reflow peak and time above liquidous, and ramp-up/heating rate. Characterization of a solder paste reflow profile by the solder paste supplier helps the end user in understanding the reflow profile and its limits, and in avoiding use of an improper reflow profile.

During reflow profiling, consideration should be given to the type of thermocouples to attach and their location at the component and across the board. The lead-free process window is narrow compared to that for SnPb, so this step takes on added importance.

Typically thermocouples are attached at low and high thermal mass locations across the board, as well as at temperature-sensitive component locations where component body and solder joint temperatures are measured.

Component temperature ratings are usually defined by the standard maximum temperature rating of that component. For non-hermetic solid state surface mount devices IPC/JEDEC J-STD-020 [35] covers the maximum component temperatures tested by the component supplier. For lead-free soldering this is from 3 × 245°C to 3 × 260°C peak temperature depending on package thickness and volume. Additionally there is testing at 1 × 260°C peak temperature for area array rework for BGA/CSP parts not rated to 260°C. The time above 217°C that the component is tested to varies in the range from 60 to 150 seconds. For non-IC components there are classification temperature ratings mentioned in IPC/JEDEC J-STD-075 [36] for components such as chip components, TH connectors, aluminum capacitors, crystals, oscillators, fuses, LEDs, relays, and inductors. For SMT reflow, these non-IC components are labeled to indicate the reflow temperature limitations and other process-related issues. The classifications range from R0 to R9, with R0 not being process sensitive, R4 classified to 3 × 260°C peak temperature, R7 classified to 3 × 245°C peak temperature, and so forth.

For components sensitive to moisture ingress, which can cause issues during reflow, IPC/JEDEC J-STD-033 [37] addresses handling, packaging, and shipping of moisture sensitive components. This standard covers dry packing of components, as well as use of moisture barrier bags, humidity indicator cards, desiccants, moisture-sensitive identification, and caution labels. For example, humidity indicator cards mentioned in this standard indicate exposure to 5%, 10%, and 60% RH, with the aim to keep component exposure to less than 5% RH. The higher SMT peak temperatures for lead-free SnAgCu solders compared to SnPb solders make it imperative to avoid moisture ingress during shipping, handling, and board assembly.

IPC/JEDEC J-STD-020 [35] also indicates component floor life classifications according to lead-free or tin-lead reflow profile with the following labeling on moisture barrier bags. The levels depend on the moisture sensitivity of the component:

- Level 1: Unlimited floor life (<30°C, 85%RH)
- Level 2: 1 year (<30°C, 60%RH)
- Level 2a: 4 weeks (<30°C, 60%RH)
- Level 3: 7 days (168 hours) (<30°C, 60%RH)
- Level 4: 3 days (72 hours) (<30°C, 60%RH)
- Level 5: 2 days (48 hours) (<30°C, 60%RH)
- Level 5a: 1 day (24 hours) (<30°C, 60%RH)
- Level 6: Time on label (TOL) (<30°C, 60%RH)

Level 1 to Level 3 components are preferred in manufacturing assembly because there is a reduced frequency/need to bake out parts before manufacture. The baking temperatures for components that are exposed to floor lives beyond their recommended limits are dependent on component part thickness and type, and the temperature rating of component trays (which are typically higher temperature rated) or component reels and component tubes (which are typically lower temperature rated). Component bake-out times at 125°C range from 5 to 96 hours. Excessive baking at 125°C for greater than 96 hours can cause solderability issues and must be avoided. For components packaged in reels or tubes, baking can also be done at 40°C or 90°C, but the baking times are longer; long bake times are detrimental to production throughput. There are also baking temperature considerations for boards that are exposed to excess moisture, since there is the potential for board delamination if reflowed or reworked. The baking temperature and times depend on the stage (as-received, after assembly, after field return) in production that the board may need to be re-baked and how much moisture the assembly has been assumed to be exposed to which will determine the baking temperature and bake time. IPC task group D-35 is developing a working draft standard (IPC-1601) guide for storage and handling of bare boards [38]. This includes development of a test method for determining the percentage of moisture in a printed board as well as determining the rate of absorption of moisture in a printed board.

2.9 EFFECT OF NITROGEN VERSUS AIR ATMOSPHERE DURING LEAD-FREE REFLOW

There has been ongoing work on the use of nitrogen instead of air atmosphere in electronics assembly. Carsac et al. [39] indicated that the use of nitrogen in reflow improved the process window. Shina et al. [40] showed SnAgCu soldered boards in nitrogen atmosphere (50 ppm O_2) had lower assembly defects on test boards compared to air atmosphere soldering (210,000 ppm O_2). Ling et al. [41] showed that 2000 ppm O_2 improved the solder spreading performance of certain lead-free pastes compared to air atmosphere soldering.

Spreading tests by Astrom [42] showed that for two out of the three lead-free SnAgCu solder pastes tested, as the O_2 concentration increased from the 100 to 1000 ppm O_2 range, there was a noticeable decrease in solder spreading performance somewhere in the 1000–5000 ppm O_2 range.

Adams et al. [43] showed that acceptable wetting times were obtained for SnAgCu solder using the wetting balance solderability test equipment on Ni/Pd SOIC leads at an O_2 concentration less than 5000 ppm. In additional work on copper coupons and wire, Hunt et al. [44] showed that good wetting times were obtained at an O_2 concentration less than 5000 ppm and for difficult-to-solder surfaces at an O_2 level of 50 ppm.

Dong et al. [45] conducted fluxless solder perform spread tests on Sn3.5Ag solder with a melting point of 221°C at 10 ppm, 100 ppm, 1000 ppm and 10,000 ppm O_2 levels. At 10 ppm O_2 the Sn3.5Ag solder spread at 230°C, at 100 ppm it spread at 238°C, and at 1000 ppm it spread at 240°C; no spreading occurred at 10,000 ppm O_2. Overall, these data show that the use of nitrogen atmosphere has benefits during lead-free soldering even more so than during tin-lead soldering.

In addition to considering the solderability/spreading of the lead-free SnAgCu alloy, consideration should be given to the board and component surfaces to be soldered. The process window on board surface finishes such as OSP, and potentially immersion tin, can be improved with the use of nitrogen. This is especially the case when considering the solderability of double-sided SMT boards or those that need to be wave soldered after reflow. In these cases there is a potential degradation of surface finish after reflow. OSP coating thickness is reduced with each reflow cycle, with the underlying copper layer becoming more oxidized and thereby making it more difficult to solder. The nitrogen reflow oven atmosphere may help preserve the OSP coating prior to wave soldering.

Marquez de Tino et al. [46] found that nitrogen levels in the reflow oven up to 2500 ppm O_2 helped preserve the solderability of the high temperature rated OSP coating so that good lead-free wave solder hole-fill could be achieved on 125-mil (3.3 mm) thick boards. The process windows for lead-free narrows compared with tin-lead, the higher lead-free reflow soldering temperatures affecting the OSP board surface finish, especially with the reduced hole-fill of lead-free wave soldering on thicker boards.

Some components may also require the use of nitrogen atmosphere for improved wetting performance. Further, the solder paste flux residue reflowed in air versus nitrogen may be more difficult to probe during in-circuit test (ICT) probe testing, leading to more false calls and more frequent probe maintenance. The higher soldering peak temperatures used for lead-free would also reduce the probeability of the solder paste flux residue, as described in Chapter 7.

There are some disadvantages to the use of nitrogen, as well. Aravamudhan et al. [47] showed that 0201 (0603 metric) components reflowed with lead-free SnAgCu solder paste had an increased tombstoning tendency with less than 50 ppm O_2 versus a higher concentration of O_2 (5000–10,000 ppm). The cause was the increased wetting performance with the higher purity of nitrogen increasing the amount of chip component tombstoning during reflow. There were no differences observed between different purities of O_2 ppm with larger 0402 (1005 metric) components in the same study.

Nitrogen also is costly. The cost of assembly increases, the higher the purity of nitrogen (lower O_2 concentration). The decision whether to use nitrogen typically depends on the assessment of benefits it would provide for SMT, wave, and rework versus the cost. Air is still used in most applications today. Large, complex boards with a need for an increased process window may justify the use of nitrogen, more so with lead-free water-soluble paste that has a tendency for increased voiding at the higher peak temperatures and times in air atmospheres. For SMT, the consumption and usage/cost of nitrogen is quite high. The benefit of nitrogen for wave soldering is the decrease in drossing rates, which are higher for lead-free solders than for tin-lead.

For SMT reflow, nitrogen levels may be characterized into different O_2 purity levels: 0 to 100 ppm O_2, 100 to 500 ppm O_2, 500 to 1000 ppm O_2, and 1000 to 2000 ppm O_2. The 0 to 100 ppm O_2 is used to enable wetting on certain components with certain solder pastes but at the highest cost. The 100 to 500 ppm O_2 purity has an intermediate cost and is sufficiently beneficial to wetting in many cases. The 500 to 1000 ppm O_2

comes at medium/low cost but has a general benefit, and the 1000 to 2000 ppm O_2 is relatively lower cost with some benefit compare to air reflow. If nitrogen is used during SMT reflow, it is typically at a purity of less than 1000 ppm O_2. Equipment manufacturers are also developing their reflow ovens so that, if nitrogen is used, the amount of usage of nitrogen can be reduced relative to older models.

Hsiao et al. [48] showed production data at different purity levels that confirm the effectiveness of using nitrogen to improve yields. Based on this information, a cost analysis could be done to compare the cost of nitrogen versus the cost for reduced yield that would lead to product repair.

2.10 HEAD-IN-PILLOW COMPONENT SOLDERING DEFECT

The head-in-pillow (HIP) component soldering defect occurs from incomplete merging of the BGA/CSP component sphere and the molten solder paste (Figure 2.5). This defect could be due to warpage of the component or board, ball coplanarity at BGA/CSP components, or non-wetting of the component based on contamination/excessive oxidation of the component coating.

An example of how a head-in-pillow defect occurs is as follows: During the preheat and soak stages of the reflow process, the BGA/CSP sphere and the flux become oxidized. At the early stages of reflow, the solder paste starts to melt, with bleed out of the flux and further oxidation of the BGA/CSP bumps. The flux is then consumed as the solder melts with no contact between the BGA/CSP solder ball and reflowed solder based on component or board warpage. During the peak reflow stage, as the BGA/CSP solder ball and reflowed solder paste come back into contact, the oxide film on the

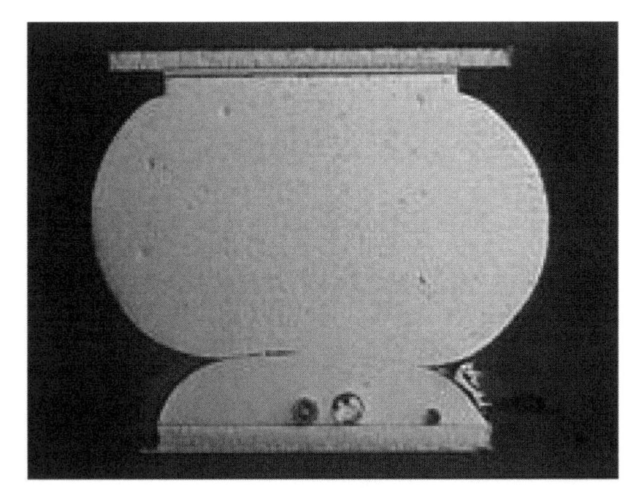

Figure 2.5 Head-in-pillow component soldering defect (Courtesy: Koki Solder/Christopher Associates)

BGA/CSP sphere surface does not melt with the reflowed solder paste, which has a flux layer with almost no activation. Additionally, the time above liquidous is limited before cooling causing the head-in-pillow defect. Fullerton [49] found that HIP defects could be produced in lead-free soldering if an extended BGA rework temperature profile is used, which exhausts the flux and promotes oxidation of the solder ball and the reflowed solder paste deposit.

As one of the above-mentioned causes of the separation of BGA/CSP sphere from the solder paste during reflow, component package warpage is especially problematic during lead-free soldering with the higher lead-free soldering temperatures. Additionally during production the solder paste screen printer can cause HIP joints if it does not deposit sufficient solder paste or if there is inaccurate print registration on the board. Even the BGA/CSP component placer can cause HIP defects if there is inaccurate x-y component placement or if there is insufficient downward pressure pushing the component spheres into the solder paste deposit. An improper reflow profile can cause HIP defects if excessive warpage of the component and/or the PCB results.

The difficultly with the HIP defect is that it is usually very difficult to detect during inspection and functional level testing. There is partial contact between the solder and the ball sphere but no real metallurgical bond. Thus, when the component is subject to mechanical or thermal stress in the field, HIP defects lead to failure.

Head-in-pillow soldering defects have also been found to be a result of a large temperature gradient (delta T) across the BGA component during reflow [50]. The HIP defect then occurs on the cooler trailing edge of the component as it passes through the reflow oven. For a delta T of around 7°C, as the hotter leading edge of the component starts to become molten and the trailing edge does not, a tilt occurs in the component that lifts the trailing edge solder joints out of the solder paste. The trailing edge paste then becomes molten but does not contact with the BGA sphere, which has oxidized. As the trailing edge of the component is dropped into the molten solder paste on the board, both surfaces are oxidized, creating a HIP defect. Some of the remedies considered for HIP have been to modify the stencil thickness and stencil apertures in order to increase paste deposition to compensate for the package and PWB warpage causing the defects [51]. In addition to solder paste inspection to ensure good paste deposits, the placement pressure in the pick and place machine was adjusted to ensure that the BGA balls were correctly seated in the printed solder paste. The reflow profile was adjusted to reduce the temperature differential between the component solder joints, and between the package top and package edge. This way, the warpage of the BGA after assembly was reduced from 0.157 mm (6 mils) to between 0.076 mm and 0.102 mm (3–4 mils). The head-in-pillow type defects were also seen during BGA rework evaluations. Rework nozzles designs were adjusted to achieve a more uniform thermal gradient across the component and reduce the defect rates.

Shadow moire techniques were also employed to understand the warpage of the component during reflow [51]. The reflow profile used in the shadow moire testing machine was adjusted to simulate the production reflow oven profile. For this specific component at room temperature, the package warpage/coplanarity was measured at 3.8 mils (0.1 mm), with the corners of the package warped upward compared to the center of the package. After heating the package, the package warpage was reduced to

3.1 mils (0.08 mm), with the corners of the package now warped downward. On cooling, the warpage increased from 3.1 mils (0.08 mm) to 6.5 mils(0.17 mm) with the most of the warpage still at the corners. In this case the warpage of the board itself was relatively minor. The warpage effect of the component during the reflow profile caused the head-in-pillow defect. It is important for component warpage/coplanarity behavior during reflow to be understood and controlled by the component supplier during component design to prevent issues occurring in production.

Vandervoort et al. [52] showed that head-in-pillow in a large sized lead-free BGA socket is caused by component and PCB warpage. The large size of the socket leads to a large temperature gradient across the part. This temperature gradient creates different melt times, with the cooler inner balls holding the component standoff height longer before the package collapses during reflow. So the outer ball contact with the solder paste is delayed, creating the defect. The time difference between the last ball to melt (which allows the part to collapse) and the first ball to solidify was found to be more important than the time above liquidous to understand the actual window for ball/paste wetting. To reduce the defect, improved heat transfer was investigated with socket designs such as cap venting to minimize the component collapse delay and expand the process window for all joints.

Increasing solder paste amount at reflow can also help reduce the defect by reducing the gap between the paste and the solder sphere at reflow. Solder paste with more robust flux activator and higher activity paste was found to clean oxides in a shorter space of time and so reduce the defect. Ball alloy composition was also found to be a factor with the absence of a specific impurity element and the presence of another impurity element, giving a combination that created a low oxide solder ball with increased wettability and reduced head in pillow.

Shea [53] showed that for lead-free solders the HIP defect is related to component warpage at the higher soldering temperatures. As the component warped, the solder ball was lifted out of the paste deposit, which resulted in oxidation of the solder paste and the component sphere. The flux in the solder paste had to be available over the whole reflow profile to remove the oxide that had formed. Printing additional paste by increasing stencil apertures helped increase the amount of solder paste and therefore the flux available to remove the oxidation. However, excessive openings of stencil apertures can lead to solder bridging, which puts a limit on this approach to solving the HIP problem.

One of the primary factors behind the head-in-pillow issue is warpage of the component and/or board. The JEDEC JEP-95 standard [54] refers measurements of component coplanarity/flatness at room temperature. The maximum package warpage at room temperature is 3 to 8 mils (0.075 mm–0.2 mm) for a ball pitch at 0.4 mm to 1.27 mm [55]. However, component coplanarity and flatness at room temperature can be different than at the SMT reflow temperatures at which HIP defects are produced. JEITA and JEDEC standards are being developed to provide guidelines on the component coplanarity and flatness specifications during SMT reflow [56,57]. Some of the suggested changes include having a maximum package warpage during reflow of 3 to 6 mils (0.075 mm–0.15 mm) dependent on the ball pitch in the range of 0.4 mm to 1.27 mm [55].

Vaccaro et al. [58] showed that large component warpage can be caused by moisture in the part and the use of higher lead-free soldering peak temperatures. This means that even though the component may be qualified to a certain moisture sensitivity level (MSL) rating for lead-free soldering according to IPC/JEDEC J-STD-020 [35], it can still cause production issues due to excessive component warpage. In tests on lead-free 1 mm pitch PBGA components ranging in body size from 23 mm × 23 mm up to 37.5 mm × 37.5 mm, component warpage ranged from 3 to 4 mils (0.075 mm–0.1 mm) for the 23 mm square parts to 5 to 9 mils (0.125 mm–0.225 mm) for the 37.5 mm square parts. For a 37.5 mm square part, an increase in die size from 5 mm × 5 mm to 11 mm × 11 mm reduced the component warpage during the lead-free reflow operation because of the additional stiffening from the larger die size.

Lathrop [59] showed that one of the main factors influencing BGA coplanarity is component laminate warpage due to underfill and overmolding operations on the package side. Lin et al. [60] conducted a study on PoP (package on package) components using the shadow moire technique to understand the influence of package warpage during lead-free reflow assembly from 25°C to 260°C and back down to 25°C. Various factors had an influence on warpage, including die size, mold compound thickness and mold compound CTE (coefficient of thermal expansion), substrate material, and the thickness and copper layer ratio, which could then be optimized to reduce package warpage. Since lead-free reflow temperatures exceed those of tin-lead, more attention should be paid to these parameters if HIP defects are to be avoided in lead-free assemblies.

For the board, current IPC standards refer to a maximum board flatness of 7.5 mils/inch [55]. For a large I/O BGA component this would allow the maximum coplanarity/warpage along the board pad diagonal to be too high. The current board standards do not scale correctly with package size and I/O count and need to be adjusted accordingly. Currently there is work underway by iNEMI to measure board land coplanarity at room and reflow temperatures [55]. Based on this work, acceptance criteria will be recommended for the relevant IPC standard groups. In addition the results of this work will be provided and input into the board and component coplanarity requirements for JEITA and JEDEC so that common standards can be developed.

Some other considerations when a head-in-pillow issue has occurred would be to check for insufficient solder paste deposit by understanding stencil aperture and stencil area/aspect ratios and non-wetting of component or board due to potential contamination or excessive oxidation of the coating. Optimization of the reflow profile also may be needed. Newly developed flux formulations with higher heat resistance throughout the reflow profile could potentially reduce the HIP defects on lead-free assemblies.

In summary, the head-in-pillow defect has many potential causes, including component and board warpage, ball coplanarity and ball oxidation, paste volume and type, component placement, and reflow profile used. Some of these, such as warpage, oxidation, and coplanarity, may be accentuated by the high temperatures associated with lead-free soldering. It typically requires a systematic approach to identify the specific cause leading to the production issue and then to focus on removing/reducing the defect rates.

2.11 VISUAL INSPECTION OF SOLDER JOINT

Assessment of solder joints have typically been done by visual inspection. The most commonly used standard for lead-free and tin-lead soldering is IPC 610 [61]. This standard indicates that even though lead-free solder joints may appear to be visually different compared with tin-lead solder joints, this is not grounds for rejection. This standard contains photos of typical lead-free soldered joints so that proper assessments can be made. It indicates that the most common difference is related to visual appearance, with the lead-free alloys more likely to have surface roughness and greater wetting contact angles versus tin-lead soldered joints. Figures 2.6 and 2.7 show pictures of SnAgCu soldered QFP and 0201 (0603 metric) components respectively. Figures 2.8, 2.9, and 2.10 show some of the differences that may appear between tin-lead and lead-free soldered joints.

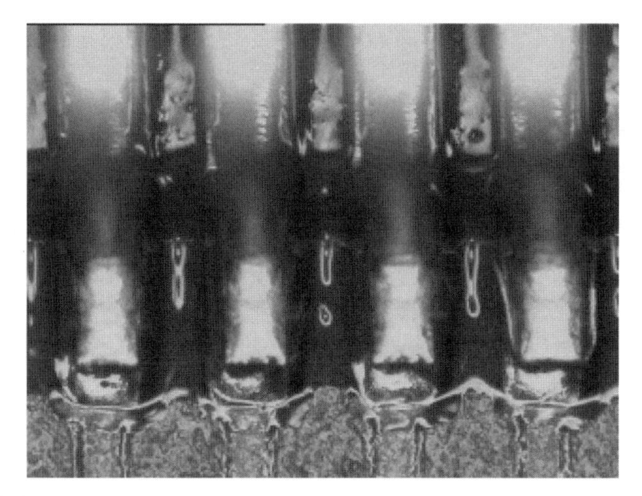

Figure 2.6 SnAgCu soldered QFP component (Courtesy: Flextronics International)

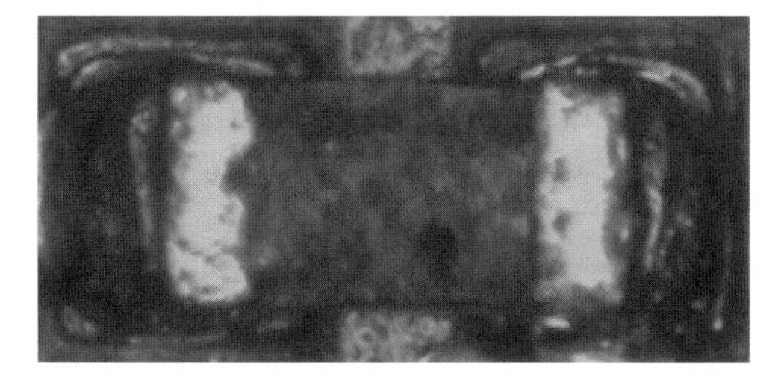

Figure 2.7 SnAgCu soldered 0201 (0603 metric) component (Courtesy: Flextronics International)

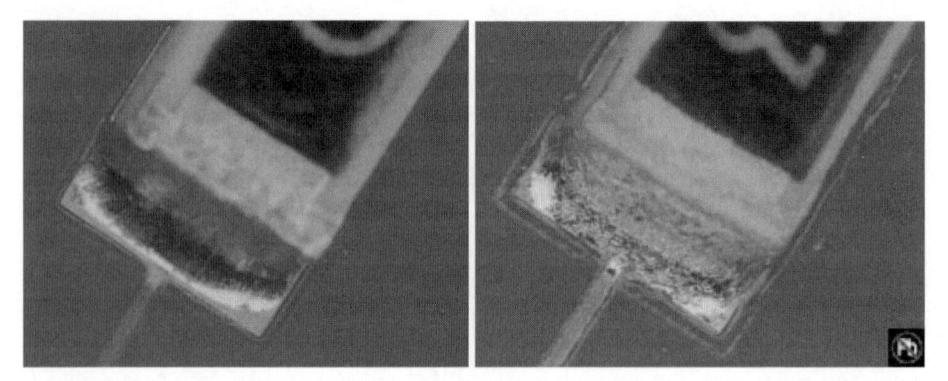

Figure 2.8 Tin-lead (left) and lead-free (right) soldered chip resistor (Courtesy: IPC, IPC 610 standard [61])

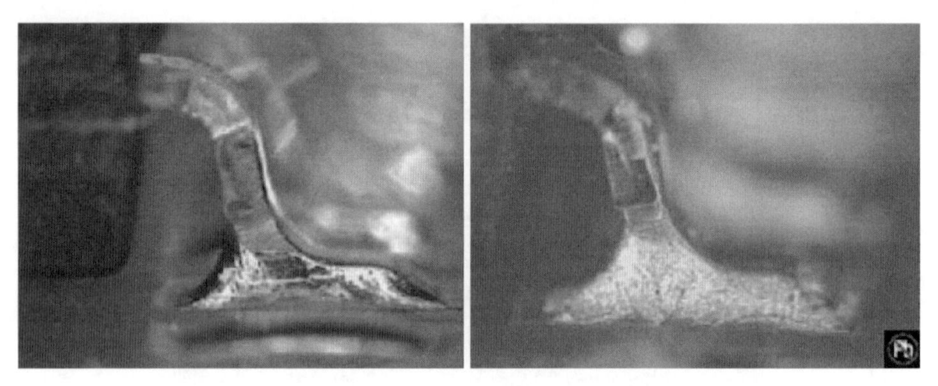

Figure 2.9 Tin-lead (left) and lead-free (right) soldered lead-frame component (Courtesy: IPC, IPC 610 standard [61])

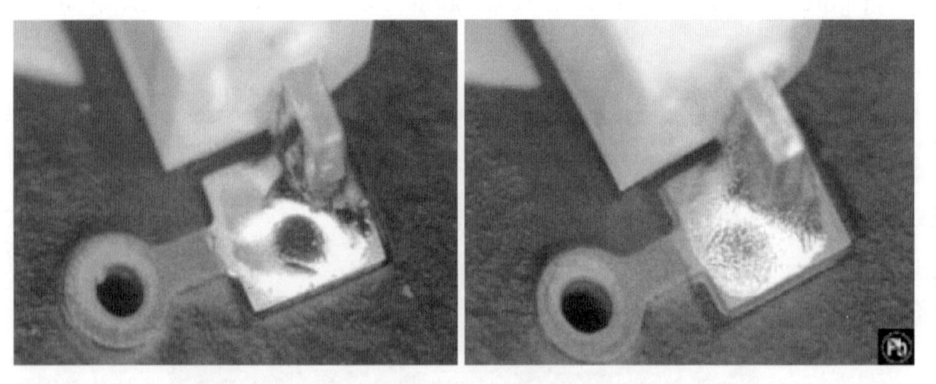

Figure 2.10 Tin-lead (left) and lead-free (right) soldered component on OSP board finish (Courtesy: IPC, IPC 610 standard [61])

2.12 AUTOMATED OPTICAL INSPECTION (AOI)

In addition to visual inspection, automated optical inspection (AOI) equipment is used in assembly operations to inspect solder joints and components. The AOI equipment is typically gauged by its performance in relation to false calls and escape rates. A false call is defined as a solder joint or component that the AOI system calls a fail but does not conform to one of the defect categories listed as a fail. An escape is a solder joint or component that has a characteristic defined as a defect but is not detected by the AOI system. Escape rates are especially a high concern for aerospace, defense, and similar industries, whereas false reject rates are a higher concern for consumer industries.

The placement of the AOI equipment on the production line can vary. For prototype/NPI (New Product Introduction) builds, the AOI can be used after reflow to capture all defects. Once the prototype/NPI builds have achieved high yields, the AOI equipment may be moved to precede the reflow stage. For high-volume production, the AOI equipment is typically used after component placement.

Beer [62] showed that converting AOI systems from tin-lead to lead-free solder inspection requires the use of equipment having flexible sensor modules, lighting, and software to adapt to changes in gray-scale values and potentially different flow behavior for lead-free solder.

Pollock [63] showed that there were no significant issues with AOI capabilities on lead-free assemblies. There was a slight increase in false fails, but this was not seen as significant and would be improved with further refining of the AOI programs for lead-free. As the lead-free joints may be less shiny than tin-lead joints, there may be some regeneration of programs that were originally developed for tin-lead products as well as operator re-training to recognize good versus potentially bad lead-free solder joints. Examples of AOI images of tin-lead versus lead-free soldered joints are shown in Figure 2.11.

Figure 2.11 Tin-lead (left) and Lead-free (right) soldered chip resistor components viewed using AOI (Courtesy: Marantz/Christopher Associates)

2.13 X-RAY INSPECTION

Since lead-free solder has less flow than tin-lead solder, there can be more problems with solder joint formation, such as reduced wetting, increased voids, and reduced wave barrel fill. Traditionally X-ray inspection has been used to detect such problems. Mayer [64] found that there are no significant differences in the X-ray images for lead-free and tin-lead soldered joints.

One consideration in using X-ray is whether to use manual or automated X-ray systems. The considerations include cost of the equipment and the number of X-ray inspections to conduct. In the transition to lead-free, there should be more concern about potentially increased defects compared to tin-lead soldering especially in the start-up/prototyping phases. In automated systems there are challenges in X-raying wave-soldered through-hole boards, as well as challenges in X-raying double-sided mirror imaged press-fit components. So, when deciding to use X-ray, the most effective way of using X-ray equipment must be determined, namely, whether it should be on a sample basis or as an in-line inspection machine during production. Typically once the lead-free assembly process was developed and optimized during the prototype/development phase, the use of detailed X-ray inspection can be scaled back as needed in full-scale production. There appears to be a trend toward using more paste inspection and AOI equipment prior to reflow, where defects can be detected earlier so that assembly issues can be resolved more quickly versus the use of X-ray inspection after reflow. X-ray inspection would still be needed for hidden solder joints such as BGA/CSP and QFN components and for wave solder hole-fill inspection, especially on thicker boards where hole-fill would be a concern.

2.14 ICT/FUNCTIONAL TESTING

Groome [65] showed that the flux residues from lead-free soldering can be harder than those associated with tin-lead soldering because of the increased soldering temperatures, and these residues can build up more on the probe tips, increasing contact resistance. Thus new probe styles may need to be considered as well as more aggressive cleaning schedules and shorter probe replacement cycles. The chisel-type probes are better than crown probes for flux residue probing, but there are limits to the probe forces used. Moreover, crown type probes can accumulate more flux residue over time. An example of flux residue accumulation on a crown type probe is shown in Figure 2.12. A nitrogen reflow atmosphere can make the flux residue more pin probeable. The solids content of the flux also affect pin probeability, with a higher flux solid content in the solder paste tending to make the flux residue remaining on the board more difficult to probe. The time between assembly and test is yet another consideration because the flux residue hardens over time. Development of a solder paste that has a softer, more probeable flux residue would also be desirable.

The board surface finish can also affect probeability. Oresjo [66] suggested covering OSP test pads with solder by paste printing over them with the stencil. For lead-free solders that do not spread well on a test pad, care must be taken to avoid hitting the

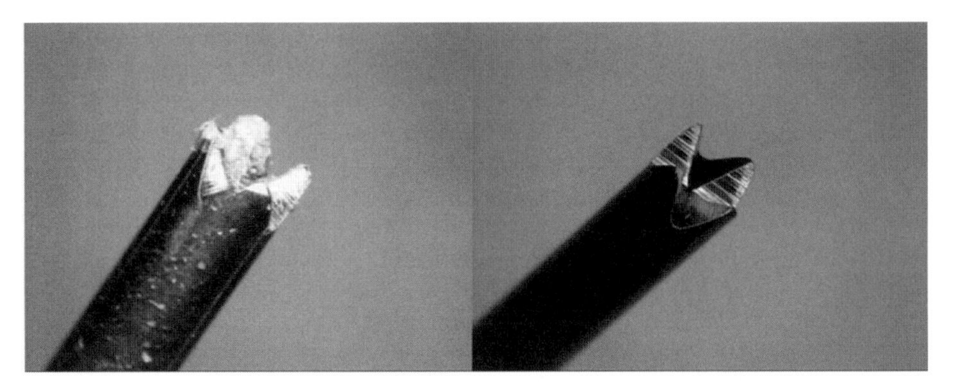

Figure 2.12 Example of flux residue accumulation of a conventional lead-free solder paste product on a crown-type probe (left) compared to a more pin probeable lead-free solder paste product (right) (Courtesy: Koki Solder)

raw copper pad with the test probe, which may result in a fixture contact problem. The same applies to test vias. For unprinted test vias or pads, the lack of solder can cause oxidized OSP pads and probe issues. One solution may be to use a soldered test pad connected to the test vias, with the test pad being probe tested instead of the test via.

In addition to considering the process (print, reflow) during the lead-free prototype builds, it is important that the test results be recorded and properly assessed to ensure that no issues remain when moving into full-scale production.

With the advances made in increasing circuit densities and speeds on modern assemblies, there are fewer electrical test access points. New methods such as boundary scan are being used to reduce the ICT/electrical access needed. However, more adoption of boundary scan and better implementation in semiconductor devices is needed before it can be widely used. There is an iNEMI project currently underway to promote the implementation of the such methods [67].

The reader is encouraged to read Chapter 7 for more information on the impact of lead-free technology on manufacturing test.

2.15 CONCLUSIONS

This chapter discussed the factors affecting the SMT process, with emphasis on the similarities and differences between lead-free and tin-lead technology. Major elements of the SMT process were described, including the solder paste used, the printing and reflow processes, inspection techniques, and some of the soldering defects that can occur and how to address them.

Halogen-free, lead-free solder pastes are being increasingly used, especially for consumer products. One of the developments needed for halogen-free pastes is improved heat tolerance to withstand long soak and reflow times at elevated temperatures. There is still a need for improved halogen-free flux activators to be used in the solder paste, however.

Solder paste handling was discussed with emphasis on the importance of transportation, usage, and storage of solder pastes. Board and stencil design were key to ensuring that the amount of solder paste printed on the board pads is optimized to reduce soldering defects and improve reliability. The board and stencil designs for lead-free and tin-lead were typically the same.

Screen printing parameters of importance include squeegee speed, squeegee pressure, and stencil separation speeds. Also discussed were some of the problems encountered when these parameters were not under control. The print results for lead-free pastes were similar to tin-lead pastes. Paste inspection equipment should be used to measure both lead-free and tin-lead solder paste deposits.

The reflow profile with the four stages (preheat, soak, reflow, cooling), all needed to be optimized to ensure a good solder joint was formed. The reflow profiles for lead-free soldering have the same four stages as tin-lead but the reflow profile was different in certain areas to account for the increased lead-free solder melting temperature.

Reflow in air compared with nitrogen atmospheres was reviewed with benefits shown for using nitrogen in both SMT assembly and the wave-soldering process. There were more benefits for lead-free compared with tin-lead to use nitrogen to improve the process window. For example, the use of nitrogen in lead-free reflow was found to protect OSP board surface finishes, especially for board locations requiring wave soldering and for thick boards where hole-fill is a concern.

Issues related to the reflow profile were discussed, including the head-in-pillow (HIP) defect. HIP has, as one of its primary causes, warpage of the component and/or the board. Lead-free soldering uses higher soldering temperatures than tin-lead soldering, so the increased probability of component warpage and BGA/CSP component solder sphere oxidation can potentially lead to more HIP soldering issues.

Some of the advantages and drawbacks of the various inspection techniques as related to lead-free soldering were reviewed. On visual inspection, lead-free solder joints tend to have less shiny and rougher appearances than tin-lead soldered joints, but this was not ground for rejection, these differences were acceptable and incorporated into the relevant solder joint inspection standards. AOI inspection of lead-free solder joints requires more initial program fine-tuning to account for a wider variety of solder joint appearances compared to tin-lead solder joints, but the changes were typically minor. X-ray inspection can be done on both tin-lead and lead-free soldered joints.

For ICT testing, because the lead-free solder paste is reflowed at higher soldering temperatures and times, there may need to be newer probe types used that can penetrate the harder lead-free soldering flux residues and more probe maintenance.

2.16 FUTURE WORK

This chapter indicated the need for improvements to be made to halogen-free lead-free solder pastes and for more optimization of board and stencil pad designs. More testsing should be done to understand the benefits of different purities of nitrogen during lead-free production. A systematic study of the parameters affecting the head-in-pillow defect would be beneficial, as would be the development of guidelines for acceptable component and board warpage levels during reflow. Standards would need to be developed to help us to understand how to control the warpage of components and boards during lead-free reflow and avoid the head-in-pillow component soldering defect.

The development of newer lead-free solder alloys, especially lower silver SnAgCu alloys such as Sn1Ag0.5Cu for BGA/CSP components, will typically affect the reflow profile needed to be used for assembly with Sn3Ag0.5Cu solder paste. Chapter 5 covers the types of low silver alloys used for BGA/CSP components in more detail. Besides the investigations in this area, there is some discussion of the use of lower silver alloy Sn1Ag0.5Cu paste (melting point of 225°C) as a replacement for Sn3Ag0.5Cu paste (melting point of 217°C). This advancement will not only mean a new reflow profile and an increase in the component temperature ratings used in J-STD-020 standard [35] but also investigations into the reliability of such a solder joint.

More ICT development, especially in the use of OSP-coated boards with lead-free solder, is needed, as well as wider adoption of boundary scan so that alternative test strategies can be used when needed.

ACKNOWLEDGMENTS

The authors would like to thank the persons and companies who contributed lead-free development papers and studies that were reviewed and discussed and formed the basis for this chapter.

REFERENCES

1. C. Poon, A. Long and J. Wang, "Halogen-Free Debate on Solder Paste: IPC Classification and Application," *EM Asia Magazine*, July 2008.
2. JPCA –ES-01-2003 Standard, *Halogen-Free Copper Clad Laminate Test Method*, 2003.
3. IEC 61249-2-21 ED. 1.0 B:2003 Standard, *Materials for Printed Boards and Other Interconnecting Structures. Part 2-21: Reinforced Base Materials, Clad and Unclad—Non-halogenated Epoxide Woven E-Glass Reinforced Laminated Sheets of Defined Flammability (Vertical Burning Test), Copper Clad*, 2003.
4. IPC-4101B Standard, *Specifications for Base Materials for Rigid and Multilayer Printed Boards*.
5. T. Jensen, Indium Corporation website, www.indium.com, 2008.
6. IPC-STD-004 Standard, "Requirements for Soldering Fluxes," 2008.
7. IPC-TM-650 Section 2.3.35, "Halide Content, Quantitative (Chloride and Bromide)," 2004.
8. IPC-TM-650 Section 2.3.28.1, "Halide Content of Soldering Fluxes and Pastes," 2004.
9. EPA SW-846 5050/9056 Standard, *Bomb/Ion Chromatography Method*.
10. EN 14582 standard, *Characterization of Waste. Halogen and Sulfur Content. Oxygen Combustion in Closed Systems and Determination Methods*.
11. JPCA ES-01-2003 Standard, *Test Method for Halogen-Free Materials*, 2003.
12. AIM Tech-Sheet, *Solder Paste Handling Guidelines*.

13. Multicore (Henkel/Loctite), *Solder Paste Handling Guidelines*, 2004.
14. Koki Solder Document, *Solder Paste Handling Guidelines*, 2002.
15. Cookson/Alpha Metals, *General Solder Paste Handling Guidelines—Americas Edition*, 2008.
16. H. Ladhar and S. Sethuraman, "Optimizing the Fear Factors in 0201 Assembly," Nepcon West/Fiberoptic Automation Expo, 2002.
17. B. Vaccaro, D. Gerlach, R. Coyle, R. Popowich, J. Manock, H. McCormick, G. Riccitelli and O. Fleming, "Influence of Package Construction and Board Design Variables on the Solder Joint Reliability of SnPb and Pb-Free PBGA Packages," SMTAI Conf. 2006.
18. IPC 7351 Standard, *Generic Requirements for Surface Mount Design and Land Pattern Standard*, 2005.
19. IPC 7525 Standard, *Stencil Design Guidelines*, 2007.
20. R. Freiberger and R. Venkat, "New Stencil Design for Epoxy Printing," Nepcon West/Fiberoptic Automation Expo, 2002.
21. I. Fleck and P. Chouta, "A New Dimension in Stencil Print Optimization," *J. SMT*, Vol. 16, no. 1, 2003.
22. R. Pandher and C. Shea, "Optimizing Stencil Design for Lead-Free SMT Processing," SMTAI Conf., 2004.
23. S. Nambiar, D. Santos, V. Shah, R. Mohanty and J. Belmonte, "From Printing to Reflow: Process Development for 01005 Assembly," SMTA Pan Pacific Conf., 2007.
24. T. Jenson and R. Lasky, "Practical Tips in Implementing the Pin in Paste Process," SMTAI Conf., 2002.
25. G. Pfennich, H. Fockenberger, L. C. Tat, T. Ho and D. Shangguan, "Board Design and Process Optimization for Paste-in-Hole Using Lead-Free Solder," SMTAI Conf., 2004.
26. B. Bentzen, *Printing of SMT Solder Paste, SMT in Focus.*
27. V. Shah, R. Mohanty, J. Belmonte, T. Jensen, R. Lasky and J. Bishop, "Process Development for 01005 Lead-Free Passive Assembly: Stencil Printing," SMTAI Conf., 2006.
28. T. Wright, "Mathematical Modeling of Solder Paste Stenciling for Process Control," SMTAI Conf., 2006.
29. Koki Solder Document, *General Information on Solder Paste.*
30. Z. Feng , A. Garcia, T. Munnerlyn, W. Meliane, S. Kingery and M. Kurwa, "Solder Paste Inspection Study with SPI and AXI," SMTAI Conf., 2006.
31. B. Gilbert, "Step 7—Soldering," *SMT Magazine*, Aug. 2001.
32. D. Barbini, G. Diepstraten and V. Marquez, "Process Considerations for Optimizing a Reflow Profile," *SMT Magazine*, July 2005.
33. P. Biocca, "Lead-Free SMT—Considerations in Developing a Reliable Soldering Process," Kester White Paper, Jan. 2005.
34. M. Sampathkumar, "Effect of Deviating from the Reflow Process Window for Lead-Free Assembly," *SMTA J.*, vol. 18, no. 4, 2005.
35. IPC/JEDEC J-STD-020 Standard, *Moisture/Reflow Sensitivity Classification for Nonhermetic Solid State Surface Mount Devices*, 2008.
36. ECA/IPC/JEDEC J-STD-075 Standard, *Classification of Non-IC Electronic Components for Assembly Processes*, 2008.
37. IPC/JEDEC J-STD-033 Standard, *Handling, Packing, Shipping and Use of Moisture/Reflow Sensitive Surface Mount Devices*, 2007.

38. IPC 1601 Working Draft Standard, *Printed Board Handling and Storage Guidelines*, 2009.
39. C. Carsac, J. Uner and M. Theriault "Inert Soldering with Lead-Free Alloys: Review and Evaluation," IPC APEX 2001 Conf., SM2-39, p. 1–10, 2001.
40. D. Abbott, R. Anderson, H. Pasquito, G. Wilkish, L. Harriman, M. Kistler, D. Pinsky, S. Shina, M. Quealy and K. Watters, "Testing and Analysis of Surface Mounted Lead-Free Soldering Materials and Processes," IPC APEX 2003 Conf., S20–29, p. 1–19, 2003.
41. H. Ling and P. Stratton, "Improving Lead-Free Solder Joint Quality with Nitrogen," *SMT Magazine*, Oct. 2006.
42. A. Astron, "The Effect of Nitrogen Reflow Soldering in a Lead-Free Process," SMTAI Conf., 2003.
43. S. Adams, P. Stratton and C. Hunt, "Atmosphere Effects on the Comparative Solderability of Eutectic Tin-Silver-Copper and Tin-Lead Alloys," SMTA/Nepcon East Conf., 2001.
44. C. Hunt, D. Lea and S. Adams, "Evaluation of the Comparative Solderability of Lead-Free Solders in Nitrogen—Part II," SMTAI Conf. 2002.
45. C. C. Dong, A. Schwarz and D. V. Roth, "Effects of Atmosphere Composition on Soldering Performance of Lead-Free Alternatives," Nepcon West 1997 Conf., 2007.
46. U. Marquez de Tino, D. Barbini and W. Enroth, "Impact of Soldering Atmosphere on Solder Joint Formation," SMTA Pan Pacific Conf., 2008.
47. S. Aravamudhan, M. Apell, J. Belmonte, G. P. Van Diep and J. Harrell, "Effect of Oxygen Concentration during Reflow on Tombstoning for Passive Resistors for Lead-Free Assemblies," SMTAI Conf., 2006.
48. H. Hsiao, J. R. Lin, E. K. Chang and S. M. Adams, "Reducing Solder Defects under Nitrogen with Varying Oxygen Concentrations," SMI Conf., 1997.
49. J. Fullerton, "Head in Pillow Defects on BGA Assemblies," EMPF, *Empfasis Magazine*, 2008.
50. B. Smith, "A Proposed Mechanism and Remedy for Ball-in-Socket and Foot-in-Mud Soldering Defects on Ball Grid Array and Quad Flat Pack Components," *SMT Magazine*, 2006.
51. M. Mehrotra, S. R. Stegura, J. R. Campbell, M. L. Scionti, E. J. Simeus and J. E. Enriquez, "BGA Warpage and Assembly Challenges," SMTAI Conf., 2004.
52. S. Vandervoort, D. Amir, R. Aspandiar, F. Hua, B. Li, G. Murtagian and R. S. Sidhu, "Head-And-Pillow Defects in BGA Sockets," SMTAI Conf., 2009.
53. C. Shea, "Reprofiling May Be the Best Containment Method for Head-on-Pillow," *Circuit Assembly Magazine*, 2008.
54. JEDEC Publication 95 (JEP 95), *JEDEC Registered and Standard Outlines for Solid State and Related Devices*, 2000.
55. M. Varnau, "Implementing High Temperature Component Requirements for Components and PWBs," INEMI presentation, www.inemi.org.
56. JEITA-ED-7306 Standard, *Measurement Methods of Package Warpage at Elevated Temperature and the Maximum Permissible Warpage*, 2007.
57. JEDEC JEP 95 SPP-024A Standard, *Reflow Flatness Requirements for Ball Grid Array Packages*, 2009.
58. B. T. Vaccaro, R. L. Shook, E. Thomas, J. J. Gilbert, C. Horvath, A. Dairo and G. J. Libricz, "Plastic Ball Grid Array Package Warpage and Impact on Traditional MSL Classification for Pb-free Assembly," *SMTAI Conf.*, 2004.

59. R. Lathrop, "BGA Coplanarity Reduction during the Ball Attach Process," SMTA Pan Pacific Conf., 2008.
60. W. Lin, A. Yoshida and M. Dreiza "Material and Package Optimization for PoP Warpage Control," SMTA Nepcon Shanghai Conf., 2007.
61. IPC-A-610 Standard, *Acceptability of Electronic Assemblies*, 2005.
62. D. Beer, "Lead-Free: AOI in High-Volume Production Assemblies," *SMT Magazine*, Jan. 2006.
63. A. Pollock, "Inspection of Lead-Free Solder Joints on Printed Circuit Assemblies Using Automated Inspection Techniques," SMTAI Conf., 2004.
64. T. Mayer, "Process Control with Automated X-ray Inspection of Solder Joints," SMTAI Conf., 2007.
65. P. R. Groome, "PCB Test and Inspection," *SMT Magazine*, Nov. 2005.
66. S. Oresjo, "Test and Inspection in Lead-Free Manufacturing," IPC APEX Conf., 2005.
67. INEMI, *Annual Report*, 2008.

3

LEAD-FREE WAVE SOLDERING

Denis Barbini (Vitronics-Soltec) and Jasbir Bath
(Bath Technical Consultancy LLC)

Wave soldering is a process, where a board with components has flux applied to it and then is preheated and moved over a molten solder wave via a conveyor system.

A good wave-soldering process will deliver soldered joints on boards that will perform without failures during the expected lifetime of the equipment that the board is part of. The surface solderability, and the thermal solderability, and the layout of all joints to be soldered, must be in accordance with the process that will be used.

Wave solder machine conveyor speed, conveyor angle, solder wave contact length, flux chemistry and application, preheating, solder temperature, solder alloy, and the use of nitrogen are all machine-related aspects of wave soldering. Board type, components, board layout, board thickness, and use of pallets are product-related aspects of wave soldering. All machine- and product-related aspects should fit well together for good wave soldering.

The following sections will discuss the various areas of lead-free wave soldering and the optimization that is necessary to aim to achieve good wave-soldering results.

3.1 WAVE-SOLDERING PROCESS BOUNDARIES

The process steps for wave soldering can be split up in two main parts. One is the soldering process with the wave-soldering machine. The other is the design of the solder

Lead-Free Solder Process Development, Edited by Gregory Henshall, Jasbir Bath, and Carol A. Handwerker

joints, including the board layout. The final result depends on optimizing both the soldering process and the board design.

For a wave-soldering process we need a machine that basically is built up from the following: a transport system, a fluxing unit, preheating unit(s), the solder pot with the wave former(s), and an exhaust system. This basic system can have several options such as extra preheating on the topside of the boards and a cooling unit to cool the board faster after wave soldering.

The transport system must be able to transport and support the boards to be soldered in a continuous flow and at a constant speed according to the speed setting. In most machines this transport system is adjustable in width so that it can handle a variety of boards or pallets in which the boards are mounted. Common transport systems use a so-called finger transport. Often the transport system can be provided with a board support system that can compensate for when there is too much bending during preheating and wave soldering.

Different conveyor finger shapes can be used if heavy boards must be handled. The transport angle is in most cases typically fixed at 7° from the horizontal but may be adjustable to a lower angle. In this case the solder wave nozzle must be designed to enable a longer contact length that is related with the lower transport angle.

The desired soldering quality level is often dependent on the product specification. In general, a good joint is a joint that is soldered in one single operation that will fulfill both its electrical and mechanical requirements within the specified lifetime conditions without failures.

The solder quality for wave or reflow is mostly controlled by visual inspection, according to common specifications like IPC J-STD-001 [1] and IPC 610 [2] standards. However, one should also use common sense when inspecting lead-free or tin-lead wave soldered joints. For example, one should not touch up a "large" joint so that it will fit to the "hollow shaped" ideal solder joint shape. The reason that the joint is large is that the "excess" solder that remains on the joint after seperation from the wave has no way to drain off to a track, for example, since these tracks are covered by solder resist/mask. The wetting of such a joint is perfect; otherwise, little or no solder would be found on that joint.

3.1.1 Fluxer

The fluxing system is the first unit that the board will pass on its way toward the wave. The flux used and the fluxing process are the most important process parameters in wave soldering next to board design. Flux is a mixture of an activator that provides the chemical cleaning reaction, a carrier that will embed and keep the activator on the board, and finally a solvent in which the activator and the carrier are dissolved. The solvent is used for easy application of the flux by a spray fluxing system.

The flux has a function to remove the oxides from the area of the joint to be soldered and from the molten solder wave that is in contact with the board. The flux must also provide a sufficient duration of flux activity so that when the board leaves the wave, it has bridge-free wave soldering. The correct fluxer setting is of the most importance.

There also needs to be considered the type of solderable board finish and solder resist (mask) used.

The spray fluxer setting is related to the conveyor speed. This setting needs to apply the correct amount of flux on the board. A guideline to how much flux should be applied can be found in the data sheets of the flux supplier. One should also measure the flux flow with different fluxer settings with that flux. It is important that not too much flux is applied, since this may cause solder splattering onto the assembly, and the remaining excessive flux residues may interfere with the electrical reliability of the soldered product.

We need the flux activity not only at the start of the solder process but also at the end of it. The formation of a single joint can only be accomplished when the solder wave separates and the solder stays only at the joint and does not create solder bridging from excessive oxide formation at the wave surface. For that, the flux should remain on the board during separation of the board from the solder wave. In that part of the process the flux activity is needed to assist by reducing or preventing oxidation of the solder wave surface.

If a flux does not have sufficient activity or if there is insufficient flux to adhere to the board surface and stay on the board during contact with the wave, soldering defects will increase, like solder bridging. Also solder icicles can be formed as a result of insufficient flux activity at the wave exit area.

The benefit of the spray fluxing system in relation to the foam fluxing system that was often used in the early history of wave machine soldering is that the flux that is applied to the board has a constant composition. With a foam fluxer there was a continuous evaporation of the solvent from the flux. This results in a continuous change of the flux consistency and excessive evaporation of VOCs (volatile organic compounds).

A disadvantage of a foam fluxer was that the flux amount on the board was difficult to control. A great benefit of the foam fluxer was the simplicity and the fact that it would bring flux within the capillary spaces between component lead and barrel wall.

With a spray fluxer or flux atomising system the filling of such gaps would be impossible during fluxing. The single droplets do not create a "reservoir" from which the flux will wick up in the gap by capillary action. To fill a capillary, one needs a liquid source that contains a sufficient amount of liquid to fill the gap. Small single droplets are not able to create such a source, even when a few drops might enter the gap during spraying.

During soldering of the board, as the board makes contact with the solder wave, part of the flux layer on the board surface will be moved by the solder flow from the wave. At the same time the flux layer is pressed between the board and the wave. This moving flux layer is capable of filling the capillary gap and will clean the metal surfaces so that the solder is able to wet the component lead and the board barrel hole.

3.1.2 Function of the Flux in Wave Soldering

The main function of the flux when the board enters the solder wave is to clean the component and board area that must be wetted by the solder. The common metals used for solder joint formation, including the liquid solder are covered with an oxide layer. As long as an oxide layer or any other layer is present between the solder and the metal to be soldered, it is difficult to create a good solder joint. Even small oxide additions on the surfaces to be wetted can cause wetting difficulties.

During solder wetting the solder will displace the flux that has cleaned the metal surfaces. Since the surface of the solder wave will instantly oxidize in air, the flux should remove those oxides continuously during soldering of the board. Although the wetting of the joint surfaces might be accomplished, the final joint formation can only take place when the board separates from the solder wave. There needs to be sufficient flux activity at the area where the board separates from the solder wave.

The solder fillets will be formed when the solder wave separates from the board. The surface tension will keep the liquid solder at the joint. However, when the solder is covered with an oxide layer, the separation of the solder will be affected. As a result solder bridging can occur between the fillets. The bridging is mainly due to a lack of flux or lack of flux activity at the end of the wave soldering process. At this stage of the process the use of nitrogen can be of help. By applying nitrogen at the point of separation between the board and the solder wave, less solder oxide will form, and the flux will have a wider process window.

Flux must adhere to the board during the soldering process. It is necessary that active flux remain at the point where the board leaves the last part of the solder wave. If the solder resist that is applied on the board is not properly cured, then the curing process may proceed in the solder wave. As a result of this a vapor film in between the solder resist and the flux will be formed. The flux will adhere to the solder resist but not be able to adhere to a vapor blanket. As a result of the friction between the solder wave and the board, the flux will be wiped off almost completely, so it will be unable to employ its full activity for the wave-soldering process.

3.1.3 Flux Chemistries

With wave fluxes there is a trend to move to no-clean fluxes. Water-soluble based fluxes are typically used if there are specific needs such as obtaining good hole-fill on thick boards or if there is a zero solder ball requirement where the solder balls present can be removed when the board is washed.

For water wash fluxes some are VOC (volatile organic compound, or alcohol based), and some are VOC-free (water-based) fluxes. The solids content for these water wash fluxes can be around 19% for VOC based fluxes to 7% to 22% for VOC-free based fluxes. Typically the higher the solids content, the more flux activity there is for the flux.

No-clean wave flux can be put into four categories. These are:

1. Low residue organic acid VOC-free water based (3–5% solids content)
2. Low residue rosin VOC (3–7% solids content)
3. High residue rosin VOC (15% solids content)
4. Rosin emulsion low VOC (2–6% solids content)

Compared with water-soluble fluxes, the no-clean wave fluxes typically have solids contents ranging from 3% up to 15%.

No-clean alcohol based fluxes facilitate good hole-fill on thick boards (>100 mils, 2.5 mm). If the solids content of these fluxes are higher than 6% or 7%, they may inhibit pin probing at ICT. Water based no-clean fluxes are more environmentally friendly, but their use makes hole-fill difficult on thick boards (>100 mils, 2.5 mm) more so than with no-clean alcohol-based fluxes.

Alcohol based fluxes tend to have some benefits over water based fluxes based on hole-fill, solder balling, solder opens, and bridging data. There are some trends on the development of wave fluxes that are halogen-free. The typical industry definition of halogen-free is <0.09 wt% (900 ppm) Cl and <0.09 wt% (900 ppm) Br, with total Cl + Br < 0.15 wt% (1500 ppm) [3]. The movement to halogen-free is mainly concentrated on consumer products presently but is expanding to include other markets. The difficulties in moving to halogen-free wave fluxes is that due to the lack of the Br and Cl containing activators in the flux, there is less flux activity and potentially less solder hole-fill.

3.1.4 Keeping the Flux on the Board during Soldering

Normally the flux is made to adhere well to the board so that it can mainly withstand the friction of the solder wave. The adhesion of the flux is also dependent on the board surface. Most boards are covered with a solder resist/mask layer. There are different types of solder resist, but most important is that the resist layer is clean and properly cured during its application so that no vapors will escape from the surface of the solder resist during soldering. Another effect that is related to solder resist is the presence of solder balls on the solder mask surface after wave soldering, which will be discussed in more detail in a later section of this chapter.

3.1.5 Correct Flux Amount

The amount of flux that is needed on the surfaces that come in contact with the solder wave is typically defined in the wave solder flux data sheet. This figure can be expressed in wet weight or dry weight. Since the flux will be applied as a liquid, it is necessary to know the wet weight. A typical wet weight figure is about 5 mg/cm^2.

Less flux may create soldering problems, while more flux may create product electrical reliability problems due to excessive flux residues and unheated or unactivated flux residues in certain areas. So it is important to keep the applied flux amount on the board within the optimal range. The flux amount on the board can be controlled and measured by weighing a test panel that is fluxed. Depending on the type of flux and the fluxing system, the flux application can be corrected to the necessary amount on the board by adjusting the fluxer settings. Flux should also be assessed by flux coverage/board penetration testing. Flux should have good flux penetration through 20 mil (0.5 mm) vias. It should have an even spray pattern across the board surface when tested using flux paper. There should also be a minimum amount of flux flooding on the trailing edge of the test fixture. Typical flux paper used to assess flux spray patterns and flux penetration includes fax type paper for alcohol based fluxes and pH/litmus type paper for water based fluxes. There are also various measurement methods and equipment available from suppliers to evaluate flux penetration on test boards.

3.1.6 Flux Application Methods

With the new fluxing systems, such as spray fluxers now commonly used in wave soldering systems, the flux amount can be well controlled. With the low solid content fluxes, this control is necessary to ensure that the specified amount of flux is applied.

The fluxing systems apply the flux as single droplets that will flow together. Current flux spraying technology, however, has difficulty in applying the flux in the gap between lead and barrel. One way of looking at it is that there is no flux "reservoir" between the board and the lead to supply capillary filling of the gap with flux. This is only possible when the board is "soaked" with flux, such as with the foam or wave fluxing systems that were used in the past. Droplet size also plays a critical role.

This is a principle drawback of these new spray fluxing systems. The fact is that good soldering results can still be obtained due to the mechanism of the solder wave. This will move part of the flux layer from the board surface into the joint gaps just before the solder is entering.

3.1.7 Preheating

After the board is fluxed the solvent from the flux must be evaporated before the flux enters the solder wave. Low solid fluxes often demand a preheat temperature that is well above the solvent evaporation temperature. There are several methods used for preheating, such as hot-air convection blowers, IR (infrared) heaters, and quartz lamps.

The hot-air convection blowers have the benefit that they can be set at a certain air temperature and kept at that temperature. This means that there is a limited risk of overheating compared to radiation heaters, which have a much higher temperature set point. The drawback of this system is that on a wet board freshly covered with flux the hot-air flow from the blower box might displace the flux and create flux "traces" on the board. In most cases this will not create soldering problems as fluxes often behave dynamically when they are in contact with the solder wave. Another aspect that should be considered is that due to the mass of the hot air convection blower box, the system can not react as fast to changes in machine preheater settings.

IR elements have a quicker response to a change in temperature settings than a hot air convection blower box, but still they do not react very fast because of the large elemental mass. Here the heat transfer to the board is a mix of radiation and convection. Often this heating system is used as the first preheater zone after the fluxing unit. It will increase the viscosity of the flux layer by partial removal of the solvent. If this unit is working in front of a hot air blower box, which works as a second preheating unit, the risk of flux "trace formation" is greatly reduced.

A third preheating unit can be quartz lamps, which are often used behind the hot air convection blower box. This unit can be used as a preheat booster, since it is a fast reacting heater. It enables the running of the preheater and the soldering machine with different recipes when a mix of boards must be soldered that each has a distinct recipe.

Preheater settings are adjusted to get optimized flux activation with more top-side preheat for better solder hole-fill. For lead-free wave soldering more heat is needed on the top side of the board to get the lead-free wave solder to flow better up the through-hole barrels compared with tin-lead wave soldering. The top-side preheater settings are raised as high as possible, while bottom side preheaters should be lowered so that flux is not exhausted. In the data sheets of the flux the typical preheat settings are given based on the flux used.

The standard position of the preheating units is that the board will be preheated from the under side. For applications using 1.6 mm (63 mil) thick boards this will be sufficient. However, when thick multilayer boards (>1.6 mm [63 mil]) are used with a high thermal heat demand, the amount of preheating is insufficient. In that case top-side preheaters are needed. Such a unit should, however, only be used if all top-side components can withstand the top-side temperatures.

One has to keep in mind that the total heat presented during the wave solder process is an addition of the heating energy from the preheater and the heat transfer by the solder wave. The total temperature inside the components must be kept below the temperature limit of the components. The allowable time–temperature load during soldering can be understood from the component data sheets. The solder process should fit in with all component temperatures involved.

The typical wave preheater board temperatures are similar for tin-lead and lead-free wave soldering and depend on the flux used. When using alcohol-based fluxes 80°C to 90°C top-side board preheat is typically needed. For water-based flux fluxes 100°C to 110°C is typically required for top-side board preheat. The typical preheat rate from 40°C to 100°C is around 1°C/s to 2°C/s. For lead-free soldering the delta T between the preheat zone and wave soldering zone may be larger than during tin-lead wave soldering as there is a higher pot temperature used for lead-free compared with tin-lead soldering. This should be monitored as there could be potential component issues due to the larger delta T between the preheat and wave soldering zones.

Relevant data about the necessary flux amount on the board that are needed, and the specified preheating temperature can be found on the data sheet of the flux or in the documentation from the flux supplier such as TGA (thermogravimetric analysis) and DSC (differential scanning calorimetry) graphs.

In the case where a shielding pallet is used, the preheat energy is mostly adsorbed in the pallet and not in the board. Boards using wave pallets, especially those with small selective openings, need to have the preheater settings adjusted so that there is sufficient heat into the board to have the required top-side board preheat temperature for wave soldering.

3.1.8 Wave Height

A board must be contacted by the solder wave for a sufficient time to make a good solder joint. The board may not enter the wave at such a wave depth that the solder will flow over the top-side of the board. The wave height should also be kept low to avoid too much solder dross formation. In general, a wave height of 6 mm to 8 mm is an appropriate setting. Lower wave height settings may give movement of the components during wave soldering as the leads may touch the nozzle rim.

A wave setting should be constant within a few tenths of a millimeter. The solder level in the pot should be monitored and corrected as needed. The correct setting of a chip wave depends on the component layout and the board slots or large apertures through which this wave may penetrate to the top side of the board. If only low wave height settings can be used, it may be necessary to reduce the conveyor speed in order to avoid skipped or open solder joints.

The solder wave setting should be such that the top of the solder wave crest is at least half the board thickness or about 1 mm lower than the topside of the board in the conveyor.

For wave height accuracy, although the solder pot is a large and robust unit, the correct wave settings and mechanical adjustment during installation and soldering need a relative high level of precision. The area in contact with the solder wave when a board is soldered can be measured and controlled with a glass test plate, which is run through the wave. This wave area profile should be parallel over the width of the wave.

3.1.9 Wave Contact Length

With a conveyor angle of 7° from the horizontal, the contact length of the main molten wave if a glass plate is used at the same level as the board would be about 25 mm. The real contact of the solder joint in the solder wave depends also on the protruding length of the leads at the solder/bottom side and on the board layout. This real contact length can be twice as much (50 mm). This is important to know when one checks the actual dwell time. Some board bending may be allowed during wave soldering. Boards can often be kept sufficiently flat during wave soldering with a board support (wire or skate) or with a wave pallet.

The dwell time in combination with the solder temperature is a process parameter that must be sufficient to give a sound joint, provided that the joint design is good. A longer dwell time will increase the thermal energy and may also assist in better solder hole-filling. A higher solder temperature will, however, provide a better solution to keeping the solder in the joint liquid. Then again, longer dwell times and higher solder temperatures will exhaust the flux more rapidly. This can give an adverse effect on the solderability/wettability. In addition component and board temperatures may be exceeded and copper barrel dissolution may occur.

The accessibility of the joint area for the solder is restricted by the use of the shielding wave pallet. Due to the natural shape that the solder will have when flowing in a nonwettable small gap, the solder has a tendency to repel or withdraw from gaps that are too small for the solder to access the joints. Also due to the curved shape of the solder surface, the edges of a gap will not be filled with solder unless a very high wave soldering pressure is used. Even when these gaps are tapered with a wedge-shaped aperture entry that is absolutely necessary, the solder has no optimal access to the joints inside such apertures. As a result of this mechanism, the contact time with the solder joints inside such a gap will be far shorter than the theoretical calculated contact time. So the heat transfer from the solder wave will be far less or even insufficient to make a sound solder joint.

3.1.10 Wave Machine Conveyor Speed

A general conveyor speed setting is in the range of 1 to 1.5 m/min (3.3–5 ft/min). The conveyor speed setting also depends on the type of board. Single-sided boards can often be soldered with a high conveyor speed because they often have a low thermal demand for the solder joint formation. Then again, a multilayer thick board may have such a high thermal demand for joint formation that a relatively slow speed such as 1 m/min (3.3 ft/min) would be too fast. Also the layout of the joints at the bottom solder side can be an important factor for the speed setting in order to prevent solder bridging.

As already indicated, the conveyor speed determines the contact time of the wave with the board, which determines how much wave hole-fill will occur. For tin-lead soldering the typical conveyor speeds are 0.6 to 1.5 m/min (2–5 ft/min), whereas for lead-free soldering the typical conveyor speeds are 0.6 to 1.2 m/min (2–4 ft/min). Slightly slower speeds may be used for lead-free wave soldering to help improve lead-free wave-soldering hole-fill.

3.2 SOLDERING TEMPERATURES ON THE CHIP AND MAIN SOLDERING WAVES

The unit where the actual soldering is performed is the solder pot with its solder wave(s). The solder pot contains the solder nozzle(s) and the pump(s). The flow of the solder is guided to a separate compartment where most of the dross that will be created during soldering will be collected and separated from the solder flow. This separated dross needs to be regularly removed to prevent the pickup of dross particles in the solder flow that may clog the chip wave and then go onto the soldered boards.

The solder pot temperature setting depends on the type of solder that is used and also the product to be soldered. In general, low temperature settings are recommended as these low temperatures will create less dross, but in most cases low temperatures can extend the lifetime of the flux so that it has a better flux activity. For tin-lead solders, 245°C to 250°C is a common pot temperature setting. If a thicker 125 mil (3.2 mm) board is used, the wave setting should be around 255°C to 260°C for tin-lead wave soldering. For SnAgCu based alloys, 260°C to 270°C is the typical pot temperature setting.

Although much can be kept under control during the wave-soldering process, there are details in the process that are beyond the process control. One such example is the wave flow dynamics when a board is soldered.

Depending on the width of the board in relation to the wave nozzle width, there will be a distortion of the flow of the wave when a board is entering the wave for soldering. Before the board comes into contact with the wave, the solder that is pumped up has a free flow over the front rim of the nozzle. As soon as a board enters the wave part the free flow is limited. Depending on the board width and the soldering depth, this will affect the wave dynamics. As the back end of the board departs from the wave, the original free flow situation of the wave will return. As the same amount of solder must still pass the nozzle over a partly blocked passage, the speed of the solder flow must inevitably increase. As a reaction the wave height increases to create the extra pressure that is needed for the higher flow rate.

Wave soldering is often done with a combination of waves. A chip wave is mostly used as the first wave to enable wetting of solder joints on chip components. This dynamic wave is necessary to get the solder to the chip component solder pads that are often surrounded by nonwettable component bodies. However, all dynamic waves should not flood the board with solder.

The chip wave has good wetting dynamics over a wide range of wave height settings. These dynamic chip waves are for most applications not suited for the smooth solder drainage that is preferred for the removal of solder bridging. That is the reason that after a dynamic chip wave a smooth main wave is used. To produce a smooth and clean wave surface during soldering, the main wave backside overflow can be optimized with an adjustable backplate.

The use of a dynamic chip wave and a second smooth main wave gives the best process flexibility and compromise. If one needs a variety of different wave height settings when soldering different boards with different demands, then this is the best option to use.

As there is a gap between these waves in that area, the board gets a thermal temperature dip. To reduce that dip, the gap between the waves is kept as small as possible.

Another consideration for a double wave system is that it might consume too much flux activity in the chip wave, leaving insufficient activity for the flux when the board leaves the main wave. This can create serious drawbacks in solder bridging. However, using the right type of flux will not cause a problem. Use of nitrogen over the wave may also be a good way to reduce solder bridging.

The maximum temperature that is allowed and the time the component would resist this temperature during soldering can be found in the component supplier data sheet. Reduction of the thermal load during wave soldering can often be accomplished by increasing the distance between the solder joint and the component body based on the design of the board. The use of temporary heat sink jigs during the soldering process may also give a solution, but in general, this is not recommended because of the extra handling and logistics of use.

When wave pallets are used, a higher solder wave pressure is needed, thus higher wave settings, to get the solder to the joints. As a result more dross will be generated. To reduce the dross, often a lower solder pot temperature is used. In the case of joints with poor thermal design, with holes ending in copper-clad (heat sinks) without thermal dissipation (thermal vias), a high solder pot temperature is needed to assist the solder flow as much as possible, so a compromise should be reached in terms of pot temperature.

The reason for having an 800-kg (1600-lb) volume solder bath is to have a more stable solder wave contact time with products as well as to have a more stable soldering temperature for the board being wave soldered. As soon as the board leaves the solder wave the solder joints will cool rapidly.

The lead at the component side, connected to the component body, has a lower temperature than the solder in the joint as the component lead has good thermal conductivity. It will produce a rapid reduction of the joint temperature. The copper tracks connected to solder pads at the solder side will also be relatively cool and will absorb the heat from the solder joint, reducing the joint temperature. These mechanisms result in a fast temperature drop (10–15°C/s) during the first few seconds after leaving the wave. This means that before external cooling can effectively be applied without affecting the solder wave, the solder in the joints have already solidified. Cooling after soldering might, however, be effective to reduce the temperature rise inside some larger components. Cooling can also allow the boards to be handled directly from the machine when lower pot temperatures are used.

All boards will bend during preheating and wave soldering. Differences in temperature between the top side and the bottom side of the board will result in different thermal expansions that will be responsible for the bending of the board. Also the weight of the board and the position of heavy components on the board will affect the bending of the board. The board must be fixed during transport in the wave machine. If finger transport is used, this system will create a force at the sides of the board that may affect the final bending of the board during the process.

Only a slight amount of bending can be allowed in wave soldering, since the solder wave should not flood the board when the board is entering the wave. To reduce board bending, board supports can be used. If such a support is necessary, the board design should allow space for the board support so that no joints will be obscured or hindered by the support system during soldering.

If solder joints are bridging after leaving the solder wave, such solder bridging is often always at the same board area or component. This is typically indicating a design layout effect: since all other joints are good, the process is relatively stable. For this reason selective debridging systems have been designed. The use of these systems is, however, limited to those situations where the solder is still in a liquid state at the moment that the joints are passing the nozzle. If the solder on the joints is already solidified, the system is not able to remove the solder bridge.

The solidification speed of a solder joint is dependent on the cooling speed of that joint. This is related to the joint design, such as board layout, lead length between solder joint and component body, and component mass. This cooling speed is therefore not the same for all solder joints.

With a temperature measuring device, a small thermocouple can be used on the joint and connected to a data recorder to show how long the solder at the joints involved will be above its solidification temperature after departing the wave. Both the wave departure point and the solder solidification temperature can be determined from the graphical presentation of the data. At the wave departure the temperature profile will show a steep drop to solidification.

3.2.1 Wetting during Wave Soldering and Wave Capillary Action

When a liquid comes into contact with a surface, there will be an interaction between the surface tension of the liquid and the surface that is in contact with that liquid. The interfacial tensions are related to the surface tensions of the materials involved.

During the soldering process we have only a limited amount of time to create the solder joint. Usually the dwell times in the wave are around 3 to 5 seconds. Too long a time can overheat and damage the components or the board, and can cause copper dissolution. Often the flux cannot withstand a long dwell time either.

Therefore the solder must be applied quite quickly to wet the joint surfaces. The solderability of the surfaces should be such that it is possible for the solder to fill the solder joint within the allowable time.

The thermal soldering property plays an important part in the design of the joint. Difficulties in soldering can be expected at multilayer barrels with connected inner layers or ground planes that act as a heat drain or heat sink. This may hamper solder flow because copper is a very good thermal conductor, causing a fast temperature drop of the solder as it tries to fill the through-hole solder joint, as can especially be the case for lead-free wave soldering.

During the wetting process of solder on the wettable metal an intermetallic layer will be formed. Such a layer can have different structures, depending on the bonding reaction(s). When a tin-lead or lead-free SnAgCu or SnCu solder wets to a gold-plated surface, intermetallic compounds such as $AuSn_4$ can be formed. On copper there can be the formation of Cu_3Sn intermetallic compounds directly at the copper surface and Cu_6Sn_5 intermetallic compounds as a layer between the Cu_3Sn and the solder. On nickel there can be the formation of Ni_3Sn_4 intermetallic compound. For lead-free SnAgCu or SnCu solders on nickel, there can also be the formation of $(Cu,Ni)_6Sn_5$ intermetallic compound.

So during soldering there is a direct exchange between tin and the wettable metal. As a result parts of the dissolved metal are able to go into the solder pot. At low levels these additions are completely soluble in the solder and will not affect the solder significantly. However, if too much metal is dissolved, separate needle shaped crystals will be formed, such as Cu_6Sn_5, that can affect the viscosity of the solder. The melting behavior of the solder is affected, since the alloy composition is affected too.

Normally these additions come to equilibrium because the solder is dragged out during wave soldering and is replaced by fresh solder during the topping up of the solder pot. When tin-lead solder is used in wave soldering, this equilibrium typically has no affect on the normal processing conditions for the process or the solder joint reliability and needs no special control.

However, when a lead-free SnAgCu- or SnCu-based solder alloy is used, where the copper is part of the wave solder alloying elements, a change in copper content based on dissolution of copper pads or leads can have a more relevant effect on the solder alloy pot composition. So the solder should be regularly checked. If copper additions increase significantly, this may increase the melting temperature of the solder in the wave solder pot.

Often the copper level in the lead-free SnAgCu or SnCu alloy in the solder pot can be kept in control, once the drift rate is known, by mixing the standard alloy solder bars with solder bar alloys such as Sn3Ag for a Sn3Ag0.5Cu solder pot.

The viscosity of molten solder is only slightly dependent on temperature. When the solder is molten, it has a high fluidity. However, if the solder has impurities, the viscosity may increase and the solder may behave more sluggishly. This again can affect the solder drainage conditions from the joints during departure from the wave and can result in unexpected solder bridging in between joints. Some of the elements that can be of concern in a lead-free wave soldering bath include copper, iron, gold, nickel, and palladium.

Besides the wetting that is necessary for solder joint formation, there are capillary forces to consider. Capillary forces give the liquid the tendency to stay in the small gaps. Although the construction of a joint between the lead and board does not look like a common capillary, it is nevertheless present between the lead and hole barrel.

The amount of liquid that will stay in a capillary depends on the surface tension of the liquid. This also depends on the shape and dimensions and position of the capillary, the density and viscosity of the molten liquid solder, and the specific gravity.

Capillary forces make it possible for a liquid to hold onto wettable parts, even when the liquid source is removed. At first the liquid solder contacts the joint area, and if that area is wetted by the solder and the board separates from the wave, the solder stays in the joint and solidifies.

Obviously the dimensions of the parts to be joined have an effect on the amount of solder that will be left after soldering. Especially on joints with thin leads this can be clearly observed. The reason is that the lead surface that is related to the lead diameter is too small to hold more solder. A longer lead will not help because the lead alone does not have enough capillary action. The capillary action that will hold the solder is just between the pad and the lead (edge).

Then again, there is a limitation to the amount of solder on larger joints. Due to the surface tension and gravity the excess solder will drain off but only if the capillary action allows that to happen.

The surface tension acts between the surfaces and it is a combination of the forces acting on the solder and the surface that must be soldered that determines how well the solder will spread out. In combination with the capillary force, these forces determine how much solder will stay at the joint.

3.3 ALLOYS FOR LEAD-FREE WAVE SOLDERING

There are two main types of lead-free wave-soldering alloys which are SnAgCu and SnCu based.

The SnAgCu-based alloys have compositions of Sn3–4Ag0.5Cu. They cost more than SnCu-based alloys because of their higher silver (Ag) content. At a melting point of 217°C, the typical wave soldering pot temperatures used are around 265°C to

270°C, with 4 to 8 seconds dwell time in the wave. When these alloys are used in wave soldering, there must be more consideration given to component and board temperature issues at the high pot temperatures used. Sn37Pb has a melting point of 183°C, and a typical pot temperature of 250°C to 260°C with contact time in the wave of 4 to 8 seconds as a comparison.

The SnCu-based alloys have compositions with little or no silver additions (0 to 1 wt%Ag). These compositions include Sn0.7Cu, Sn0.7CuNi, Sn0.3Ag0.7Cu, Sn0.3Ag0.7CuNi, and Sn1Ag0.7Cu. The lower silver amount lowers solder cost, but the melting point of these alloys increases into the 227°C melting point range. Alloys such as Sn1Ag0.7Cu and Sn0.3Ag0.7Cu are general wave solder alloys; alloys such as Sn0.3Ag0.7Cu0.03Ni also have the benefit of helping reduce copper dissolution during lead-free wave soldering. Wave soldering pot temperatures for these SnCu-based alloys are slightly higher than SnAgCu-based alloys, with typical pot temperatures of 270°C and 4 to 8 seconds dwell time in the wave.

If we consider the delta T between solder melting point and pot temperature for the three alloys, SnPb solder has a delta T of 67°C that is greater than SnAgCu, which has a delta T of 58°C which is greater than SnCu, which has a delta T of 53°C. Based on these facts there should be better flow and wave hole-fill for SnPb compared with SnAgCu compared with SnCu wave solder alloys. If we used the example of a 100-mil (2.5 mm) thick board with a 260°C pot temperature for all three alloys, the SnPb solder would typically give 100% hole-fill, the SnAgCu solder would give around 75% hole-fill, and the SnCu solder would give around 65% hole-fill.

3.4 FUNCTION OF NITROGEN IN WAVE SOLDERING

The use of nitrogen can extend and support the flux activity during the separation of the board from the solder wave. During this separation process the solder should stay at the joint location but not in between the joints. Solder oxide formation at this stage increases the risk of solder bridging considerably. That is why the flux must have sufficient flux activity left to reduce these solder oxides and prevent solder bridging. The application of nitrogen in the soldering area can displace the air (oxygen) at that region and so assist in better drainage conditions. Nitrogen can also help improve hole-fill by maintaining flux activity and preventing reduced solderability of the board and component surface finishes.

The nitrogen should be applied at the area where the solder wave separates from the board. The amount applied and the necessary purity of the nitrogen depend on the need. In most cases one can solder well without the use of nitrogen. However, the use of nitrogen will increase the process window especially during lead-free wave soldering.

There are issues to using OSP (organic solderability preservative)—and potentially immersion silver or immersion tin coated boards—if nitrogen is not used as the solderability of double-sided SMT boards or boards to be wave soldered can be an issue (based on the degradation of the board surface finish). With OSP, the coating thickness reduces with each reflow cycle. The underlying copper layer becomes more oxidized making it more difficult to solder. The nitrogen atmosphere can help preserve the OSP coating somewhat.

There are other considerations that help determine whether nitrogen should be used. Nitrogen use may be justified for larger complex boards with a need for an increased process window. There should also be a cost/usage comparison of the nitrogen benefit for the SMT and wave. For SMT reflow the amount of nitrogen used would typically be more compared to wave soldering, so the benefit of using nitrogen in wave soldering based on cost would be higher. Whether nitrogen is necessary depends mainly on the product and the flux used. A positive side effect of nitrogen use is that it can reduce solder dross (oxide) formation in the wave machine.

When dross forms, the layer is floating on flowing solder partly underneath in the liquid. It is for that reason that the dross layer should be removed at regular intervals. If these intervals become too long, the dross layer will grow and inevitably dross particles will be picked up by the solder flow and these particles will show up in the solder wave.

Tests were done by exposing board test vehicles to lead-free reflow profiles under controlled atmospheric conditions, followed by the secondary wave soldering process where through-hole components were soldered. Lead-free wave through-hole penetration was measured and used to determine the impact of atmosphere on joint formation. As already mentioned, some of the common problems associated with the use of OSP board finishes are related to reduced solderability. Work quantified the relationship between the OSP thickness and the through-hole penetration observed.

Figures 3.1 and 3.2 show the percentage loss of OSP thickness when subjected to the full tunnel (preheat and reflow zones) atmosphere and reflow zone atmosphere only configurations, respectively. Notice that the use of nitrogen produces less degradation on the thickness of the OSP, resulting in better integrity of the surface finish during the manufacturing process, whereas air produces more degradation. Nitrogen with different oxygen levels does not show a significant difference in OSP thickness with the full tunnel atmosphere. It was observed that where the nitrogen was applied affected

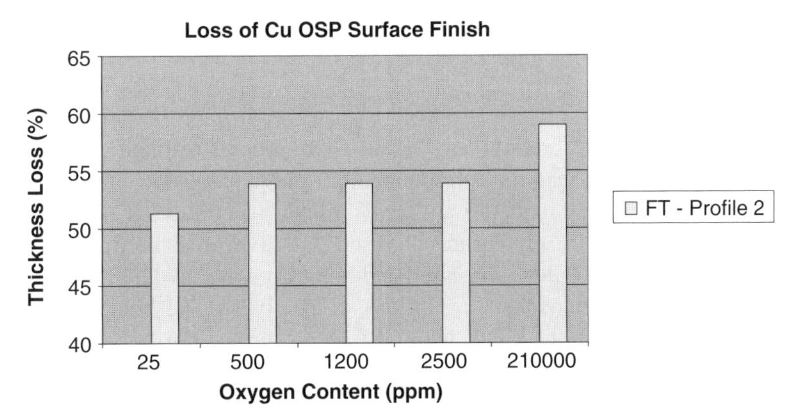

Figure 3.1 Percentage of OSP degradation for "full tunnel" atmosphere reflow processes

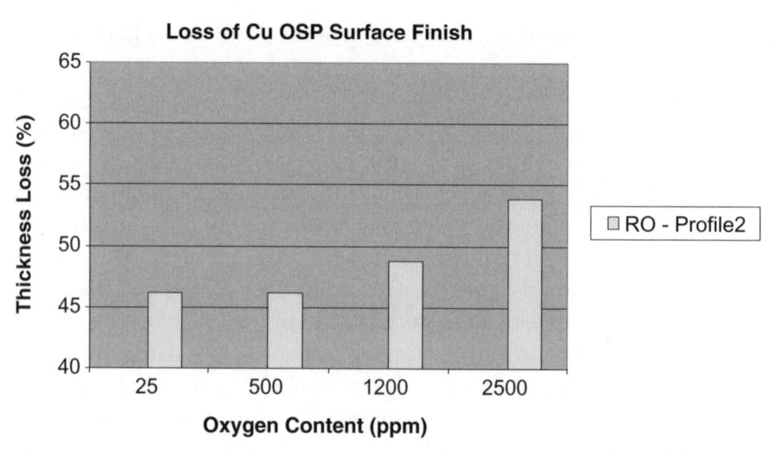

Figure 3.2 Percentage of OSP degradation for "reflow only" atmosphere reflow processes

Figure 3.3 OSP board reflow processed and then wave soldered in nitrogen atmosphere using SnAgCu solder on a DIP component

the OSP thickness result. A "reflow only" atmosphere process resulted in less degradation in the OSP thickness than the full tunnel atmosphere process.

This phenomenon may explain why poor through penetration of complex boards occurs when assembled in air. Figures 3.3 and 3.4 give cross-sectional images of DIP connectors that were reflow processed and SnAgCu wave soldered under nitrogen (2500 ppm O_2) and air atmosphere conditions, respectively. The board thickness for this illustration was 125 mil (3.2 mm) with OSP board surface finish, and Sn3Ag0.5Cu alloy was used in the wave-soldering process. Poor through-hole penetration was observed for 125-mil (3.2 mm) thick boards when they were processed in the air environment.

Figure 3.4 OSP board reflow processed and then wave soldered in air atmosphere using SnAgCu solder on a DIP component

3.5 EFFECT OF PCB DESIGN ON WAVE SOLDER JOINT FORMATION

The fact that most defects are related to the design layout can simply be proved by comparing the majority of sound joints to those joints that give a defect. One should thereby keep in mind that all these joints are soldered with the same wave machine settings.

Another important aspect to keep in mind is that some joints may give a bi-stable behavior. This means that with the same setting one board may give a perfect result without a specific defect, while at the next board a specific defect appears, but always in the same area.

One could easily draw the conclusion that the process is unstable. This is, however, seldom the case. It is the bi-stable character of such joints that gives a false impression. The main reason for the process instability is the "random" drainage condition at the point where the board separates from the solder wave when there is a concentration of joints. By optimizing the process settings, one often has the possibility to bring the balance in the right direction, so that the process becomes more robust. This is, however, in most cases only possible when a series of boards are soldered under defined test conditions. If only a small number of boards are available for a first development to determine optimum process settings, such process optimization may be too difficult to achieve. The best advice one can give in such situations is to try to optimize the board design layout to enable good wave soldering.

3.5.1 Solder Bridging

In the case of a concentration of joints and/or long protruding leads from the board, the remaining flux that is present on the board during the departure from the wave may not be sufficient to prevent bridging. The use of nitrogen atmosphere, which helps to prevent oxidation, could help reduce bridging. A "drawback" of nitrogen, however, is that it will increase the surface tension of the solder, so more solder will be left on each joint, appearing as a larger fillet. When joints are close together, this might then also give solder bridging. Increasing the space between joints, by reducing the board solder pad size, would give the best solution in this case.

Increasing the flux amount may also help, but this can have a negative effect in the form of skipped solder joints and violate board cleanliness requirements. If the design of neighboring joints is such that the wetting forces between such joints are high, solder bridging can often hardly be avoided. One has to bear in mind that the same force mechanisms that create the solder joints are also acting on the wetting between such joints. In such unfavorable conditions solder bridging is likely to occur.

Some other factors for solder bridging include the pitch of the component and shadowing effects and orientation of components. The optimization of the flux amount and the type of flux used would be important. If too high a preheat temperature and time are used, there may be an issue with VOC-free water-based no-clean fluxes. As increase in the conveyor speed may reduce the bridging.

3.5.2 Large Fillet Formation

Another aspect related to the board design is the difference in the conditions at which the wave will separate from the joints (drainage conditions). If joints are relatively wide apart from each other, the solder will smoothly separate from such joints, giving all the joints the same joint shape (solder cone shape).

If joints are close together such as with a connector, the solder will separate more slowly. This is because the solder has a tendency to hold at such clustered joints due to the concentration of wettable surfaces. When the solder separates from such clusters, it happens stepwise and not in a smooth draining process. This layout related behavior would therefore often result in the formation of a few "large joints" in such areas. The contour of the lead would still be visible so there would be no doubt as to the quality of such joints.

3.5.3 Lead-Free Wave Hole-Fill

Provided that the joints are fluxed well and the solder wave makes contact with the joint for the set period of time, the usual cause for reduced hole-fill is poor thermal solderability. If the "heat sink effect" of the soldered joint is such that the solder will solidify during the penetration of the joint gap, capillary action will stop because the solder freezes. This effect is often intensified by poor surface solderability as well. The filling of the gap between component lead and barrel wall with solder is based on capillary action on the cleaned metal surfaces. Wetting and capillary action occur with molten solder. If solder is solidifying, it is no longer liquid and the wicking process stops.

The thermal layout of the solder joint is related to the component as well as to the board layout. If the component body or the barrel absorbs too much heat during the process, the liquid solder will solidify before filling the hole completely. If the barrel ends in massive copper cladding (heat sink), it will be difficult to get good hole-fill.

To make a reliable solder joint within the specified process settings, both the surface solderability and the thermal solderability must be in accordance with the process.

For some components, such as can capacitors, they should be mounted on a spacer to provide the correct soldering distance. If that distance is not maintained, the lead may transfer so much thermal energy to the component body during the soldering process that the solder will solidify before the joint is completely filled with solder. This effect blocks the flow of the solder even with ideal surface solderability.

For lead-free soldering, the temperature difference (delta *T*) between the melting point of the SnAgCu or SnCu alloy and the pot temperature is smaller than that for tin-lead soldering. So there is more difficulty in getting good hole-fill with these lead-free solders.

Mendez et al. [4] and Boulos et al. [5] conducted work on a 12-layer, 93-mil-thick OSP surface finish board. Both SnAgCu and SnCu alloys were tested using an open wave pallet design by Mendez et al. [4]. SnAgCu and SnPb wave alloys were tested by Boulos et al. [5]. The SnAgCu pot temperature used was 270°C, whereas the SnPb pot temperature used was 250°C [5]. No-clean alcohol-based flux was used with a top-side board preheat temperature which ranged from 80°C to 100°C. A selective wave pallet was used with openings for through-hole components [5]. SMT pad to PTH pad spacings were also tested in the evaluation.

In evaluations using a 40-pin power connector that was run parallel to the direction of the conveyor (perpendicular to the wave solder) where the pin diameter was 25-mil (0.6 mm) with a pitch of 100 mils (2.5 mm), the pin-to-hole area ratio of 0.25 gave the best hole-fill result where the pin-to-hole area ratio was defined as the component pin cross-sectional area divided by the through-hole barrel cross-sectional area. Pin-to-hole area ratios of 0.51 gave the worst hole-fill result. Over 0.35, the pin-to-hole area ratio (42 mil [1.05 mm]) hole) gave significantly worse hole-fill results especially for SnAgCu. It was also found that increasing the number of ground planes reduced the hole-fill. More hole-fill defects were observed with SnAgCu versus SnPb for all pin-to-hole area ratios evaluated.

For DIMM168 connectors that were run perpendicular to the direction of the conveyor (parallel to the wave solder) with a 50-mil (1.27 mm) pitch and 30-mil (0.75 mm) pin diameter, increasing the number of ground planes gave increased hole-fill issues:

- 4 ground planes directly connected to the through-hole barrel had very poor hole fill results.
- 4 spoke thermal relief connections had better hole-fill results than a direct solid ground connection.

Boulos et al. [5] found that hole-fill for SnAgCu was worse than for SnPb. Hole-fill for both SnAgCu and SnPb improved with increased SMT to PTH pad to pad spacing when the selective wave pallet was used. For SnPb wave soldering, previously a minimum 200-mil (5.1 mm) gap between the edge of the nearest SMT component and the edge of PTH barrels was used with a selective wave pallet that ensured good solder flow and hole-fill. Based on the test results this reduced to a 150-mils(3.8 mm) gap [5]. For SnAgCu wave soldering, previously a minimum 250-mil (6.4 mm) gap between the edge of the nearest SMT component and the edge of PTH barrels was used with a selective wave pallet that ensured good solder flow and hole-fill. Based on the test results the 250-mil (6.4 mm) gap was the minimum gap required for SnAgCu wave soldering. It was recommended that the gap be larger than 250-mils(6.4 mm) for SnAgCu wave soldering when possible.

For bottom-side 1608 [0603 imperial] and 2012 [0805 imperial] chip components that were run perpendicular to the direction of the conveyor (parallel to the wave solder) [4], pad to pad spacings of 20-mils(0.5 mm) or 25-mils(0.62 mm) increased the number of visual defects during lead-free wave soldering for these components especially for the 2012 [0805 imperial] components. A pad to pad spacing greater than 25-mils (0.62 mm) was recommended.

Marquez de Tino et al. [6] conducted work on 93-mil (2.3 mm) and 125-mil (3.2 mm) thick OSP surface finish test boards with SnAgCu solder. VOC-free and VOC-containing wave fluxes were used. By using a pin connector component with a pin diameter of 25 mils(0.62 mm), the best lead-free hole-fill results were obtained with a 70-mil (1.75 mm) pad diameter and a maximum of two connected copper barrel layers during wave soldering of the component, which was run parallel to the direction of the conveyor (perpendicular to the wave solder). Pin-to-hole area ratios of 0.51 to 0.34 could be used but not a pin-to-hole area ratio of 0.28 on the 93-mil (2.3 mm) thick boards. For the 125-mil (3.2 mm) thick board, a pin-to-hole area ratio of 0.51 was preferred.

By evaluating an axial resistor on the same board with a pin diameter of 22-mils(0.55 mm), the best hole-fill results were obtained with a maximum of one connected copper barrel layer during lead-free wave soldering, which was run perpendicular to the direction of the conveyor (parallel to the wave solder). A pin-to-hole area ratio of 0.32 could be used but not 0.40. For a DIP16 component on the board with a rectangular cross-sectional pin of 20 mils (0.5 mm) × 11 mils(0.27 mm), lead-free wave soldering on 93-mil (2.3 mm) thick boards was acceptable in terms of hole-fill but unacceptable on 125-mil (3.2 mm) thick boards because there was no component lead pin protrusion from the bottom side of the board.

Hubbard [7] conducted tests with SnAgCu and SnPb solder on 93-mil (2.3 mm) and 125-mil (3.2 mm) thick OSP surface finish 12-layer boards. All pin-through-hole locations were connected to the planes by 4-spoke thermal relief connections with 10-mil (0.25 mm) spoke widths using one of the following four types of copper plane configurations:

A: 3 top planes connected (layers 2, 4, 6)

B: 3 bottom planes connected (layers 7, 9, 11)

C: 6 planes connected (layers 2, 4, 6, 7, 9, 11)

D: 1 single plane connection (layer 2)

Evaluations were done with selective pallets and without wave pallets. Two component types were used. The first was a 24-mil (0.6 mm) lead diameter 2 pin electrolytic capacitor with 5 mm (200-mil) pitch. The pin-to-hole area ratios varied from 0.38 to 0.08 for this component. The second component was a 31-mil (0.77 mm) lead diameter 2 pin electrolytic capacitor with 7.5 mm (300-mil) pitch. Pin-to-hole area ratios varied from 0.31 to 0.11.

For SnPb (93-mil [2.3 mm] and 125-mil [3.2 mm] thick boards) and SnAgCu (93-mil [2.3 mm] thick board), wave soldering, as the pin-to-hole area ratio was reduced (hole size increased), wave solder hole-fill improved for both component lead diameters.

For SnAgCu (125-mil [3.2 mm] thick board) on the 24-mil (0.6 mm) lead diameter component, as the pin-to-hole area ratio was reduced from 0.38 to 0.14, the hole-fill improved but not to a significant extent. For SnAgCu (125-mil [3.2 mm] thick board) on the 31-mil (0.77 mm) lead diameter component, as the pin-to-hole area ratio was reduced from 0.31 to 0.17, the hole-fill improved, but again not to a significant extent.

For SnPb (93 mil [2.3 mm] and 125 mil [3.2 mm] thick boards) and SnAgCu (93 mil [2.3 mm] thick board), hole-fill generally improved from copper plane connection C (6 copper planes connected [top and bottom]) to A (3 top planes connected) to B (3 bottom planes connected) to D (single plane connected).

For SnAgCu (125-mil [3.2 mm] thick board), hole-fill generally improved from copper plane connection C (6 copper planes connected [top and bottom]) to B (3 bottom planes connected) to A (top planes connected) to D (single plane connected).

The improvement in hole-fill for copper planes connected to the top planes versus the bottom planes for the 125-mil [3.2 mm] thick SnAgCu soldered board compared with the results using SnPb (93-mil [2.3 mm] and 125-mil [3.2 mm] thick boards) and 93-mil (2.3 mm) thick SnAgCu soldered boards indicated the hole-fill reduction that can occur during lead-free soldering on thermally challenging boards. So there is need to consider and investigate more carefully optimization of board design, wave fluxes, heat transfer into the board, and the equipment and process.

Hole-fill was generally reduced with the use of the selective pallet versus no wave pallet for both tin-lead and lead-free wave soldering. The hole-fill results were reduced more with lead-free wave soldering than tin-lead soldering with the selective pallet. The pallet reduces the amount of heat transfer into the board. This was particularly critical for lead-free soldering as it used a relatively small difference between pot temperature and the melting temperature compared with that for tin-lead soldering.

For lead-free SnAgCu or SnCu wave soldering, typically water based fluxes could be used for less than around 100-mil (2.5 mm) thick OSP coated boards. For greater than around 100-mil (2.5 mm) thick OSP coated boards, it would be advisable to use alcohol based fluxes with SnAgCu solder, but there could still be hole-fill issues on difficult multiple thermal planes or large thermal mass components.

Some of the considerations when having hole-fill issues during lead-free wave soldering would be to assess the wave preheat: too high a preheat may not be good for water-based no-clean fluxes compared with alcohol-based fluxes. Solder pot temperature could be considered. Too high a pot temperature can cause component or board temperature issues or copper dissolution. For wave contact time, a contact time that is too high can cause component or board temperature issues or copper dissolution. Other factors affecting hole-fill are board thickness (number, type, and distribution of copper layers) and board design (pin-to-hole area ratio, pin shape, board and component surface finish, and flux type including the use of alcohol versus water-based fluxes and their solids content). The type of equipment used would play a part, including wave machine, preheater, and spray fluxer type.

3.5.4 Open/Skipped Solder Joints

Open or skipped solder joints during wave soldering can be caused by board pads that are too small in relation to the component dimensions. Another cause may be that small components are positioned too close to larger components, where the latter is shadowing the smaller one, so the wave cannot get into contact with the solder pads.

There is further the possibility of too much flux collected under or in between the components. The flux will then evaporate faster as the wave touches the component. This will cool the wave surface and may form a bubble that prevents the solder wave from contacting the joint area.

A limitation of the dynamic chip wave height setting, caused by the risk of solder overflow in gaps or holes in the board, will reduce the wetting capacity of the waves and therefore increase the risk of skipped solder joints. A lower conveyor speed setting may be a solution for some of these components, but in general, the best solution would be to improve the board design pad layout.

One should also keep in mind that the general layout rules given by the designer of the components are typically valid only for the types of components used. As soon as components with different dimensions are mixed, the general board layout rules may not apply. In the case where components with different dimensions are placed next to each other, one should always try to follow the rules on keep-outspacing between components vaild for the largest component used.

Some of the things to check for when observing open/skipped solder joints are insufficient solder or flux during soldering, the design of the board, shadowing, orientation of components, incorrect time in the wave and wave temperature. Additionally the conveyor speed has the potential to increase solder skips. Incorrect wave pallet design may create non-optimal clearance of the pallet around the through-hole component parts.

3.5.5 Solder Balls

The formation of solder balls in a wave soldering process is unavoidable because it is part of the physical behavior of the molten solder separating from the formed solder joint. Normally these molten solder droplets formed during separation will fall back into the solder wave, unless there is a strong "adhesive" bond between the solder resist/mask and the solder. In that case the solder droplets may remain on the board surface, often in a reproducible pattern, between individual adjacent solder joints.

Solder balls adhering to the board surface depend on the interactions among between the solder resist(mask), the flux, and the solder. Some of the considerations when solder balls are observed include solder mask type (shiny versus matte). Typically a glossy solder mask gives more solder balls than a matte mask because of the smaller contact area on matte mask for the solder balls to adhere to. The flux type also typically creates fewer solder balls with alcohol-based fluxes, as compared to water-based no-clean fluxes. Increased conveyor speed can help to reduce solder balls with alcohol-based no-clean fluxes.

3.6 STANDARDS RELATED TO WAVE SOLDERING

There are various standards used for wave soldering. They are gradually being updated to reflect lead-free soldering developments.

JEDEC JESD22-B106-D standard [8] gives an overview of the temperature rating for the wave soldered through-hole components. For SnPb wave soldering, a component should be rated to a wave soldering temperature of 260°C with a 10 second wave-soldering dwell time. For SnAgCu soldering, a component should be rated to a wave-soldering temperature of 270°C with a 7 second wave-soldering dwell time.

JEDEC JESD22-A111 standard [9] gives an overview of the temperature that a wave-soldered immersed bottom side small body SMT component would be rated to. Currently it indicates a wave pot temperature of 260°C for 10 seconds but it does not take into account the typical temperatures used in lead-free wave soldering which is an issue.

J-STD-075 standard [10] covers identification of both wave and SMT components. It gives an assembly classification for chip components, through-hole connectors, aluminum capacitors, crystals, oscillators, fuses, LEDs, relays, inductors and other non-IC devices. For wave soldering, it provides classification of the non-IC wave-soldered components by identifications ranging from W0 to W9. W0 is determined not to be process sensitive, whereas W1 is sand to be able to withstand 2 × (2 times) 275°C wave pot temperature. A W4 component is classified as able to withstand 2 × (2 times) 260°C wave pot temperature. Additional labeling can be used to indicate wave dwell time limitations and other process-related issues.

In terms of the wave hole-fill, the IPC-A-610 standard [2] indicates 75% hole-fill is acceptable for non–ground plane connections and 50% for multiple ground plane connections—according to IPC Class 2 requirements. The challenge is that thicker boards especially those soldered with lead-free wave solder will have more difficulty obtaining this amount of hole-fill especially when there are multiple ground plane connections and there are limitations on the pot temperatures that can be used due to current component and board temperature rating limits.

Telcordia/ Bellcore GR-78 standard [11] has been updated to allow the 47-mils (1.2-mm) pin wetted length on PTH (pin-through-hole) components with ground or power plane connections. The IPC-A-610DC Telecom Addendum Standard [12] has also adopted these criteria. The IPC-A-610DC standard requires that where large areas of copper or other heat sinks are connected to PTHs in PCBs resulting in insufficient vertical solder fill hole-fill be sufficient to ensure that the minimum pin-wetted length within the barrel is at least 0.047-inch (47 mils [1.2 mm]) regardless of board thickness.

3.7 CONCLUSIONS

Various aspects of wave soldering were reviewed in this chapter, including the need to optimize wave height, wave contact time and length. Fluxes used and preheater types also play an important role in wave soldering. Solder pot temperatures and times for lead-free wave alloys, especially for SnAgCu and SnCu were also discussed compared with tin-lead soldering. Board design was found to be a factor in reducing bridging, solder balls, opens and improving hole-fill. Nitrogen was also found to benefit the wave-soldering process, not only during wave soldering but in maintaining the solderability of board finishes such as OSP after reflow, especially during wave soldering of thicker boards with lead-free wave solder.

For lead-free wave soldering, all process parameters must be balanced. For example, sufficient hole-fill and few solder skips/opens usually require lower conveyor speeds and reduced solder balls and solder bridging usually require increased conveyor speeds.

For hole-fill on thick boards (>100-mils [2.5 mm]) lead-free wave soldering is more of a challenge. There are various ways to improve the process window, including the development and use of alcohol-based fluxes. Standards were reviewed that would help reduce the hole-fill requirement and increase board and component temperature ratings which would be complemented by component and board design optimization. Consideration was also needed for potential changes in board surface finish. Using a combination of the mentioned factors would help reduce the difficulty of lead-free wave soldering for thicker boards but challenges still remain. Development of the correct pin-to-hole area ratios to use for lead-free wave soldering, especially for thicker boards, is still a work in progress.

The importance of optimizing all parts of the wave process was stressed, and this includes the equipment used such as the spray fluxer, preheater, and wave machine.

3.8 FUTURE WORK

Standards relevant to lead-free wave soldering are still being developed in terms of component and board temperature ratings and hole-fill requirements. More work is needed in this area. Data needs to be collected and provided to the standards group on temperatures needed for lead-free wave soldering on complex and large thermally challenging boards. Hole-fill standards for lead-free wave soldering also need further work.

More flux development for lead-free wave soldering, especially for thicker boards, and optimization of board designs should help improve lead-free wave hole-fill.

ACKNOWLEDGMENTS

The authors would like to thank the persons and companies who contributed lead-free development papers and studies that were reviewed and discussed in this chapter.

REFERENCES

1. IPC J-STD-001 Standard, *Requirements for Soldered Electrical and Electronic Assemblies*, 2005.
2. IPC-A-610 Standard, *Acceptability of Electronic Assemblies*, 2005.

3. IEC 61249-2-21 ED. 1.0 B: 2003 Standard, *Materials for Printed Boards and Other Interconnecting Structures. Part 2–21: Reinforced Base Materials, Clad and Unclad—Non-Halogenated Epoxide Woven E-Glass Reinforced Laminated Sheets of Defined Flammability (Vertical Burning Test), Copper Clad*, 2003.
4. R. Mendez, M. Moreno, G. Soto, J. Herrera and C. Hamilton, "Design for Manufacturability in the Lead-Free Wave Solder Process," *IPC APEX Conf.*, 2008.
5. M. Boulos, C. Hamilton, M. Moreno, R. Mendez, G. Soto and J. Herrera, "Selective Wave Soldering DOE to Develop DFM Guidelines for Lead and Lead-Free Assemblies," *SMTAI conf.*, 2008.
6. U. Marquez de Tino, D. Barbini and W. Enroth, "Impact of Soldering Atmosphere on Solder Joint Formation", *Pan Pacific Conf.*, 2008.
7. K. Hubbard, "A Comparative Analysis of PTH Holefill Performance between SnPb and Pb-Free Using Design of Experiments on Complex PCBAs," *SMTAI Conf.*, 2008.
8. JEDEC JESD22-B106D Standard, *Resistance to Solder Shock for Through-Hole Mounted Devices*, 2008.
9. JEDEC JESD22-A111 Standard, *Evaluation Procedure for Determining Capability to Bottom Side Board Attach by Full Body Solder Immersion of Small Surface Mount Solid State Devices*, 2004.
10. ECA/IPC/JEDEC J-STD-075 Standard, *Classification of Non-IC Electronic Components for Assembly Processes*, 2008.
11. Telecordia/Bellcore GR-78 Standard, *Generic Requirements for the Physical Design and Manufacture of Telecommunications Products and Equipment*, Issue 2, 2007.
12. IPC-A-610DC, *Telecom Addendum Standard*, 2009.

4

LEAD-FREE REWORK

Alan Donaldson (Intel)

4.1 INTRODUCTION

Lead-free rework has many different aspects that need to be examined because of the complexity involved in assembling product boards. The operation of reworking a defective product is one that requires a great deal of skill. The goal is always to get the board back to a state as good as when it was first assembled. Training and practice is an important part of any rework program. Good operators usually have been reworking circuit boards for a number of years and have used scrap boards to practice and develop their rework skills.

With any rework process, safety should be the number one priority. Safety glasses should always be worn unless working under a microscope. When reworking any board with a battery, it should be removed prior to any heat being applied to the board as the battery can explode when heated. Through-hole electrolytic capacitors should never have tweezer tip soldering irons touch both leads at the same time also because of the explosion potential. A solder extraction tool should be used to clear one barrel at a time to avoid an explosion. Hot air rework near electrolytic capacitors can carry the same risks, so they should be shielded with heat-resistant tape or be removed.

Lead-Free Solder Process Development, Edited by Gregory Henshall, Jasbir Bath, and Carol A. Handwerker

4.2 SURFACE MOUNT TECHNOLOGY (SMT) HAND SOLDERING/TOUCH-UP

4.2.1 Passive Component Rework

Passive components include components such as capacitors, resistors, inductors, transistors, diodes, and LEDs (light-emitting diodes). Passive component rework can entail adding solder to an empty pad (that was skipped) or the removal and replacement of a component. It can also include adding solder to an insufficient joint, removing excessive solder, or simply reflowing a cold, disturbed, dewetted or nonwetted solder joint. Adding solder to a passive component should be done with the lowest temperature soldering tip that will allow proper solder flow. When too high a temperature soldering tip is used, it can cause thermal shock to the component and cause micro cracks. Micro cracks that occur within the component may not fail when tested but can fail out in the field over time. Capacitors are extremely susceptible to this phenomenon. When adding the core solder to a passive component, it is best to start on the pad and let the heat transfer to the component to complete the solder joint. Small-diameter core lead-free SnAgCu solder from 10 to 20 mil (0.254 mm–0.508 mm) should be used for this operation.

For passive component removal, a soldering iron tip should be used that fits on both ends of the component (i.e., a bifurcated tip). Some soldering equipment allows the use of tweezer type of solder tips to remove the component. After the component has been removed, two methods for pad/land cleaning are available. The first method is to use a soldering iron and solder wick, and the second method is to use vacuum extraction. Soldering iron and/or extractor tips are selected based on land width. Replacing the component is accomplished by pre-filling one land with solder, using a soldering iron and core solder. Both lands are fluxed, and the new component is carefully hand soldered, starting at the preloaded pad, while held with tweezers or a wood stick. The solder iron tip should be in contact with the pad and allow the solder to flow onto the passive component lead termination. The other land is then soldered starting at the pad and allowing the solder to flow on to the component termination [1].

Iron tip temperatures can vary a lot depending on the type of passive component that is being soldered. Typical iron tip soldering temperatures range from 260°C to 427°C (500–800°F). The dwell time should be no more than 3 to 5 seconds spent at each solder joint location.

4.2.2 QFP/Gull Wing Rework

There are four types of gull wing components: small outline integrated circuits (SOIC), small outline transistor (SOT), quad flat packs (QFP), and thin small outline packages (TSOP). The methods described herein are for QFPs.

The two methods for removing QFP components are by a hand solder iron or by hot air convection rework equipment. The more common method is using a hand solder iron that has the exact shaped tip that fits the component package. Typical soldering iron tip temperatures range from 316°C to 427°C (600–800°F).

Figure 4.1 Hand solder iron tip for a QFP package

There are two methods of prepping the component leads before applying the QFP hand-soldering iron tip. The first method is to bridge all of the leads with additional solder using a broad surface soldering iron tip and lead-free solder. The second method is to apply a ring of core solder that is the exact size of the soldering iron tip. This is done by solder tacking the core solder at one end of the component leads and wrapping it completely around the package until tacking it at the last lead. Additionally a tacky flux or liquid flux should be applied to all of the component leads before starting the removal process. Figure 4.1 shows a typical hand solder iron tip for a QFP component package.

Removing these types of component packages with hot air rework machines typically falls into two categories. The first one is a manual hot air removal operation that requires the operator to watch through magnification to see when the solder starts melting and remove the defective package when all the solder on the leads has melted. Temperature ranges are from 230°C to 250°C (446–482°F) and rework times can range from 2 to 8 minutes. It is helpful to apply tacky or liquid flux to the leads, but it is not required.

The second method is through a semi-automatic rework machine that involves creating a removal/replacement heat profile so that the defective package can be removed during this rework cycle. No additional flux needs to be applied to the package. Figure 4.2 shows a typical removal/replacement profile for a QFP component.

After the package is removed from the board, there is non-uniform solder left on the pads that is removed by a hand soldering iron and solder wick. Flux should be applied to the pads to aid in the solder wicking process. This operation requires skill in order to not lift or tear off the board pads, especially since lead-free rework temperatures tend to increase this risk. Typical soldering iron tip temperatures range from 316°C to 427°C (600–800°F) and no more than a few seconds on the pad should be needed.

The process of placing a new component package on the board can be completed by two methods. The first method is to hand solder the new package on the board using a soldering iron with a small solder iron tip and small diameter lead-free SnAgCu cored rework solder wire of 10 to 20 mil (0.254 mm–0.508 mm). Usually some additional flux is applied to help with the soldering process. Typical soldering iron tip temperatures range from 316°C to 427°C (600–800°F).

The procedure starts by soldering one corner lead of the QFP component, making sure that all the other leads are properly aligned on their respective pads. After the first

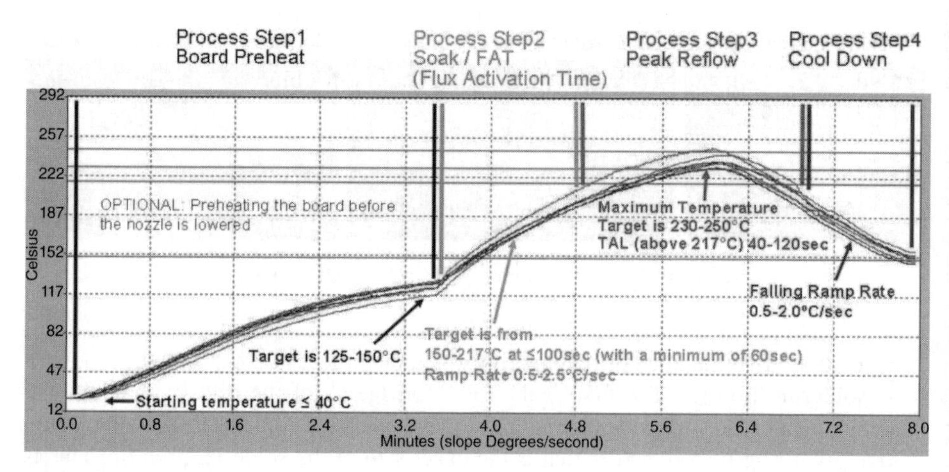

Figure 4.2 Typical lead-free QFP temperature profile ranges using a semi-automatic rework machine [2]

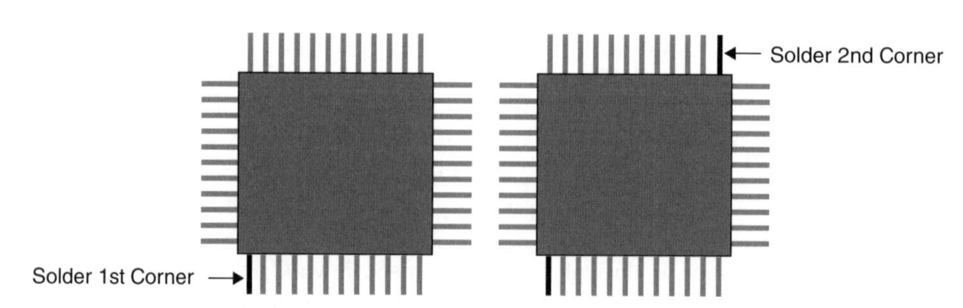

Figure 4.3 QFP manual soldering process for proper alignment to the pads

lead has been soldered the opposite diagonal corner lead is soldered to the pad [1]. Figure 4.3 shows the soldering process for the QFP component leads.

All other leads now can be soldered one at a time. It is typical to use the smallest diameter core solder that allows a good solder joint to form, and the soldering iron should contact the connection for a few seconds for each joint.

The second method is to use a mini-stencil that matches the pad pattern on the board, and then to print the lead-free SnAgCu solder paste through the stencil apertures onto the component board pads. Typically the stencil apertures match the pads at a 1 : 1 ratio and thicknesses can vary from 4 to 8 mils (0.1016 mm–0.2032 mm). Most stencil thicknesses match what is being used in the SMT solder paste printing process. This also applies to the type of solder paste that is used, which is typically a type 3 or 4 solder paste.

Using a mini-stencil requires careful removal after printing to keep from smearing the solder paste. The board is taken to the hot air rework machine, and the new package is placed on the board. Whether it is a manual or semi-automatic hot air rework machine does not matter because the hot air will melt the solder paste to form a new solder joint. The same temperature profile parameters are used as those used to remove the package.

If there are multiple packages to be reworked with a lot of leads, then the solder pasting method is the faster method. If the package only has a few leads, then hand soldering is the preferred method.

After the new package has been reworked, it is important to visually inspect and x-ray the rework before sending the board back to test.

4.2.3 ECO Wiring

Engineering change order (ECO) are another area for rework/repair. This procedure requires the running of wires from one component or board location to another.

ECO wiring is often required when changes to the board design are made after the circuit board has been assembled. This can require lifting component leads, soldering to pads, cutting traces, and so forth, to fix or upgrade the board so that it becomes fully functional.

Lifting component leads can be a very delicate operation, requiring a skilled rework operator. The soldering iron tip is placed on the board pad to melt the solder that is holding the component lead in place. When solder melting is observed, the lead is slowly lifted up with the aid of fine tweezers or a dental pick. The ECO wire can now be soldered to the component lead and routed to the other location on the board to complete the new connection. The wire should be tacked down with tape or glued along the way to make sure it does not get damaged. The other end of the wire could be connected to a cut trace, another component lead, board via, and so forth, and then needs to be soldered. Figure 4.4 shows two ECO wires on the back of a board routed from one via to another via.

Typical soldering iron tip temperatures range from 316°C to 427°C (600–800°F). Small diameter core lead-free SnAgCu solder from 10 to 20 mil (0.254–0.508 mm) should be used, and the soldering time should be no more than a few seconds at each location.

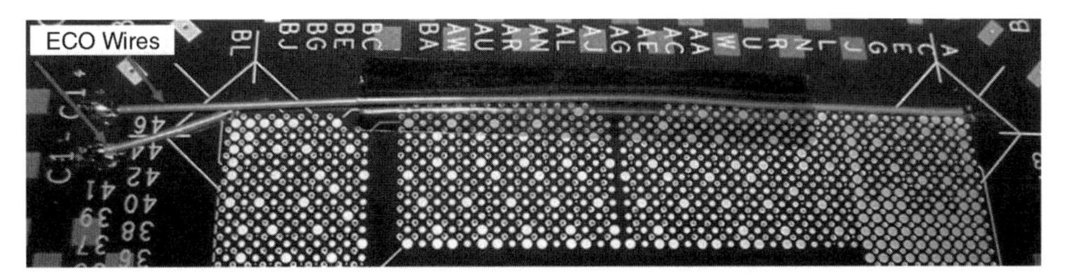

Figure 4.4 ECO wires routed from via to via on the back of a test board

4.2.4 Lead-Free Solder Pad Repair

Repairing a solder pad on the printed circuit board can be an involved process and can occur during a BGA rework removal or a solder wicking process. This requires that a new pad be placed in the location where one is missing. One removes the replacement pad from a similar location of a scrap board, using a knife blade and a solder iron tip to ease off the pad from the board. It is important that the pad have enough length to provide maximum conductor overlap for soldering to the existing conductor on the damaged board.

The board area that will receive the new pad with the trace needs to be thoroughly cleaned. The cleaning process ensures that there is no flux residue or other contaminants left that could interfere when gluing the new pad to the board. Solder mask needs to be scraped off for a minimum twice the conductor width. The exposed area will need to be fluxed and tinned after this [1]. Once the pad and trace are ready to be placed in the empty pad location, glue that can survive the high lead-free hand soldering temperatures can be applied and the pad placed. It is important that glue not get onto the pad or the solderability of the pad will be negatively affected. Placing the new land into position with a piece of polyimide tape can help in the alignment. The glue will need to be completely cured before attaching the pad trace to the trace on the board. Once the pad trace is soldered to the board trace, the area should be cleaned and a protective coat of glue put over the exposed trace area. The glue helps give strength to the trace and also protects it from environmental conditions.

The glue curing times vary by the type used and can be up to an hour or more. Some cure quickly with accelerators, and others require two parts to be mixed together for an exothermic reaction to occur.

Typical soldering iron tip temperatures range from 316°C to 371°C (600–700°F) for soldering the traces together, and small diameter core lead-free SnAgCu solder from 10 to 20 mil (0.254 mm–0.508 mm) should be used. Pad repair can be time-consuming and take over an hour to complete.

Lead-free boards are more prone to pad damage than tin-lead soldered boards because of the higher soldering temperatures used. The BGA pad solder wicking process during rework is where most of the board pad damage can occur and with pad sizes getting smaller, this will only get worse.

4.3 BGA/CSP REWORK

To remove a defective BGA package from the circuit board requires some type of rework machine (typically hot air convection). When a first time removal of this package is required, a proper profile should be developed using a profile board. If a profile board is not available, then a profile that is close to the board type and package size is selected.

For BGA rework there may be a heat sink that needs to be removed before any rework can begin. Figure 4.5*a* shows a board with two heat sinks over BGA packages, and Figure 4.5*b* shows the same board after both heat sinks have been removed.

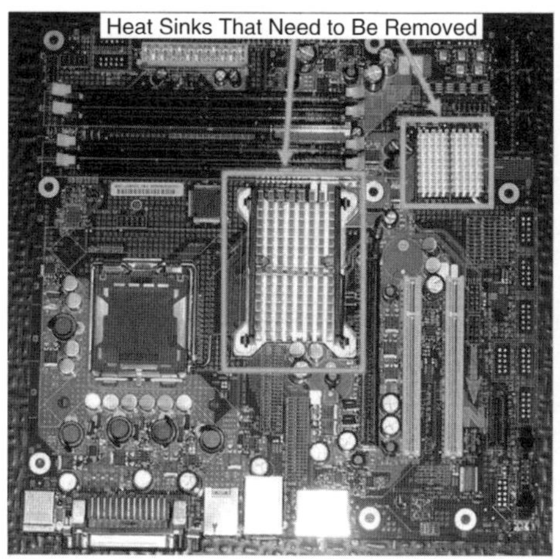

Figure 4.5a Heat sinks that need to be removed

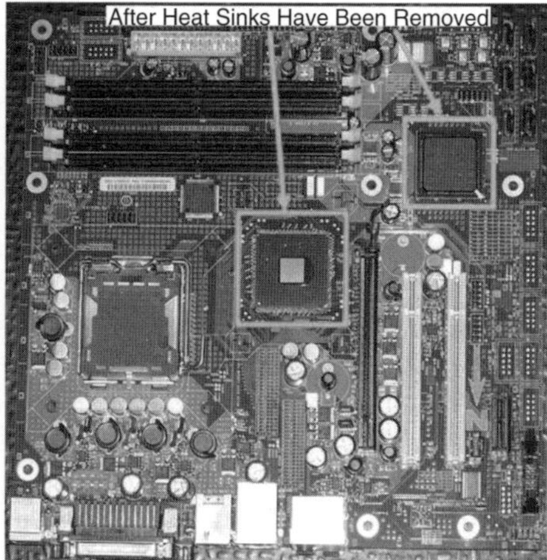

Figure 4.5b Heat sinks that have been removed

Figure 4.6a Connector close to a BGA package

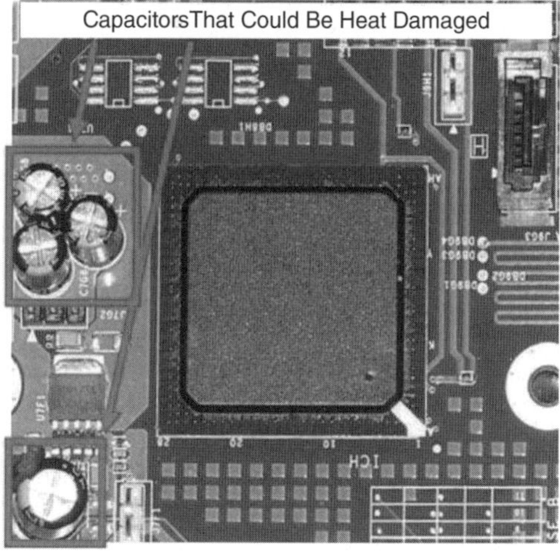

Figure 4.6b Electrolytic capacitors close to a BGA package

There also may be areas near the defective BGA package that should be heat shielded with thermal tape or a heat sinking device. These temperature sensitive parts can include plastic connectors and electrolytic capacitors. Electrolytic capacitors can explode if they get overheated, so this is a safety requirement. Figures 4.6*a* and 4.6*b* show examples of two BGA packages with one having a PCI ExpressX16 connector and the other BGA package having many electrolytic capacitors near it.

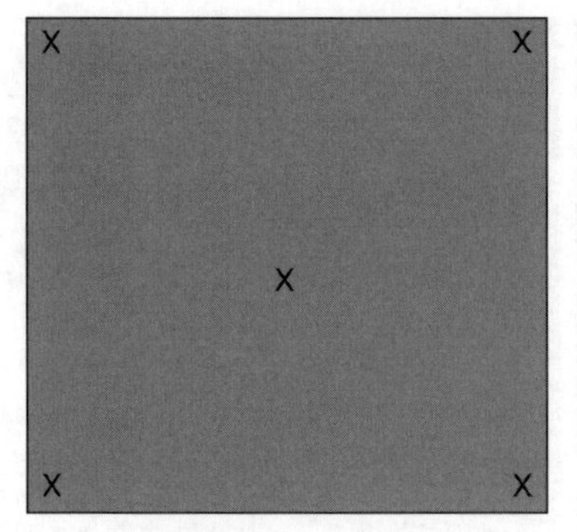

Figure 4.7a Full BGA array with thermocouple locations marked

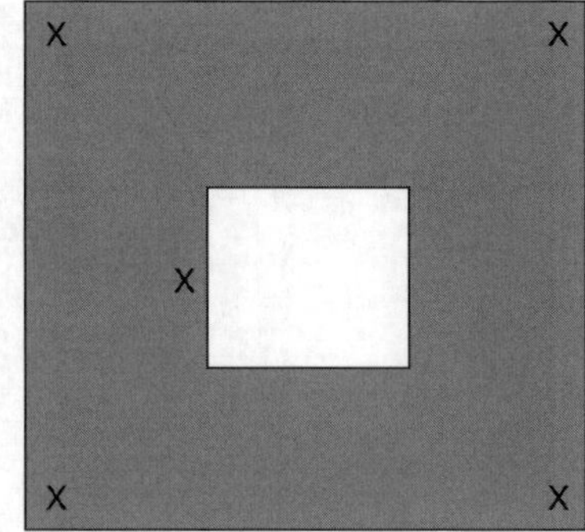

Figure 4.7b Partial BGA array with thermocouple locations marked

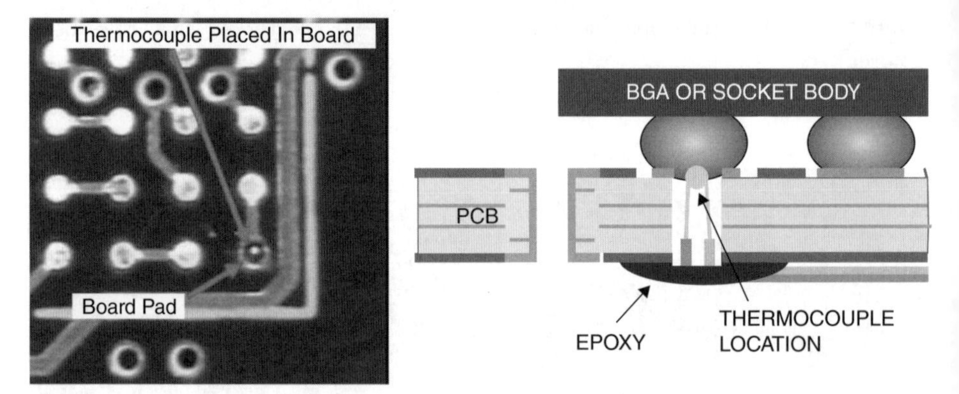

Figure 4.8 Properly centered thermocouple wire and the correct position to place it vertically

Thermocouple placement is important in creating a good rework profile. BGA packages that are 10 mm and larger in body size usually have thermocouples placed at all four outer corners and one in the center solder joint. If there are no solder balls in the center, then the thermocouple needs to be placed as close as possible to a center solder joint location. Figures 4.7*a* and 4.7*b* shows typical board thermocouple placement locations as marked by the Xs for a full and partial BGA array.

Type 36 gauge or smaller thermocouples should be used for the profile board. A hole will need to be drilled through the board at the center of the board pad. An appropriate drill size is one that is slightly larger than the thermocouple tip, which can be measured using calipers or a micrometer. The thermocouple tip should be positioned right at the board surface. Figure 4.8 shows a thermocouple wire positioned in the center of the board pad along with a drawing of the proper vertical placement.

The thermocouple wires should be secured in their correct locations using a thermally conductive epoxy that can survive temperatures up to 260°C (500°F). This epoxy

Figure 4.9 Thermocouples attached at the center of the package (die) and at the corner

also needs to be able to survive repeated profiling runs. Usually the same epoxy that is used for thermocouple wire attachment in the SMT reflow process is also used for rework profiling.

For smaller BGA/CSP packages (less than 10 mm body size) usually one to two thermocouples are enough to create a good profile. One thermocouple should be placed in the center of the small BGA/CSP package array. If an additional thermocouple will fit, then it can be placed on an outside corner pad location.

During BGA/CSP profiling, thermocouples attached to board pads connected to vias that run to power or ground planes can create temperature profiling issues. These pad connections can make the solder joint much lower in temperature than the other solder joints nearby, causing soldering issues. Care needs to be taken during profiling for these types of boards.

After a BGA/CSP package has been soldered on to a board with thermocouples placed at the board pad locations, it is helpful to attach a thermocouple to the top of the component package. The typical location is at the center or at the corner of the package. For plastic bodied BGAs, a thermocouple should be placed through a drilled hole in the center of the device down to the top surface of the die. Metal lid BGAs need to be drilled through the bottom of the PCB, up through the BGA substrate, to the bottom surface of die. Figure 4.9 shows two thermocouples attached to the top of a BGA package.

If there are any temperature-sensitive components nearby, then additional thermocouples can be attached on or near them to monitor their temperature during rework. A lead-free BGA/CSP rework profile should fall into the following temperature

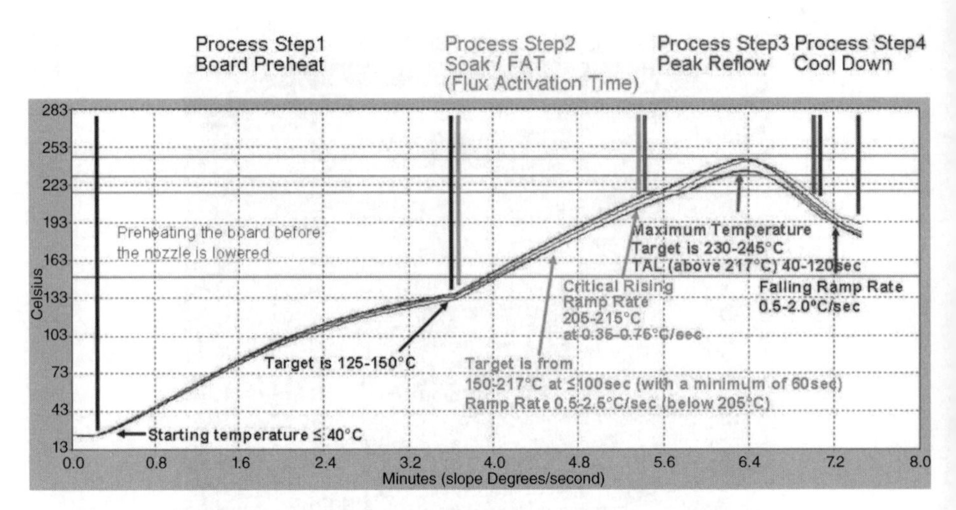

Figure 4.10 Typical lead-free BGA/CSP solder joint temperature profile range [2]

TABLE 4.1 Summary of lead-free BGA/CSP profile requirements

BGA/CSP Profile Description	Parameters
Solder joint peak temperature range	230–245°C (446–473°F)
Time above 217°C (422.6°F)	40–120 s
Maximum package body temperature	Not to exceed component supplier specifications
Delta temperature difference across all solder joint thermocouples	10°C (50°F)
Soak time at 150–217°C (302–422.6°F)	100 s with a minimum of 60 s (soak time and temperature can vary based on SnAgCu solder paste supplier recommendations)
Rising ramp rate below 205°C (401°F)	0.5–2.5°C/s (32.9–36.5°F/s)
Critical rising ramp rate at 205–215°C (401–419°F)	0.35–0.75°C/s (32.6–33.4°F/s)
Falling ramp rate	0.5–2.0°C/s (32.9–35.6°F/s)

ranges as seen in Figure 4.10. Table 4.1 summarizes the lead-free BGA/CSP profile requirements.

Once a good rework profile has been created for the BGA package location in need of repair, then it can be consistently removed and replaced. Usually the same profile that is used to remove the package is also used to replace it.

After a defective BGA package is removed the warm board should be immediately taken to an area where the excess non-uniform solder can be wicked off the board pads,

Figure 4.11 Solder spikes after the BGA package has been removed

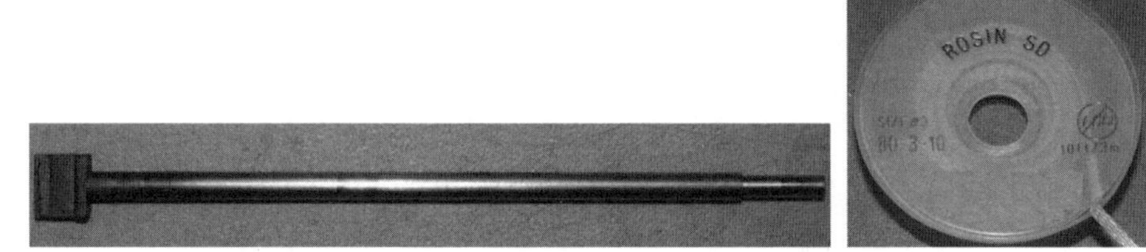

Figure 4.12 Typical blade solder tip with solder wick

leaving a uniform solder-coated board pad surface. Figure 4.11 shows board pads with solder spikes left over from the removal process.

The wicking process is usually accomplished with a blade tip soldering iron and solder wick, as shown in Figure 4.12. The solder wick contains some flux, but additional flux should be applied to the area requiring cleaning and also to the solder wick. Typical blade temperatures range from 371°C to 427°C (700–800°F).

Some BGA/CSP rework machines have built-in solder suction/scavenging features that can smooth and remove excess solder from the board pads using the same machine that removed the package. These can work quite well, but they cut down on through-put time because the same machine that is being used to remove and then subsequently replace the BGA package is now also used to clean the board pads.

The wicking process is one of the most delicate and critical parts of reworking a BGA package. Pad damage can occur easily when a hot solder tip with solder wick comes into contact with the area that needs cleaning (Figure 4.13).

Lead-free reworked boards are more susceptible to damage because of the higher rework temperatures. Pressure and time on the pad are the key elements in causing the damage. Light pressure (similar to writing with a pencil), continuous movement in one direction, and consistent/even blade temperature are essential in accomplishing good board pad cleaning.

The solder wick with the heated soldering iron should be placed on the board in front of the pads to preheat the solder wick before it touches the first row of pads. Vias will be encountered as the solder wick and iron travel over the pads. Tented vias will not cause many issues, but untented vias can be a major problem if the solder mask

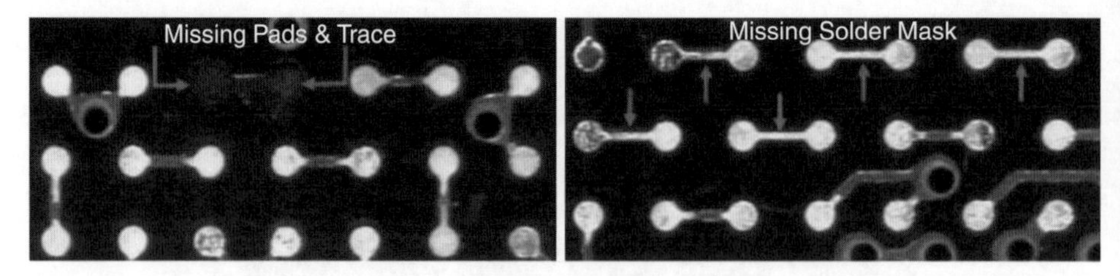

Figure 4.13 After lead-free solder wicking there can be missing pads, trace and solder mask damage

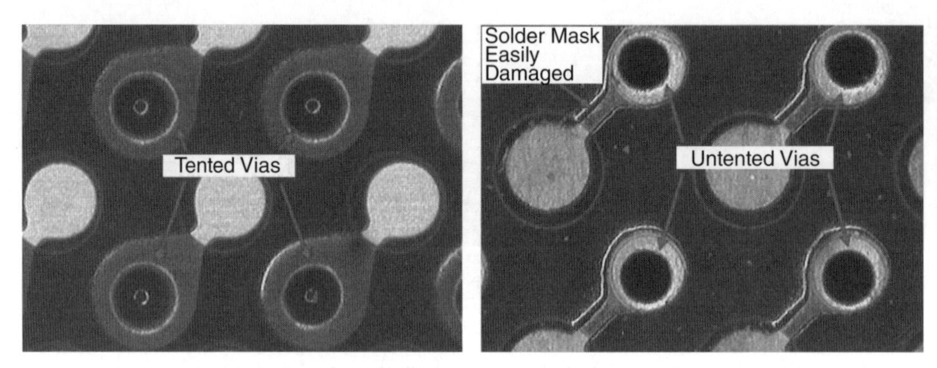

Figure 4.14 Visual differences between tented and untented vias

between the via and pad gets broken or removed. This can allow the solder from the new BGA component to wick down the via hole during the replacement process. If an untented via-to-pad solder mask location gets damaged, then it will need to be repaired before the new BGA component is placed. Figure 4.14 shows examples of both tented and untented vias.

After the board has been solder wicked and inspected to ensure that no board pad or solder mask damage has occurred, it is ready to have the new BGA package placed on it. But first either solder paste or flux must be applied to the package or board pads. If solder paste is being used, then the easiest method is to use a clam shell device with a stencil that matches the BGA package pattern. The BGA package is placed into the clam shell device. The solder paste is placed on the stencil and, using a metal blade, printed so that the solder paste goes through and fills the stencil openings while the remaining solder paste is removed. Figure 4.15 shows the clam shell stenciling process, and Figure 4.16 shows a package with lead-free SnAgCu solder paste applied to the solder balls.

The solder paste volume is calculated by using the following formula: $p \times$ (Stencil aperture radius)^2 $\times$ Stencil thickness. A good solder paste volume is what is used in the SMT process but can be adjusted up or down depending on the situation. Aperture size for a clam shell stencil is usually calculated at 80% to 85% of the ball size.

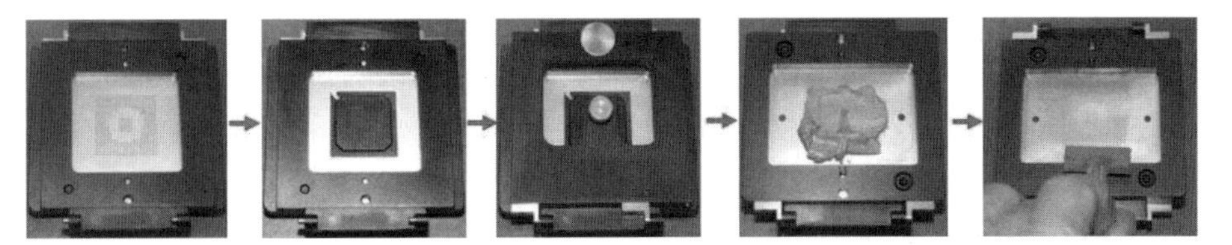

Figure 4.15 Typical clam shell package loading and solder paste printing process

Solder Paste On Package Solder Balls

Package

Solder Ball

Solder Paste

Figure 4.16 An example of a package with lead-free SnAgCu solder paste printed on the solder balls

Solder paste can also be applied directly to the board using a mini-stencil that matches the BGA pads, but this is a much more difficult process. Some BGA packages have thousands of board pads, and this can make removing the stencil from the board without smearing the solder paste difficult.

Using a flux-only method requires a liquid flux or tacky flux to be applied directly to the pads before starting the replacement cycle. The flux should be very mild, such as a no-clean flux type, and should not cause corrosion issues over the life cycle of the board. Some liquid fluxes can get trapped in areas of the board where they do not receive enough heat to activate them. If this happens and the flux is too active, then the board will fail prematurely in the field due to corrosion or dendrite growth.

A water-soluble flux (usually an organic acid type) can solder highly oxidized surfaces better than no-clean fluxes. If a water-soluble flux is used, then a good water cleaning process is needed to make sure all flux residues are removed. If any flux residue is left, then it will chemically react with the board and solder balls. In most cases this will eventually damage the board.

Use of flux is the easiest rework replacement method to use for both BGA and CSP packages because it requires the least amount of time to apply, unlike solder paste. If the BGA/CSP package has coplanarity issues, then solder paste can help increase yields significantly.

Once the solder paste or flux has been applied to the board, it can be placed on the rework machine. With the proper rework profile used and the new BGA package loaded in the rework machine, the rework process can be started. After the rework machine has completed its rework cycle, the board can be taken to x-ray and check for solder bridges, opens, voids, or other abnormal solder joints. If there are no issues, the board can be sent back to test.

4.4 BGA SOCKET REWORK

BGA socket rework is very similar to BGA package rework except that the solder profiles are slightly different and solder paste has to be used. The reason solder paste has to be used is that coplanarity of the socket can vary significantly from lot to lot or component supplier to supplier. Figure 4.17 shows a typical BGA socket.

If there is a heat sink on the socket, then it needs to be removed before any rework can be done. It is also important to make sure that no adjacent components will be damaged by the rework heat profile before beginning. If there are heat-sensitive components, then they need to be shielded or removed.

A profile will need to be created for this package, and it will typically have a higher time above liquidus range versus other BGA/CSP packages. Figure 4.18 shows a typical lead-free BGA socket rework profile. Table 4.2 summarizes the typical lead-free BGA socket profile requirements.

The higher time above liquidus range (up to 200 s) is because many BGA sockets have a lot of metal contained in the BGA socket body. So the socket retains heat in the center for a significant amount of time during the rework cycle in the reflow and cool-down stages.

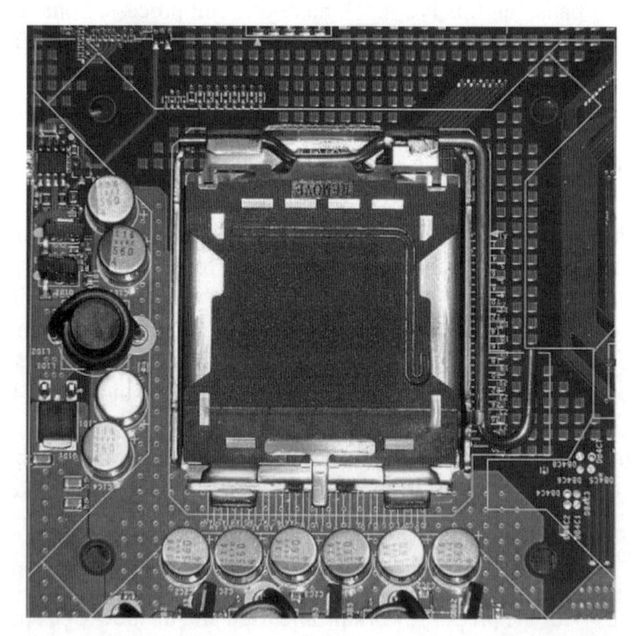

Figure 4.17 Example of a typical BGA socket requiring rework

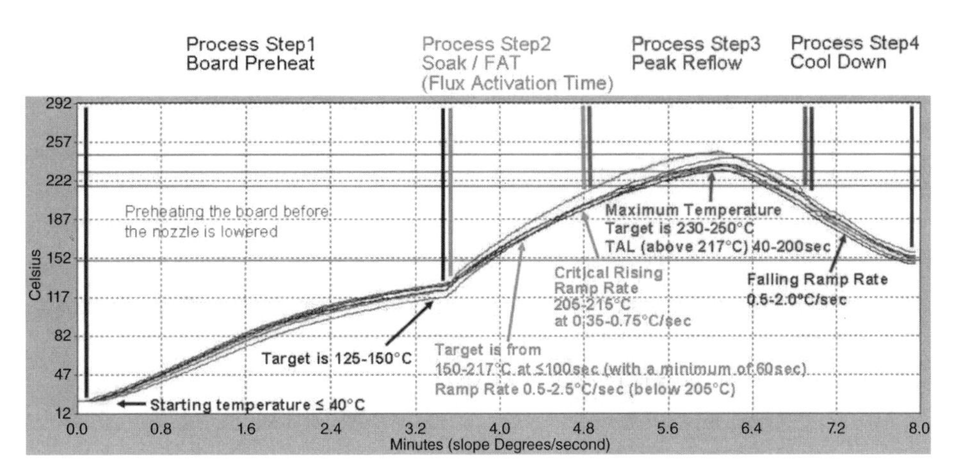

Figure 4.18 Typical lead-free BGA socket profile requirements [2]

TABLE 4.2 Summary of lead-free BGA socket profile requirements

BGA Socket Profile Description	Parameters
Solder joint peak temperature range	230–250°C (446–482°F)
Time above 217°C (422.6°F)	40–200 s
Maximum package body temperature	Not to exceed component supplier specifications
Delta temperature difference across all solder joint thermocouples	15°C (59°F)
Soak time at 150–217°C (302–422.6°F)	100 s with a minimum of 60 s (soak time and temperature can vary based on lead-free SnAgCu solder paste supplier recommendations)
Rising ramp rate below 205°C (401°F)	0.5–2.5°C/s (32.9–36.5°F/s)
Critical rising ramp rate at 205–215°C (401–419°F)	0.35–0.75°C/s (32.6–33.4°F/s)
Falling ramp rate	0.5–2.0°C/s (32.9–35.6°F/s)

This rework process follows the same pattern as regular BGA rework but differs in that solder paste has to be used in the replacement of the socket. If solder paste is applied to the socket using a clam shell fixture, then flux should be applied to the board pads too. This significantly increases yield on the rework process.

Solder paste can also be applied directly to the board pads using a mini-stencil. Typically the mini-stencil thickness and aperture sizes follow what is used in the SMT paste printing process but can be modified to help increase yields. Manual printing is difficult and requires a steady hand during the mini-stencil removal to keep from smearing the solder paste on the board pads.

After the rework has been completed the board is x-rayed and, if no issues arise, submitted to testing.

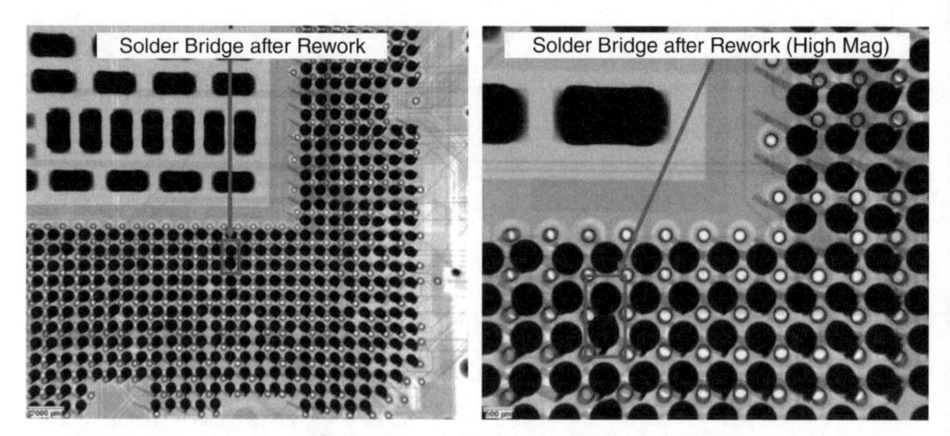

Figure 4.19 Lead-free solder bridging at low and high magnification after rework

4.5 X-RAYING

X-raying is necessary to check how good the rework is after it has been completed. When X-raying BGA packages, it is important to look for solder bridges, missing balls, or shifted balls. Figure 4.19 gives an example of a lead-free solder bridge after rework. X-raying is an important procedure because there can be unnecessary testing of a board that was incorrectly repaired at the rework stage.

4.6 THROUGH-HOLE HAND-SOLDERING REWORK

Only through-hole component types with up to 10 I/O pins should be repaired by hand soldering. If there are more than 10 pins, then it makes more sense to use a solder mini-pot/fountain to do the rework because of the increased time and risk of damaging the through-hole copper barrels.

Boards should be pre-heated to a minimum of 75°C for a 0.62-inch (1.5748 mm) thick, 4-layer board and to 125°C for a 0.93-inch (2.3622 mm), 10-layer board for through-hole lead-free hand soldering [3]. This makes it much easier to remove the solder from the holes using a heated solder sucker. For boards that are 0.125 inch (3.175 mm) and thicker, through-hole hand solder rework can be very difficult to achieve and may require much higher preheating temperatures. If an unheated solder sucker is being used to clean out the holes, then the board will have to be above its solder melting point during the whole rework operation. If the boards are being handled, then gloves are required to keep the rework operators from burning their hands.

After all the holes have been cleaned out, the part can be removed. The new part is then placed in the board, and both the board and part are preheated for through-hole soldering. A soldering iron with the appropriate tip (usually a blunt or conical tip) at a temperature range of 371°C to 427°C (700–800°F) is used with a core solder diameter range of 15 to 25 mil (0.381 mm–0.635 mm) to solder the through-hole leads. The board preheating ensures that there is good hole-fill during the solder process.

In the through-hole rework area, boards without thermally relieved through-hole power or ground planes can be the source of many problems while the components are removed. The heat from a soldering iron tip can be sucked away from the through-hole and require extended contact with it, which can cause excessive stress. Barrel damage has been known to occur because of this. The same issues also apply to through-hole solder fountain rework soldering.

There should always be a visual check to see whether there are any solder bridges or skipped soldered pins after rework. The board should also be x-rayed to verify that there is good hole-fill. After checking the X ray, the board can be sent back for testing.

4.7 THROUGH-HOLE MINI-POT/SOLDER FOUNTAIN REWORK

Typical through-hole components that are reworked on a solder mini-pot/fountain are connectors, USB ports, sockets, headers, and any component greater than 10 I/O pins. These components are much easier and faster to rework on a solder mini-pot/fountain than by hand soldering rework. Figure 4.20 shows some of typical components that require mini-pot/solder fountain rework.

4.7.1 Board Preparation

Through-hole components need to have the bottom of the board prepared before any rework can begin. The area around the component that is going to be reworked needs to be covered with heat-resistant tape, Kapton® tape, or other masking material. Figure 4.21 shows an example of a connector taped with heat-resistant tape. Use of tape is critical to keep from reflowing adjacent areas near the component needing rework. There are also retention pins on some of the through-hole components that need to be made nonfunctional so that these components can be removed as easily as possible.

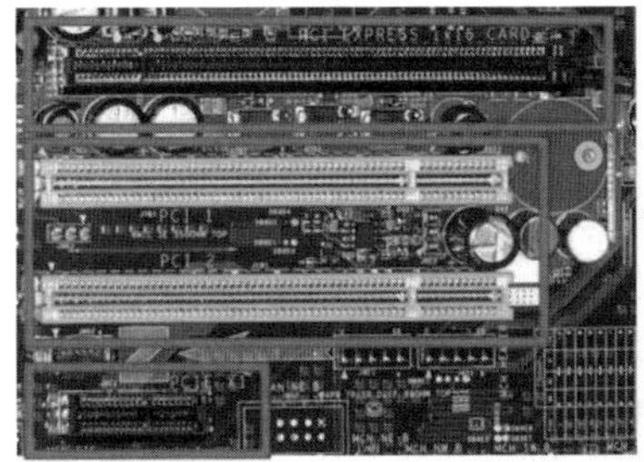

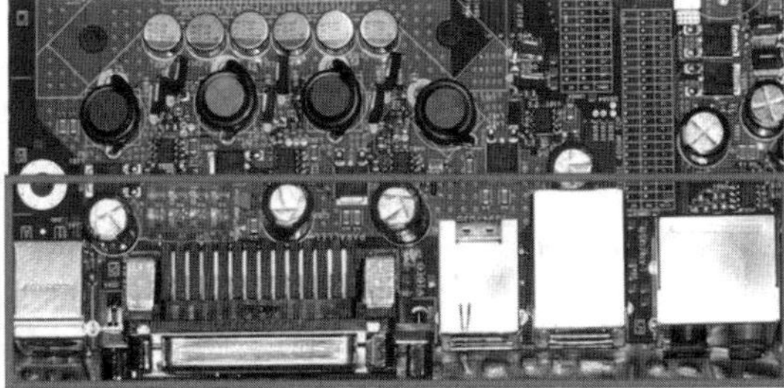

Figure 4.20 Typical through-hole component packages that require mini-pot/solder fountain rework

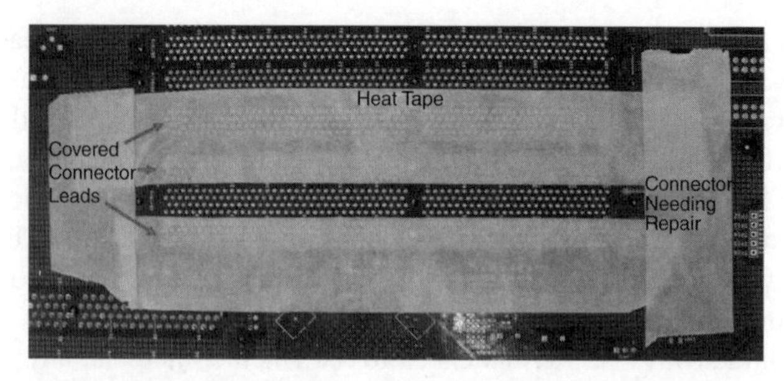

Figure 4.21 Area around the component covered with heat-resistant tape prior to lead-free solder fountain rework

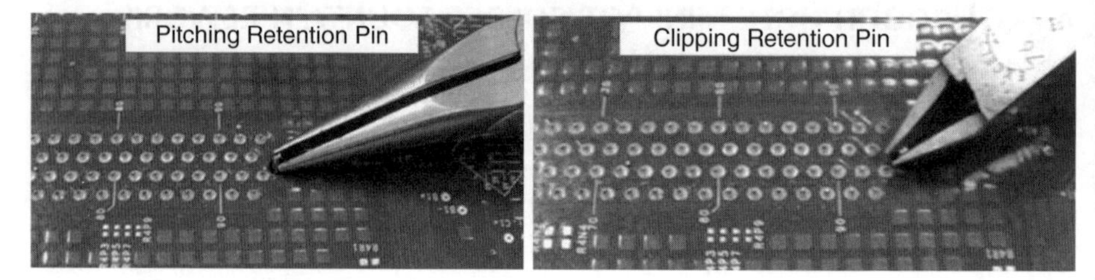

Figure 4.22 Disabling component retention pins by pinching or clipping prior to lead-free mini-pot/solder fountain rework

There are two ways to disable retention pins: by pinching the pins with needle nosed pliers or by clipping them with cutters. Figure 4.22 shows examples of both procedures disabling component retention pins.

4.7.2 Component Removal

The pot temperature can be set to varying temperatures depending on the solder alloy that is in the pot. Sn3Ag0.5Cu (SAC305) alloy pot temperatures with a melting temperature of 217°C (422.6°F) typically vary from 260°C to 270°C (500–518°F) and other solder alloys like Sn0.7Cu or Sn0.7CuNiGe with a melting temperature of 227°C (440.6°F) typically vary from 265°C to 275°C (509–527°F). The Sn3Ag0.5Cu alloy allows the solder pot temperature to be set lower, but copper barrel erosion/dissolution can occur at a faster rate with this alloy. The Sn0.7Cu alloy has a higher melting point than Sn3Ag0.5Cu and solidifies very quickly making it more difficult to work with. The Sn0.7CuNiGe alloy also has a higher melting point similar to Sn0.7Cu, but with the addition of the nickel and germanium. This alloy does not solidify as quickly as Sn0.7Cu and has a lower copper barrel erosion/dissolution rate [4–8]. There are other unique alloys from different solder suppliers that can help both keep the solder pot temperature low and reduce the copper barrel erosion/dissolution rate.

According to the JEDEC JESD22-B106 standard, the maximum temperature a lead-free through-hole component is rated to during wave soldering and mini-pot rework operations is 270°C, so attention is needed in selecting a pot temperature that conforms to this standard [9]. This standard was developed with limited data at the time and can be updated in terms of component temperature ratings and times with more lead-free production from wave solder and mini-pot/fountain rework data.

Too high of a pot temperature can cause copper barrel erosion/dissolution and barrel cracks as well as potential board or component damage. Copper barrel damage usually cannot be seen without cross-sectioning unless there is no copper board pad left. A minimum of 12.7 micron (0.5 mil) of copper at the barrel knee should be left after rework has been completed [10–12]. Too low a pot temperature can make removing a defective component very difficult and leave many pins in the board when the component is removed.

4.7.3 Component Replacement

Once the site under repair has been prepared, the mini-pot fountain rework nozzle is selected for component replacement that extends approximately 0.25°C to 0.5 inch (5.08 mm–12.7 mm) beyond the rework area on each end. The nozzle width should be approximately 0.10 to 0.25 inch (2.54 mm–5.08 mm) beyond the rework area [2]. This allows for easy alignment and good solder flow through out the whole length of the component being reworked.

The rework nozzle is then placed on the solder mini-pot fountain, and the board is aligned to the component that needs to be repaired. If the mini-pot has an incorporated heating unit, then the board is moved into the heating unit until it gets to a top side board temperature of 110°C to 150°C (230–302°F). A thermocouple should be used to determine the time it takes to get to the appropriate board preheat temperature. If there is no incorporated heater, then a stand-alone batch oven can be used to preheat the board. A profile will need to be created on a setup board and thermocouples attached to the board to ensure that the board is properly preheated to 110°C to 150°C. Gloves need to be used when handling any board that will be preheated in an oven.

The replacement component should have its component pins pre-fluxed by dipping or brushing on flux. This component should also be preheated with the board that needs repair. While the board and replacement component are being preheated, it is important to preheat the rework nozzle by cycling the lead-free solder through it for approximately 3 minutes. This should be done just before the board is ready to be reworked so that the rework nozzle is still hot. A hot rework nozzle will keep the solder from cooling down at the ends of the nozzle.

After the board has been properly preheated, it can be moved toward the solder pot; then the bottom side area should be quickly fluxed and positioned over the nozzle. Solder flow should be just enough to have it flowing out of each end of the nozzle at a steady low volume. This is considered to be low solder flow. The component leads should be approximately 0.0254 mm–0.0762 mm (0.001–0.003 inch) away from the solder mini-pot fountain nozzle [2].

Depending on the size of the component being reworked, it can take 5 to 30 seconds before all the component solder leads start to reflow. Low pin count jumper blocks usually are in the short rework time frame (5 seconds), whereas FBDIMM connectors can take up to 30 seconds to remove. To determine when the component is ready to be removed, a gentle rocking motion is applied to it. When the component rocks easily, it can be removed, with a new component immediately replacing it. The lead-free solder is allowed to flow for a few additional seconds in the nozzle after receiving the new component before the solder flow is shut off. The board is allowed to cool down for at least 1 minute before removing from the mini-pot. This allows the solder to properly solidify and the excess flux fumes to be extracted by the fume hood.

The bottom of the board can now be cleaned with isopropyl alcohol to remove the excess flux, after which it can be x-rayed and then tested.

A three-step through-hole rework process instead of the one-step process previously described can be used. This process involves removing the defective component using the solder mini-pot/fountain and then immediately cleaning all the through-holes manually. After all the through-holes have been cleaned with a solder sucking device, the board and new component can be preheated, fluxed, and placed on the solder mini-pot/fountain to be re-soldered. This can be a very time-consuming process for components with hundreds of through-holes. The advantage is that the time on the solder mini-pot/fountain can be reduced by 5 to 20 seconds, which will reduce the copper barrel erosion/dissolution rate and improve board reliability. Table 4.3 summarizes the different lead-free solder mini-pot/fountain alloys and the pot temperatures and pot times used.

According to the JEDEC JESD22-B106 standard, the maximum time that a through-hole component can withstand rework in the mini-pot/solder fountain is 17 seconds [9]. This specification applies only to the replacement of a component and not to its removal. As always, it is a good idea to keep the replacement time as short as possible.

TABLE 4.3 Summary of different lead-free solder mini-pot/fountain alloys temperatures and pot times [2]

Solder Alloy	Solder Alloy Melting Point	Pot Temperature Range	Component Removal Times	Component Replacement Times
Sn3Ag0.5Cu (SAC305)	217°C (422.6°F)	260–270°C (500–518°F)	5–30 s	5–20 s
Sn0.7Cu	227°C (440.6°F)	265–275°C (509–527°F)	5–30 s	5–20 s
Sn0.7CuNiGe	227°C (440.6°F)	265–275°C (509–527°F)	5–30 s	5–20 s
Other lead-free alloys	217–227°C (422.6–440.6°F)	260–275°C (500–527°F)	5–30 s	5–20 s

4.8 BEST PRACTICES AND REWORK EQUIPMENT CALIBRATIONS

Rework equipment calibration keeps the yields as high as possible. A daily check of the temperature function of solder iron tips and heater elements for BGA rework machines can accomplish this. Rework machines that are not performing at their proper temperatures can cause open solder joints, component package damage, or board damage.

Other issues that can affect rework equipment performance are air flow fluctuations. When the air flow to the rework equipment varies, this causes temperature profile problems, and in some cases equipment malfunction. Certain BGA rework machines can have issues with component placement when the component pickup tube is bent, and this will cause the yield to be abnormally low.

4.9 CONCLUSIONS

SMT hand soldering/touch-up covers a wide variety of rework, with passive component rework being the most common. This rework can be done using fixed or tweezer type tips and requires some level of manual dexterity. Capacitors can be damaged very easily during this rework, and proper soldering procedures need to be followed.

Another area of SMT hand soldering rework is QFP/gull wing packages that can be reworked either manually or using a semi-automated process. The choice of which type of rework should be done depends on equipment availability and operator skill. Solder wicking skills to redress the board pads are needed for this type of rework.

The final areas of SMT hand soldering/touch-up cover ECO wiring and solder pad repair. ECO wiring can call for component pad lifts, trace cuts, and soldering to vias. Solder pad repair entails pad gluing and trace soldering. This is an area where highly skilled operators are needed because of the fine detail of rework that has to be achieved.

BGA/CSP rework requires good solder wicking skills to redress the board pads. Many BGA packages have in excess of 1000 pads on a board. Many BGA/CSP pads are from 10 to 24 mils (0.254 mm–0.610 mm) in diameter. Wicking lots of small pads requires a good training program. Mini-stencil solder paste application skills may also be needed for this type of rework.

BGA socket rework also requires good solder wicking skills to re-dress the board pads because most component processor sockets have nearly 1000 board pads to clean. Fortunately, the board pad sizes for these components are usually a little larger than the BGA/CSP pads. Mini-stencil pasting skills are needed for this type of rework. This could include paste printing on the board or using a clam shell device to deposit the solder paste. Keeping the processes as simple as possible significantly improves the yield.

Through-hole hand soldering requires board preheating for removing lead-free solder from the holes. The preheat allows the holes to be cleaned more easily and reduces the chance of plated through-hole barrel damage. Re-soldering through-hole components also requires preheat to achieve good hole-fill. Properly maintained solder sucking devices is critical to the success of this rework.

Through-hole mini-pot/solder fountain rework is an area that seems straightforward enough to do but has the most unseen potential for damage of all rework processes. Excessive time on the mini-pot/solder fountain can cause significant copper barrel erosion/dissolution. This damage can be hard to see because the solder joint can still form from the solder barrel pad to the lead. Using board preheat and solder alloys that are more resistant to copper barrel erosion/dissolution can greatly reduce this risk.

After any rework procedure it is a good practice to x-ray the reworked location for any defects as this reduces the chance for any defective rework reaching test.

4.10 FUTURE WORK

New lead-free solder alloys will continue to be developed over time to help with rework. As these new alloys are developed for the different rework areas, more process margin should be obtained. Through-hole rework should be the greatest benefactor from new alloys that target reducing copper barrel erosion/dissolution. Lower melting temperature lead-free alloys may also aid in reworking QFPs/gull wing and BGA type packages.

Board pad (peel) strength may be improved over time with new materials or process changes that aim to reduce pad lift during the solder wicking rework process of BGA packages.

REFERENCES

1. IPC-7711B/7721B Standard, *Rework, Repair and Modification of Electronic Assemblies*, 2007.
2. Intel Corporation, *Lead-Free Rework Specification*, Mar. 2008.
3. A. Donaldson, "Hand Solder Rework of Lead-Free Through-hole Components and Leaded Surface Mount Packages," SMTA, *2005 SMTA Int. Conf. Proc.,* Rosemont, IL, Sept. 25–29, 2005, pp. 127–132.
4. C. Hamilton, "A Study of Copper Dissolution during Pb-Free PTH Rework," CMAP, *Proc. Int. Conf. Lead-Free Soldering* (CMAP) Toronto, May 2006.
5. C. Hamilton, P. Snugovsky, and M. Kelly, "A Study of Copper Dissolution during Lead-Free PTH Rework Using a Thermally Massive Test Vehicle," SMTA, *2006 SMTA Int. Conf. Proc.,* Rosemont, IL, Sept. 24–28, 2006, pp. 177–182.
6. C. Hamilton, P. Snugovsky, and M. Kelly, "Have High Cu Dissolution Rates of SAC305/405 Alloys Forced a Change in the Lead-Free Alloy Used during PTH Processes," SMTA, *Proc. Pan* Pacific Microelectr., Jan. 2007.
7. C. Hamilton, M. Moreno, R. Mendez, G. Soto, J. Herrera, T. Ng, J. Fangkangwanwong, M. Kelly, and J. Bielick, "A Study of Alternative Lead-Free Wave Alloys: From Process Yield to Reliability," APEX Conf., Apr. 2008.
8. A. Donaldson, R. Aspandiar, and K. Doss, "Comparison of Copper Erosion at Plated Through-hole Knees in Motherboards Using SAC305 and an SnCuNiGe Alternative Alloy for Wave Soldering and Mini-pot Rework," APEX Conf., Apr. 2008.

9. JEDEC JESD22-B106D Standard, *Resistance to Solder Shock for Through-hole Mounted Devices*, 2008.
10. M. Kelly, M. Cole, J. Wilcox, J. Bielick, T. Younger, and D. Braun, "Qualification of a Lead-Free Card Assembly and Test Process for a Server Complexity PCBA," *2007 SMTA Int. Conf. Proc.,* Orlando, FL, Oct. 7–11, 2007, pp. 806–815.
11. L. Ma, A. Donaldson, S. Walwadkar, and I. Hsu, "Reliability Challenges of Lead-Free (LF) Plated–Through-hole (PTH) Mini-pot Rework," IPC/JEDEC Lead-Free Reliability Conf., Boston, April 2007.
12. I. Williams, "Copper Removal from Plated Through-hole Barrels during Lead-Free Mini-pot Rework," IPC Midwest Conf., Chicago, IL, Sept. 28, 2007.

5

LEAD-FREE ALLOYS FOR BGA/CSP COMPONENTS

Gregory Henshall (Hewlett-Packard)

5.1 INTRODUCTION

5.1.1 Limitations of Near-Eutectic SnAgCu Alloys

The electronics industry adopted "near-eutectic" Sn–Ag–Cu (SnAgCu) alloys as the standard lead-free alloys to replace eutectic Sn-37Pb during the RoHS transition. Those in common use include Sn-3.8 wt%Ag-0.7 wt%Cu (SAC387), Sn-4.0 wt%Ag-0.5 wt%Cu (SAC405), and Sn-3.0 wt%Ag-0.5 wt%Cu (SAC305). These alloys were selected by industry consortia [1–3], balancing many factors, notably their relatively low melting point and good thermal fatigue reliability in accelerated testing. The SAC family of alloys has proved to be an acceptable working solution for a broad range of applications.

The relatively short transition period to lead-free solders did not allow sufficient time for systematic alloy design and complete testing to identify optimal SnPb replacements. The selection of near-eutectic SnAgCu alloys was done prior to understanding their impact on mechanical robustness and other factors, making them less than an ideal lead-free solution. Other concerns include poor barrel hole-fill on thick boards for some surface finishes, copper dissolution, hot tearing, and other surface phenomena that create inspection issues and possibly unnecessary rework. Another potential drawback of near-eutectic SnAgCu is the expense of silver (Ag), which was $200 per lb in January

Lead-Free Solder Process Development, Edited by Gregory Henshall, Jasbir Bath, and Carol A. Handwerker

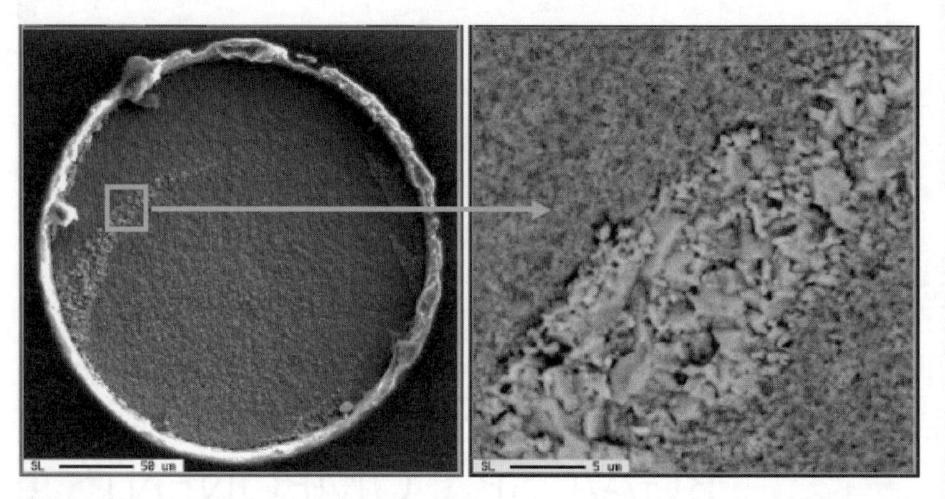

Figure 5.1 Micrograph of BGA brittle fracture of SAC405 solder joints on Ni/Au pads. After Gregorich et al. [5]

2009 while that of tin was about $5.30 per lb. This cost difference has driven the desire to reduce silver content in wave solder bar, and some of these alloys are now being considered for ball grid array (BGA) ball alloys [4].

One of the key drawbacks of near-eutectic SnAgCu for BGA components is its poor mechanical shock performance, especially on Ni/Au surfaces. An early indication of this problem was published by Gregorich et al. [5] for SAC405 joints on Ni/Au. Figure 5.1 illustrates the brittle appearance of such fractures, which are unrelated to the well-documented phenomenon of "black pad" on electroless nickel/immersion gold (ENIG) surfaces. Early work by Garner et al. [6] also pointed to the fact that high (3–4 wt%) Ag concentrations result in the high stiffness and strength of near-eutectic alloys, which in turn affect mechanical performance. Soon significant investigations into the mechanical shock resistance of SnAgCu BGA joints and the impact of alloy composition were undertaken [7–15].

Due to the limitations of near-eutectic SnAgCu, the industry has seen the development of a wide range of new alloys, as documented by Holder et al. [16]. The increasing number of lead-free alloys available provides opportunities to address the issues described above. At the same time these alloy choices present challenges in managing the supply chain and introduce a variety of risks. For example, the high melting point of low silver alloys (e.g., Sn1Ag0.5Cu) represents a risk in the reflow process if not managed properly. Low silver alloys may also present risks for thermal fatigue failure of solder joints in some circumstances, though much is unknown about the impacts of silver (Ag), copper (Cu), and dopant concentrations on thermal fatigue resistance.

Another concern is that the perceived needs of the high volume consumer segment are a key driver for the use of new alloys. These needs do not necessarily match those of the low-volume "high-reliability" segment (medical, aerospace, IT, and communications infrastructure, etc.). Many "high-reliability" original equipment manufacturers (OEMs) have not switched to lead-free technology because of various exemptions in substance restriction legislation, and have rigorous requirements for evaluation and qualification of lead-free materials and processes. For these companies the deployment of new alloys may represent a "moving target" for their transition to lead-free solders.

A general concern is that the acceptability of alternate alloys will vary from product class to product class, and possibly from company to company. Similarly BGA package suppliers want to minimize the number of alloys they have to deal with while still meeting customer needs, which may drive them to have multiple alloy solutions. Thus having such a wide range of solder alloys in use adds complexity and risk that has not existed within the supply chain up to now.

At the same time OEMs and electronics manufacturing services (EMS) suppliers cannot stop solder and component suppliers from innovating and bringing new alloys to market. The development of new alloys is a natural part of lead-free technology maturing, and may provide improvements over the long run. Therefore, as discussed later, efforts are required to manage supply chain complexity so that the benefits of new alloys can be achieved while minimizing the risk of unintended consequences.

The industry is currently very active in performing technical investigations of various alloys. Some information is proprietary, but more information is becoming publicly available. In making alloy choices, the industry is considering a wider range of alloy properties and is considering the performance of the alloy in a wider range of situations than was done in the first round of lead-free alloy selection. The market will ultimately decide which alloy or alloys "win," and whether alloy choice remains large or a few alloys become de facto standards.

5.2 OVERVIEW OF NEW LEAD-FREE ALLOYS

Because of the concerns discussed above, efforts have been made to develop and deploy new lead-free alloys with improved characteristics. This section provides an overview of these new alloys, while details about behavior are given in later sections. The scope of this chapter is limited to alloys based on the SnAgCu system, with $0 \leq [Ag] \leq 4.0$ wt% and $0 \leq [Cu] \leq 1.2$ wt%. Alloys based on the SnBi [17], SnZn [18], or other systems are not addressed, largely because they have not found widespread use in the electronics industry.

One obvious approach to addressing some of the concerns with near-eutectic SnAgCu alloys, particularly poor mechanical shock performance, has been to decrease the silver content. Lowering the silver concentration had some effect but not as much as was hoped. Thus the addition of other alloying elements, typically at levels of 0.1 wt% or less ("microalloy additions"), was investigated to provide further improvement. Table 5.1 summarizes how these strategies have been used to address specific drawbacks of near-eutectic SnAgCu [19].

TABLE 5.1 Alloying strategies for improving lead-free solder alloys

Problem	Reduce or Remove Silver	Micro-Alloy	Other Solutions
High flow stress	X	X	
Brittle joint failure	X	X	
Low impact strength	X	X	
Shrinkage defects	X	X	Move closer to eutectic complosition
Copper erosion	X	X	
Cost	X		

Source: Courtesy: Nihon Superior.

TABLE 5.2 Listing of some new lead-free, SnAgCu-based alloys

Alloys	Investigators [ref.]
Sn1.0Ag0.5Cu (SAC105)	Pandher [8], H. Kim [9], D. Kim [11], Syed [10], Kobayashi [12], Liu [25]
SAC205	Zhang [58]
Sn-3.5Ag	Cavasin [13], Darveaux [45], Liu [25]
Sn0.3Ag0.7Cu + Bi (SACX)	Pandher [8]
Sn0.3Ag0.7Cu + Bi + Ni + Cr (SACX)	Pandher [8]
SAC 305 + 0.05Ni + 0.5In	Syed [10]
SAC255 + 0.5Co	Syed [10]
SAC107 + 0.1Ge	Syed [10]
SAC125 + 0.05-0.5Ni (LF35)	Syed [10], D. Kim [11], Kobayashi [12], Darveaux [45]
SAC101 + 0.02Ni + 0.5In	Syed [10]
Sn-3.5Ag + 0.05-0.25La	Pei & Qu [15]
Sn-0.7Cu	Darveaux [45]
Sn0-4Ag0.5Cu + Al + Ni	Huang [14]
SAC305 + 0.019Ce	Rooney [59], Song [24]
Sn-2.5Ag-0.8Cu-0.5Sb	Song [24]
Sn-0.7Cu-0.05Ni	Song [24], Che [23]
Sn-0.7Cu-0.05Ni + Ge (SN100C)	Nogita [60]
SAC105 + 0.02Ti	Liu [25]
SAC105 + 0.05Mn	Liu [25]
Sn-3.0Ag-1.0Cu	De Sousa [61]

5.2.1 Range of New Lead-Free Alloys

The number of lead-free alloy choices is expanding as various approaches are developed to address the limitations of near-eutectic SnAgCu alloys. The range of alloys investigated is large. Table 5.2 gives a partial listing of new alloys being investigated, and in some cases being used in production.

The wide range of alloy choices is both an opportunity and a risk for the electronics industry. Clearly, the new alloys perform better in some respects than near-eutectic SnAgCu. However, many of these alloys have not yet been fully characterized over the full range of behavior relevant to the industry as a whole.

5.2.2 Alloys in Current Use for Area Array Packaging

Given all the activity in the industry to develop new alloys, in 2007 Hewlett-Packard Co. (HP) performed a survey of its top BGA/CSP suppliers. Some of the key objectives of the survey included identification of (1) lead-free ball alloys and surface finishes currently in production, (2) the intended applications for each of these alloys, (3) motivation for selecting alloys, (4) methods of customer notification, and (5) the resulting cost. Responses were provided by 13 of the 15 suppliers contacted, including some of the industry's largest BGA/CSP producers. The following describes the major findings of this survey.

Low-silver SnAgCu alloys, specifically those based on Sn1Ag0.5Cu (SAC105), Sn1.2Ag0.5Cu (SAC125), and Sn1Ag0.1Cu (SAC101) made up 45% of total of supplier responses, as shown in Figure 5.2. Typically these low-silver alloys contain "micro-alloy" additions or "dopants," which refers to intentionally added elements in concentrations of about 0.1% or less. Such additions may be present to improve specific properties, such as oxidation resistance or mechanical reliability. Many suppliers provided multiple alloys, depending on package type and application. Of the suppliers surveyed, 77% were providing "alternate" lead-free alloys (not near-eutectic SnAgCu) for at least some of their devices as of mid 2007.

Another finding was that different types of surface finish were used in combination with different alloys, as shown in Figure 5.3. Eighty-five percent of suppliers were using SAC405 or SAC305 on Ni/Au. Using high-silver ball alloys on Ni/Au could be a concern. As mentioned earlier, mechanical shock studies have concluded that this combination of ball alloy and surface finish would be a poor choice because of an unacceptably high propensity for brittle fracture [5,7,20]. Fortunately, most suppliers will not be using this material combination on devices intended for mobile products, where the concern over brittle fracture is greatest.

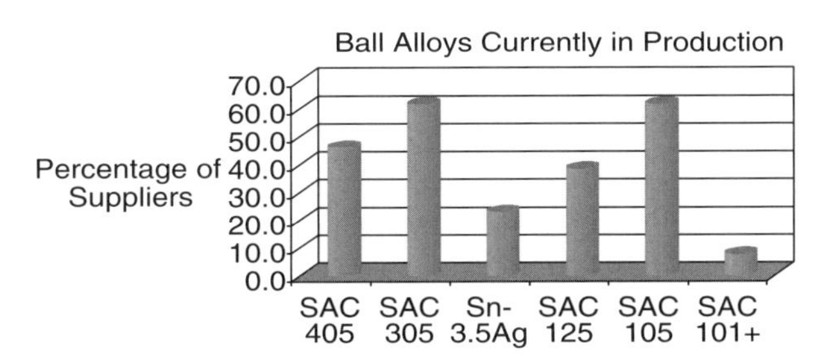

Figure 5.2 Results of company BGA/CSP component supplier survey: BGA ball alloy use in production of BGA/CSP devices

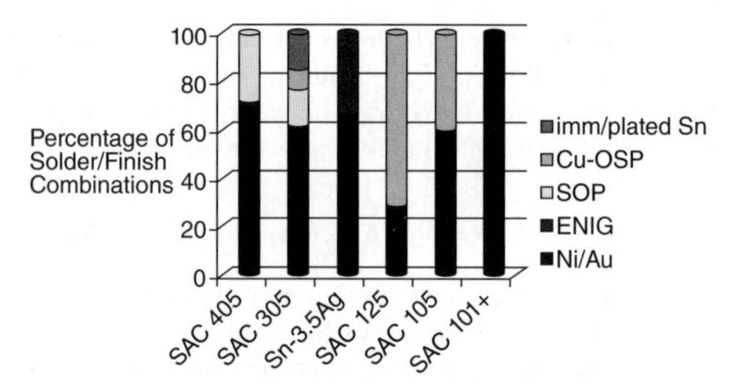

Figure 5.3 Results of company BGA/CSP component supplier survey. For each BGA ball alloy, the percentage use of each surface finish is indicated

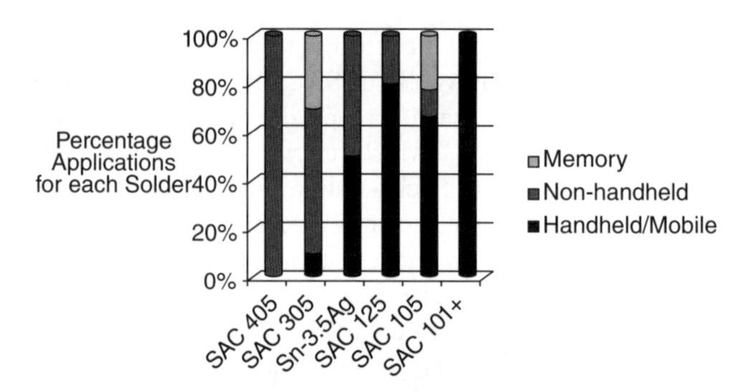

Figure 5.4 Results of the company BGA/CSP component supplier survey. The percentage of intended application of packages for each BGA ball alloy type is indicated

The survey also gave insights into how component and alloy suppliers were targeting different solutions for different applications. Intended uses for each of the lead-free solder ball alloys were categorized into three main applications (Figure 5.4): (1) handheld/mobile, (2) non-handheld, and (3) memory. Devices intended for use in mobile products had ball alloys that were mostly low silver; SAC305 or Sn3.5Ag were rarely used. Mostly high-silver ball alloys were used for devices intended for non-handheld products. Memory applications were split between SAC305 and SAC105. Many suppliers had their alloy selections relatively well aligned with mechanical or thermal concerns for a specific application: mechanical shock resistance for mobile applications (low silver), and thermal fatigue resistance for other applications (high silver).

These solutions were consistent with supplier reports for the motivation to select a specific lead-free solder. Improved thermal fatigue resistance was usually cited as the reason high-silver alloys were chosen. Improved mechanical shock performance was usually cited as the reason for choosing low-silver alloys. Cost reduction was almost never cited as a selection criterion for BGA ball alloy, in contrast with wave solders where alloy cost would play a major role. In fact, decreasing silver content generally did not lower costs for BGA/CSP producers.

5.3 BENEFITS OF NEW ALLOYS FOR BGAS AND CSPS

5.3.1 Mechanical Shock Resistance

Low-Silver and Doped Alloy Performance Studies consistently show that low-silver alloys generally perform better in mechanical shock than high silver alloys [8–11, 20–23]. For example, Figure 5.5 shows the data of Syed et al. [10] who compared mechanical shock data for four alloys. The 3%Ag SAC305 performance was significantly poorer than that of the other alloys, each of which had a silver concentration of 1%. The data of Kim et al. [11] provided in Figure 5.6 show a similar tendency. In this case the mechanical shock performances of SAC105 (1% silver) and LF35 (Sn1.2Ag0.5Cu + Ni) are far superior to that of the 4%Ag SAC405 alloy.

Many studies have shown that micro-alloying additions improve the mechanical shock performance of SnAgCu alloys on copper surfaces [8,10,21–25]. For example, the data of Kim et al. [11] given in Figure 5.6 comparing the performance of undoped SAC105 (Sn1Ag0.5Cu) with that of LF35 suggests that the addition of small amounts of nickel has a positive influence on mechanical reliability, at least on copper surfaces. Another example is provided in Figure 5.7 for the addition 0.1% nickel to SACX (Sn0.3Ag0.7Cu + Bi) and the addition of 0.03% chromium to a SAC105 alloy with 0.1% nickel. Other elements that have been added to improve properties include bismuth, cobalt, indium, and germanium [8].

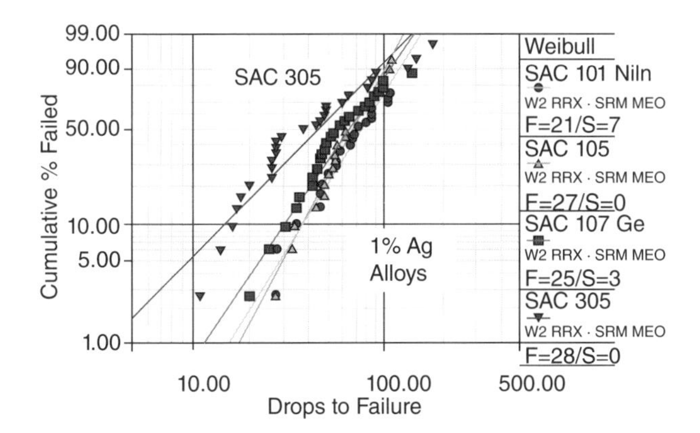

Figure 5.5 Data of Syed et al. [10] showing improved mechanical shock performance of three 1% silver alloys compared to SAC305

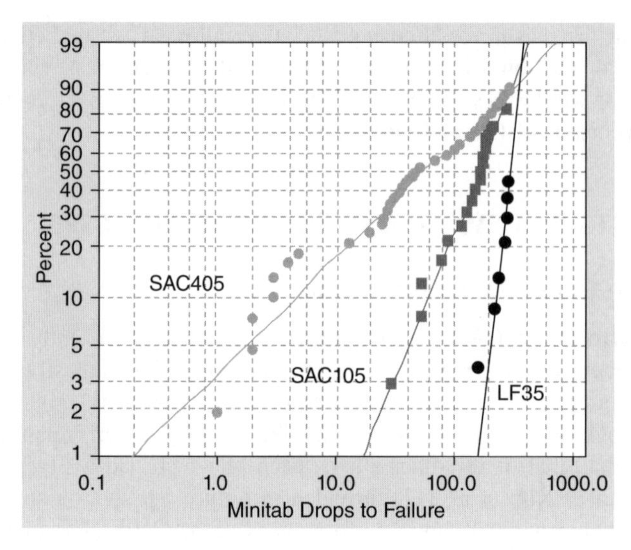

Figure 5.6 Data of Kim et al. [11] showing the cumulative failure percentage versus the number of drops in mechanical shock testing for SAC405, SAC105, and LF35(Sn1.2Ag0.5Cu + Ni) balled components on Cu-OSP substrates

Reports of improved mechanical shock resistance for alternate alloys are consistent with observations of changes in failure mode. The findings of Kim et al. [9] are shown in Figure 5.8. In their study the majority of cracking in SAC405 solders was through the IMC layer on the package side of the joint. In contrast, cracking in SAC105 joints was more complex, with cracks going through the bulk solder near the IMC layer and in the IMC. Similarly Pandher et al. [8] found that when small amounts of chromium and nickel were combined in the low-silver SACX alloy, the occurrence of brittle fractures was reduced by 80% relative to that of undoped SACX. Such changes in failure mode may explain the differences in Weibull slopes between high-silver and low-silver alloys apparent in Figures 5.5 and 5.6 during mechanical shock testing.

Theories Accounting for Improved Mechanical Shock Performance A number of theories have been proposed to account for the improved performance of low-silver and micro-alloyed SnAgCu solders in mechanical shock. Some focus on the impact of alloy composition on the mechanical properties of the bulk, while others focus on impacts to the composition, crystal structure, growth rate, and mechanical properties of the IMC layers at pad interfaces. It is likely that the composition of the new alloys has multiple effects, so a number of theories could help account for the overall improvement in mechanical shock resistance for low-silver, doped SnAgCu alloys. In this section a summary of a few theories is presented.

A reduction in silver concentration below the eutectic has been stated to decrease the strength and elastic modulus of the bulk solder. For a given amount of strain imposed on the joint, low-strength and stiffness alloys transfer less stress to the

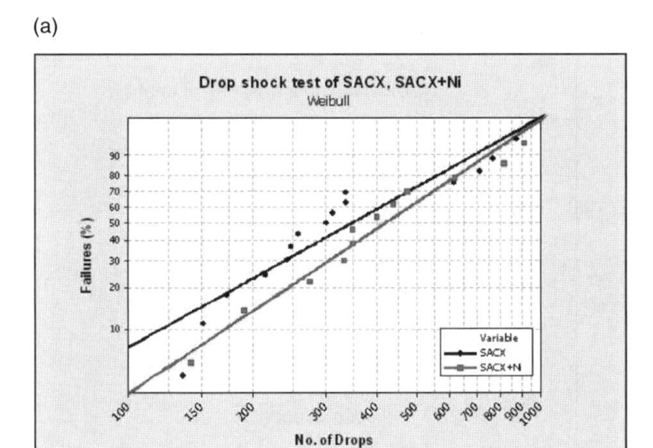

Figure 5.7 Mechanical shock data of Pandher et al. [8]: (a) SACX and SACX with 0.1% Ni addition and (b) SAC105 with 0.1% Ni and SAC105 with 0.1%Ni with a further addition of 0.03% Cr

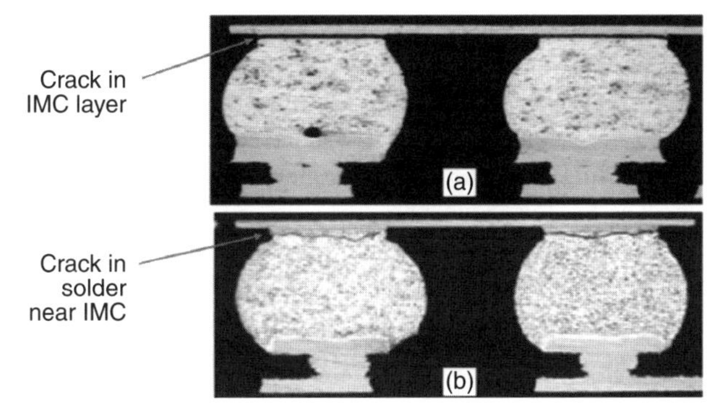

Figure 5.8 Micrographs comparing the drop failure modes of solder joints using ball alloys of (a) SAC405 and (b) SAC105. After Kim et al. [9]

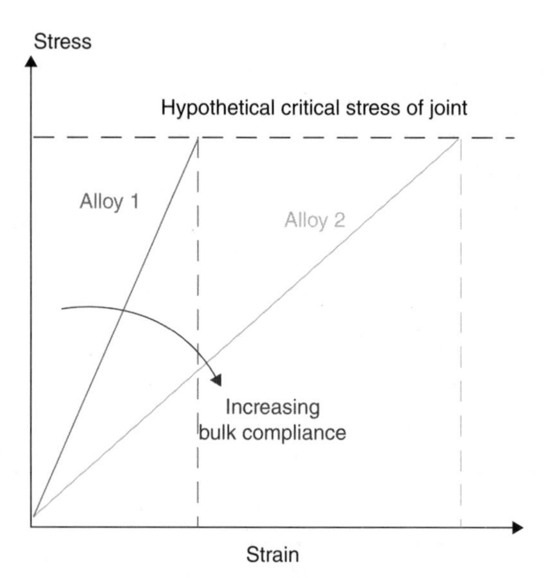

Figure 5.9 Schematic stress–strain behavior of solder joints for two hypothetical alloys with different elastic compliance. After Kim et al. [11]

solder/substrate interface [8,9,11,14]. Thus, as illustrated in Figure 5.9, low elastic modulus (and low yield strength) improves the mechanical shock performance of solder joints with low-silver alloys. Kim et al. [11] also pointed out that optimization of these properties required increasing the amount of primary Sn relative to the Ag_3Sn and Cu_6Sn_5 phases in the alloy, which was accomplished by reduced silver concentration. Note that the alloy failure strain or "ductility" as measured by elongation to failure was not an issue for any SnAgCu alloy. They are all ductile alloys. It would be their high strength and elastic stiffness that would result in "brittle" failure elsewhere in the system (e.g., the IMC layers or PCB laminate).

One role of micro-alloying additions would be to affect the IMC layers at the solder/pad interfaces. For example, Pandher et al. [8] showed that micro-alloying additions, particularly nickel, slowed inter diffusion, thus reducing IMC thickness (Figure 5.10) or the propensity for void formation. Reduced Cu_3Sn growth, in particular, improved mechanical shock performance. Kim et al. [11] also noted the general observation that interfacial strength or joint integrity inversely scales with the overall thickness of IMC layers on copper base metal. These authors concluded that the key to establishing a thinner IMC layer was to slow down the diffusion process. Through TEM and X-ray measurements of the intermetallic layers, they found that even a minimal amount of nickel substitution for copper in Cu_6Sn_5 or Cu_3Sn could effectively suppress the flow of tin atoms and retard IMC growth, especially of the Cu_3Sn phase.

In the case of nickel additions to joints on copper surfaces, another theory was that the nickel stabilizes the hexagonal close packed (hcp) form of the Cu_6Sn_5, ensuring the integrity of the intermetallic layer [26]. In the absence of nickel, hcp Cu_6Sn_5 transformed to the monoclinic form upon cooling below about 190°C. This transformation

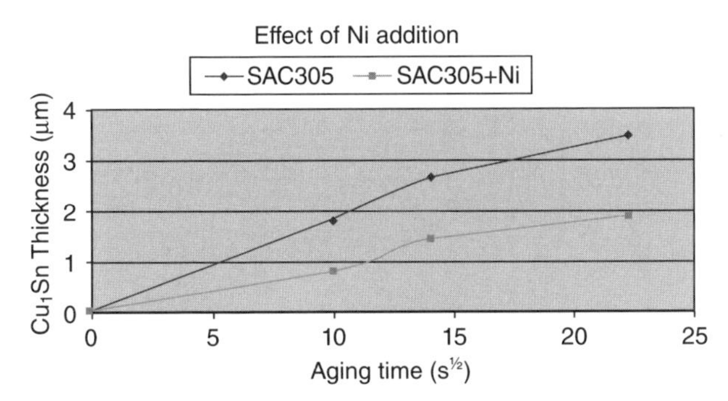

Figure 5.10 Intermetallic growth of SAC305 and SAC305 + 0.1% nickel soldered onto OSP copper pads. After Pandher et al. [8]

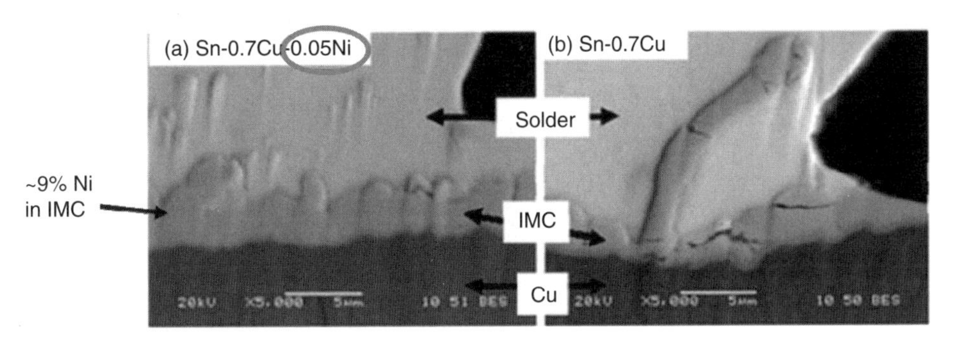

Figure 5.11 The impact of small nickel additions to Sn0.7Cu on crack formation in the Cu_6Sn_5 layer. After Nogita and Nishimura [26]

was accompanied by a 26% volume change that could lead to cracking in the IMC, as shown in Figure 5.11*b*. Small additions of nickel in the solder alloy concentrated in the IMC would stabilize the hcp form of Cu_6Sn_5, leading to an IMC without mechanical damage, Figure 5.11*a*.

Dependence of Mechanical Reliability on Pad Finish One complicating factor when assessing the impact of alloy composition on mechanical reliability would be the impact of surface finish, both on the package and the PCB sides of the joint. An illustration of the complexity comes from the work of Syed et al. [10,21]. Unlike the studies mentioned earlier, Syed et al. did not always observe a clear improvement in drop/shock performance for low-silver alloys with dopants compared to similar non-doped alloys. Specifically, their data on OSP-coated PCBs showed that SAC125 + nickel did not produce a significant drop/shock performance improvement over undoped SAC305 for a Ni/Au package finish. However, SAC125 + nickel was the best performer for Cu-OSP package finish.

Tanaka et al. [20] studied the mechanical shock resistance of solder joints formed using SAC305 and SAC125 + Ni (LF35) on both Cu-OSP and Ni/Au surfaces. As

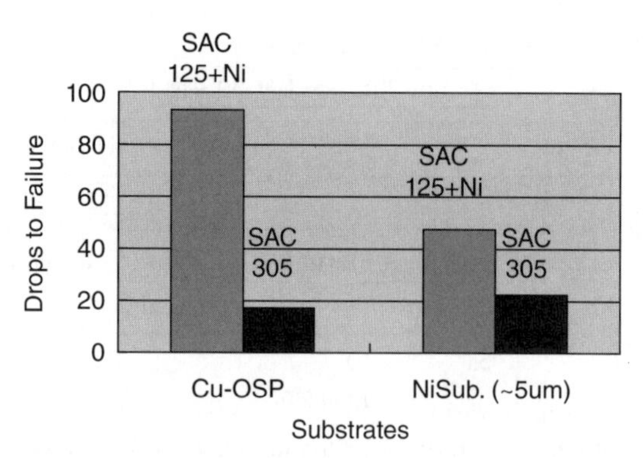

Figure 5.12 Effect of surface finish on the mechanical shock reliability of SAC305 and SAC125 + nickel joints. After Tanaka et al. [20]

shown in Figure 5.12, the LF35 performed better than SAC305 in this study for both surfaces. However, the LF35 performed much better on the Cu-OSP surface than on the Ni/Au, while the performance of SAC305 was virtually the same on both surfaces. Thus the data of Tanaka et al. suggest that doped low-silver alloys are more sensitive to surface finish than near-eutectic SnAgCu.

These data indicate the need for researchers to clearly state the PCB and package pad surface finishes when presenting mechanical shock data. Furthermore users must select alloys with a full understanding of which surface finishes will be present in the PCAs they manufacture.

5.3.2 Resistance to Bend/Flex Loading

BGA solder joints may be subjected to significant strains during in-circuit test (ICT), handling during manufacturing, connector or card insertion, and so forth. A recent investigation by the iNEMI "Alloy Alternatives" project team led to the conclusion that the impact of alloy composition on the resistance of BGA solder joints to bending or flexural strains was virtually unknown at this time [19,27]. However, the potential for significant impact of alloy composition on bend response was suggested by the decrease in strain limits due to the transition from SnPb to lead-free solders. An earlier iNEMI study [28] found that both load and deflection to failure were significantly reduced for four-point bend test parts soldered with near-eutectic SnAgCu compared to those soldered with SnPb paste. One would expect the lower elastic modulus and strength of low-silver alloys relative to near-eutectic SnAgCu to be beneficial, but there is little data to support this supposition. Song et al. [24] provided limited data suggesting that some "alternate alloys" may improve bend/flex performance. However, their investigation was limited in scope, and much remains unknown regarding the effect of alloy composition on bend/flex behavior. Clearly, understanding the impact of alloy composition on solder joint reliability under flexural strains at moderate strain rates is an area for future work.

5.4 TECHNICAL CONCERNS

5.4.1 Surface Mount Assembly

Impact of Low-Silver Content The potential benefit of alternate alloy balled BGA components would bring some associated impact to the manufacturing process, particularly when they are incorporated into a thermally challenging assembly. To understand the issue, the impact of the alloy composition on the melting point must be understood. Figure 5.13 shows the liquidus temperatures (the equilibrium temperature at which the full liquid phase exists) of several common SnAgCu alloy compositions. Notice that the reduction of silver concentration from the eutectic value of nearly 4% down to 1% has a significant impact on the liquidus temperature and (though not explicitly shown) the pasty range. This fact was one of the key reasons that near-eutectic SnAgCu alloys were chosen for the initial transition to lead-free solders [29]. Reduction of silver content could increase the melting point of the solder ball by as much as 10°C over that for SAC305 or SAC405 ball compositions. The addition of other alloying elements meant to affect undercooling, formation of various intermetallics, matrix properties, and microstructure may also affect the melting kinetics of the alloy, even if they do not significantly alter the equilibrium melting temperature.

OEMs and EMS providers are concerned about potential impacts of composition and melting point on proper solder joint formation. The causes for concern are (1) not

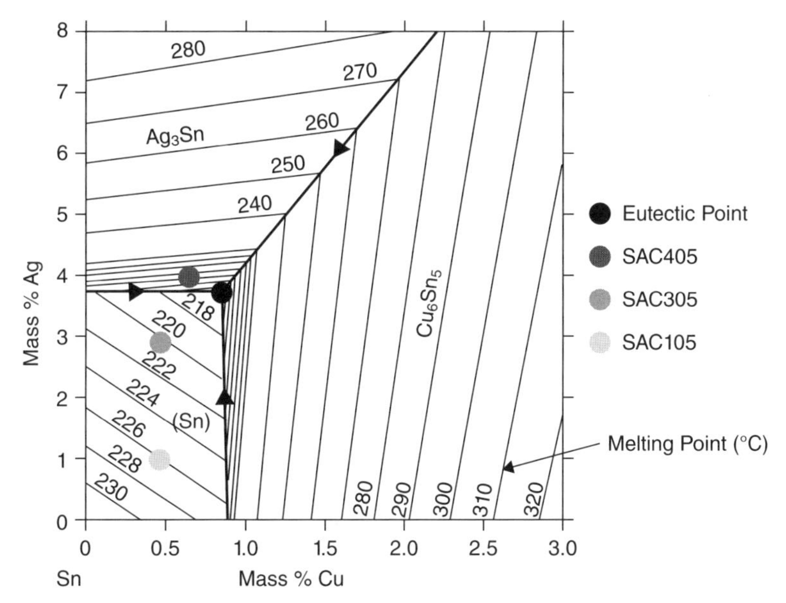

Figure 5.13 Liquidus projection for the Sn-rich portion of the SnAgCu system. After NIST [30]. The ternary eutectic composition at the 217°C eutectic temperature is indicated as the "eutectic point." For each alloy indicated, the pasty range is the difference between the liquidus temperature shown in the plot and the 217°C solidus temperature.

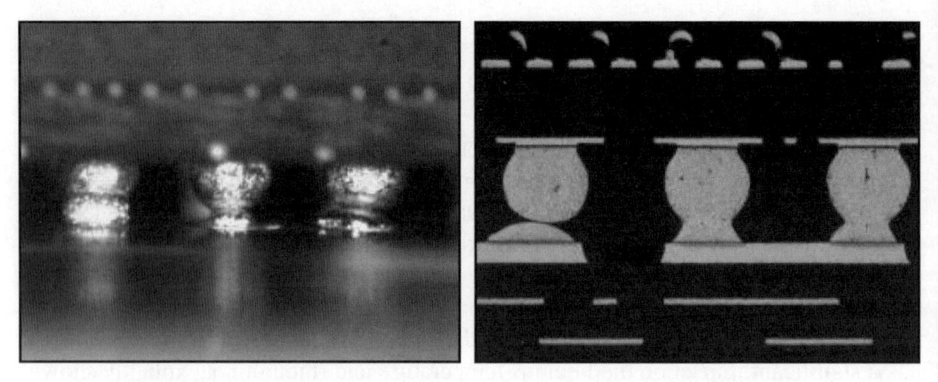

Figure 5.14 Unmelted solder balls and unacceptable solder joints [31]

all BGA/CSP component suppliers on the approved vendor list (AVL) for a component have the same ball composition and (2) a BGA supplier could make a change in the ball alloy composition and not change either the marking or the part number of the package to indicate this change. These situations can impact the assembly yields, or worse yet, create unacceptable solder joints because the assembly was soldered at too low a temperature. Improperly assembled components, such as those shown in Figure 5.14 [31], would pose a significant reliability risk because they may pass electrical test but fail more rapidly in the field than a properly formed solder joint.

The obvious solution to the increased melting temperature of low-silver alloys is to simply raise the assembly temperature of all lead-free printed circuit assemblies (PCAs) from the current minimum peak temperature of 230°C to 232°C by approximately 5°C to 7°C. This might be practical for simple, less thermally challenging PCAs where the range of package types would drive a low thermal gradient and none of the parts would exceed the maximum allowable body temperatures, as specified by the IPC/JEDEC J-STD-020 standard [32]. However, raising the soldering temperature on thermally challenging assemblies could be next to impossible without running the risk of overheating certain packages on the board.

Increasing temperatures will also put more strain on the PCB laminate, possibly causing excessive warpage or increasing the likelihood of pad cratering. Early profiling studies [31] suggested that 1% silver alloys may be incompatible with current industry lead-free assembly specifications that require a minimum peak reflow temperature of 230°C and time above liquidus (TAL) of 60 seconds. This limitation could preclude the use of low-silver alloys on thermally massive components and/or on thermally challenging assemblies.

Backward Compatibility with SnPb Solder Not all PCAs produced today are manufactured using lead-free solders. Those OEMs that qualify for exemptions specified in the RoHS Directive are still assembling many of their products using SnPb solder pastes. The challenge for these OEMs has been a shrinking supply of SnPb-balled BGAs, especially if the part was also used for consumer applications,

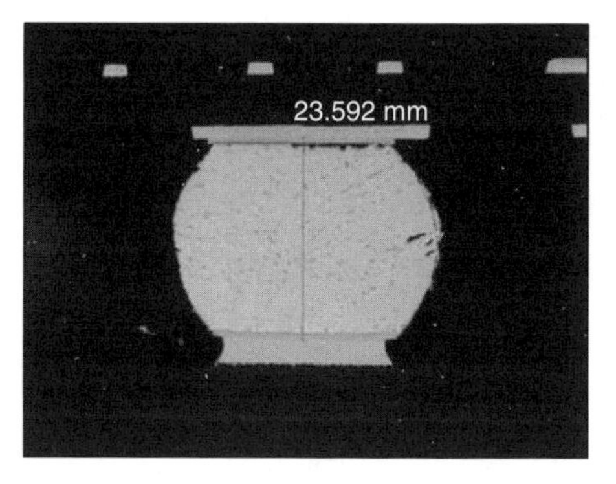

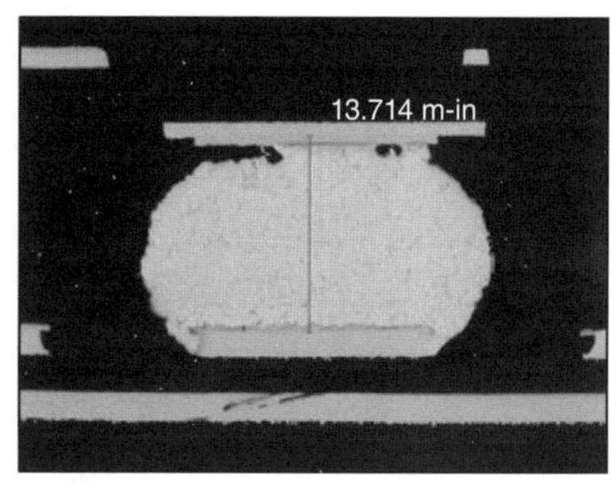

Figure 5.15 Shrinkage voids observed after rework of a SAC105 BGA using tin-lead paste [39]

where no exemption for SnPb solder exists. In some cases the use of lead-free balled BGAs is the only available option. In this situation manufacturers are forced to use the lead-free BGA in a SnPb soldering process. Previous reliability studies on backward compatibility, conducted using SAC305 and SAC405 BGAs soldered using SnPb paste, demonstrated that by soldering at peak temperatures in excess of 217°C, SnPb solder paste and SnAgCu BGA balls would mix completely to form a homogeneous microstructure with adequate reliability performance for most electronic applications [33–38].

However, the introduction of BGAs with low-silver content balls has changed this situation, particularly for rework operations. Unlike SAC305 or SAC405 BGAs, when SAC105 BGAs are reworked with SnPb solder paste, shrinkage voids can occur at the component-to-ball interface, as shown in Figure 5.15.

As described by Snugovsky et al. [39], when a SAC105 ball was mixed with SnPb eutectic solder paste, the resulting alloy has a very wide "pasty" range of approximately 45°C (177–224°C) compared to only about 30°C for the SAC305 ball/SnPb paste combination. The extra 15°C would be the result of the additional tin in the SAC105 relative to SAC305, driving up the overall melting point of the alloy and resulting in more time being required for the solidification of certain constituents. In production soldering situations the presence of additional alloy dopants (e.g., nickel, manganese, bismuth, cerium) could lower the final solidification temperature even further. As shown in Figure 5.16 and discussed in detail by Smith et al. [40], contraction forces during solidification over a large pasty range can lead to incomplete solder joint formation.

Although primarily a rework issue, it is possible that such voiding occurs during the assembly of high-density, thick boards or whenever uniform heating of the assembly is a challenge, although this author has not been made aware of this problem being observed at this time.

In summary, the new alloy alternatives add complexity to the manufacture of thermally challenging PCAs and mixed SnAgCu/SnPb alloy assemblies.

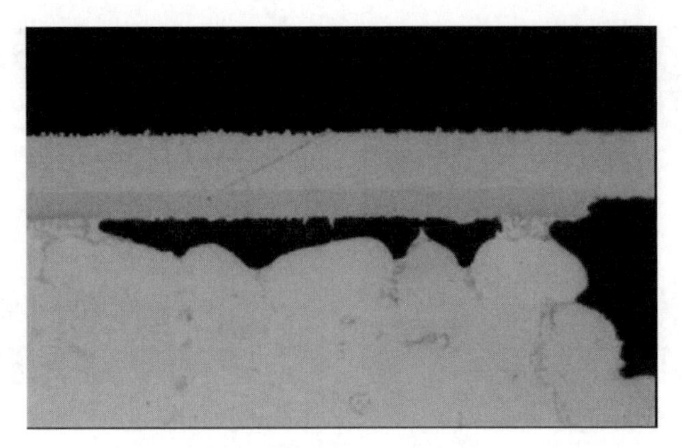

Figure 5.16 Shrinkage voids from repairing a BGA with SAC105 balls using tin-lead paste [40]

5.4.2 Thermal Fatigue Resistance

For the types of mobile devices that have motivated the push toward low silver and doped alloys, thermal fatigue is typically not a concern. For this reason alloy and component suppliers did not initially assess thermal fatigue performance for this class of material. As these alloys gain wider use, however, thermal fatigue performance will likely become a concern. A case in point is memory devices. As shown in Figure 5.4, low-silver alloys are being introduced into such components. As these memory devices begin to be used in the next generation of lead-free servers, telecommunications infrastructure equipment, and the like, understanding thermal fatigue behavior will be critical for OEMs to assess long-term reliability. Further, if low-silver alloys make their way into larger, higher power components than is currently the case (Figure 5.4), thermal fatigue will become a bigger issue to address.

A recent assessment by the iNEMI Alloy Alternatives project team found a major gap of knowledge about these new alloys to be lack of understanding of thermal fatigue performance [19,27]. Some of the key findings from that assessment, and also more indepth discussion, are provided in this section.

Thermal cycling studies for low-silver content and micro-alloyed SnAgCu solders are extremely limited and sometimes conflicting. Some accelerated thermal cycling (ATC) results show better performance for low-silver content alloys, while others show better performance for high-silver content alloys. For example, Kang et al. [41] investigated three low-silver alloys (Sn2.5Ag0.9Cu, Sn2.1Ag0.9Cu, and Sn2.3Ag0.5Cu0.2Bi) and compared their behavior to that of a Sn3.8Ag0.7Cu (SAC387) control. Various joint cooling rates and thermal cycle profiles were studied. Overall, despite expectations that hypoeutectic alloys would eliminate Ag_3Sn platelets in the microstructure and lead to improved performance, a clear and systematic improvement in thermal fatigue life with decreased silver concentration was not established.

Conversely, as shown in Figure 5.17, Terashima et al. [42] found a systematic increase in thermal fatigue performance in SnXAg0.5Cu alloys with increasing silver

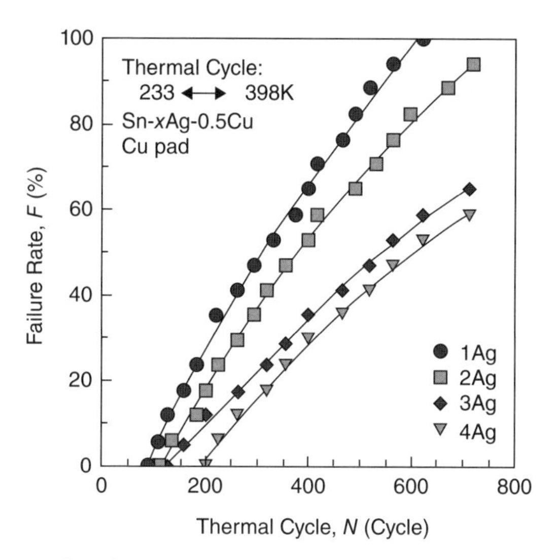

Figure 5.17 ATC test data from Terashima showing the direct relationship between silver content and thermal fatigue life [42]

concentration, with X taking values of 1, 2, 3, and 4%. The Terashima work was limited to flip chip interconnects (not BGA balls) and the ATC cycle was –40°C/125°C with approximately 15-minute dwells. Based on a detailed failure analysis, Terashima surmised that higher silver content inhibited microstructural coarsening and prolonged fatigue life. Similarly the data of Henshall et al. [43], shown in Figure 5.18, suggest that low-silver alloys performed worse than high-silver alloys in thermal fatigue.

The performance of low-silver alloys relative to the historical case of eutectic SnPb is not clear, since no studies have directly compared the results to a SnPb control. As pointed out by the iNEMI group and by Henshall et al. [43], further studies directly comparing the thermal fatigue performance of SnPb joints to those made using low-silver ball alloys are required in order to directly benchmark the lead-free alloys against the industry's historical solution.

To date there have been limited published studies regarding the impact of micro-alloy additions on thermal fatigue performance. Fortunately, this type of data has begun to be generated. Pandher et al. [44] published data that suggest the addition of about 1% bismuth (Bi) to low-silver alloys significantly improves temperature cycling performance, while other additives, such as nickel, had little to no effect in improving temperature cycling performance. Better understanding of how micro-alloy additions affect thermal fatigue performance is needed to bridge a major gap in our knowledge as an industry.

Finally, the impact of significant alloy changes on the acceleration factor that would relate field life to accelerated test life is unknown. Darveaux and Reichman [45] performed hysteresis loop predictions for various lead-free alloys based on measured mechanical property data. The shape and area of the loops, and their dependence on

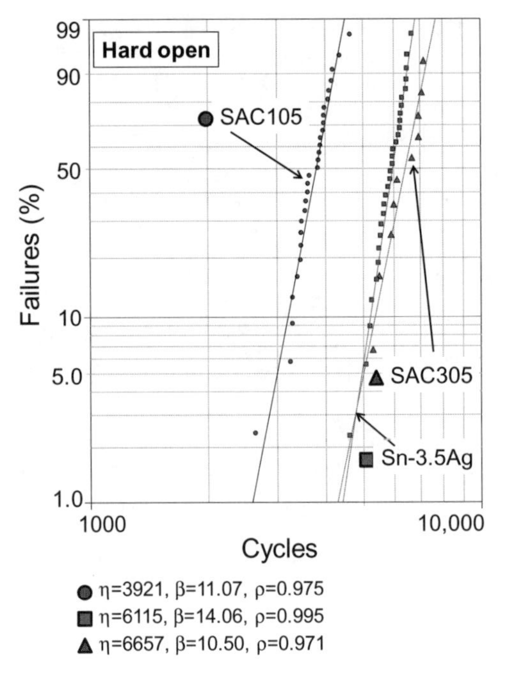

Figure 5.18 Weibull plot of failure life for the three different ball alloy joints using an electrical failure criterion of a hard open. Data of Henshall et al. [43]

thermal cycle parameters, varied significantly for different alloys. This led the authors to conclude that the acceleration factor would also vary by alloy. However, direct measurements of the acceleration factors from ATC testing have yet to be published for any of the new alloys.

The iNEMI Alloy Alternatives team also concluded that thermal fatigue reliability appeared to be dependent on process, microstructure, and micro-alloy content, and those dependencies had yet to be characterized fully and understood. Overall, the impact of ball alloy composition on thermal fatigue life in the field was difficult to judge at this time.

5.5 MANAGEMENT OF NEW ALLOYS

5.5.1 Overview of Concerns

Management of alloy choices presents a challenge for OEMs. Product portfolios can be broad, supply chains are complicated, and business models range from no-touch to in-house design. Assurance of supply concerns exist if only one patented solder with limited licensees is to be used on multiple product lines and/or factory locations. The management of multiple alloys at a single or multiple factory sites can be complicated, as we have seen in the original lead-free transition. Product support and repair also become more complex when multiple solder alloys are used. A part number change for BGA/CSP components when switching from high-silver to low-silver ball alloys can help in managing the reflow process but may have logistical and cost implications for component suppliers. A variety of industry standards still need to be updated to account for new alloys. Finally, establishing methods to evaluate the acceptability of various alloys for different products remains a challenge. This section discusses some of these concerns and possible ways to address them.

5.5.2 Assessment of New Alloys for Acceptability

One situation that has created uncertainty in the industry regarding new alloys, and that may slow the adoption of improved materials, is the lack of defined information requirements for alloy acceptance. The acceptability of any alloy may vary from product class to product class, and possibly from company to company. However, the methodology and data requirements may be largely the same, regardless of product or company requirements.

A significant obstacle to useful, data-driven assessments of alternate lead-free alloys has been the inconsistent testing that has been performed on new materials. The data from experiments conducted, while valid on their own, are often not comparable in any meaningful way to data from other equally valid experiments due to differences in the choice of test conditions, controls, or other parameters. Also alloys formulated to meet specific goals, such as improved mechanical shock resistance, have not always been tested to determine suitability for general use by assessing other performance aspects, such as thermal fatigue life [46–48].

Holder et al. published the outline of a company test protocol designed to provide a balanced assessment of potential alloys [16]. Since then, a specification has been released for company suppliers to use in qualifying new alloys for company products [49]. If widely adopted and standardized, such a test protocol would help bridge the gap between the specialized data generated on materials optimized for particular attributes and the data required for assessing the suitability of alloys for general use. Efforts to standardize alloy test requirements are under way [50].

One key feature of the protocol would be to have an alloy or combination of alloys specified as controls. These controls would allow for ordered ranking against currently accepted materials. By bounding the performance of new materials with the current near-eutectic SnAgCu alloys and the historic solution of eutectic SnPb, appropriate pass/fail criteria would often be more apparent.

The required tests would be divided into three major areas:

- Reliability
- Manufacturability
- Material properties

The set of tests and parameters is summarized in Table 5.3.

TABLE 5.3 Highlights of a proposed test requirement for new BGA/CSP ball alloys

Test and Method	Test Parameters and Controls	Report
Accelerated thermal cycling: IPC-9701A standard	• TC1 (0–100°C, 10 min dwell) • 6000 cycles or failure of all 32 components • New alloy ball and paste • Controls: Sn37Pb, SAC305	• Weibull failure curve • Tabulated failure data
Mechanical shock: JEDEC JESD22-B111 standard	• Condition B–1500 G, 0.5 ms half-sine pulse • Failure of all 32 components • New alloy ball and paste • Controls: Sn37Pb, SAC405	• Per JEDEC JESD22-B111 standard • Weibull failure plots
Stress–strain response	• Bulk new alloy • Strain rate of 10^{-3} s^{-1} • SAC305 control • Defined specimen geometry	• Engineering stress–strain curve • True stress–strain curve • 0.2%-offset yield stress • Ultimate tensile stress • Elongation to failure
Dynamic elastic modulus: ASTM 1875-00 standard	• Sn37Pb control	• Young's modulus • Shear modulus • Poisson's ratio
Impact on reflow: company protocol	• Peak reflow temperatures from 230–265°C; TAL 15–120 s. • Controls: SAC305	• Micrographs of solder joints and joint cross sections. • IMC layer thickness
Wetting balance: IPC J-STD-003 standard	• Test F1 (lead-free solder) • Controls: SAC305	• Wetting balance curves

Note: Details are provided in [49].

5.5.3 Updating Industry Standards

There are a number of key industry standards that require updating or modification to address the new lead-free alloys. Updates to these standards could help manage alloy change and choice, and reduce confusion throughout the industry. The three standards discussed here address labeling of components and PCAs, change notification, and solder alloy requirements.

Recently the iNEMI Alloy Alternatives project team suggested to IPC and JEDEC that clarifications to account for new lead-free alloys be made within IPC/JEDEC J-STD-609 standard, "Marking and Labeling of Components, PCBs and PCBAs to Identify Lead (Pb), Pb-Free and Other Attributes." This standard requires so-called e-code labels on components, PCBs and PCAs to alert EMS and repair suppliers of the alloys used on component terminations and attachment solders. For mainstream lead-free BGA ball alloys, two e-codes are relevant:

- e1 = SnAgCu.
- e2 = tin (Sn) alloys with no bismuth (Bi) nor zinc (Zn), excluding SnAgCu.

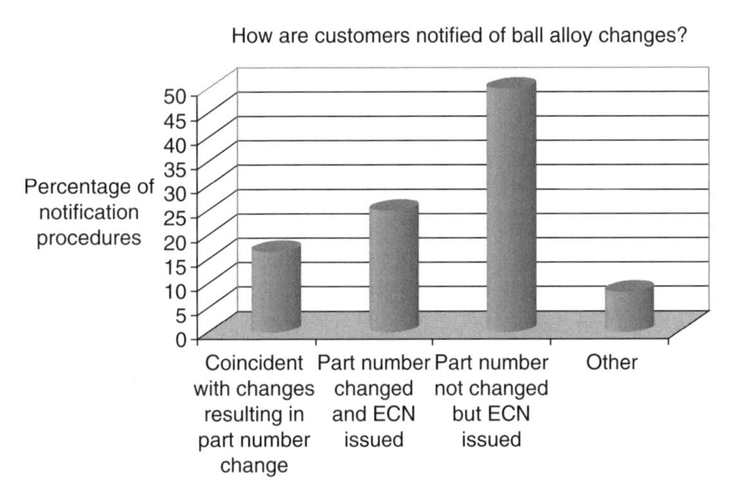

Figure 5.19 Results of the 2007 BGA/CSP component supplier survey on the approach taken to notify customers of ball alloy changes

The confusion here is how to label low-silver and micro-alloyed materials. For example, some large OEMs have learned that different BGA suppliers use different e-code labels for the same doped, low-silver alloy. Based partly on the iNEMI input, the committees responsible for the IPC/JEDEC J-STD-609 standard have updated the document to provide clarification on labeling for the new alloys.

As discussed in Section 5.4, a lead-free BGA ball alloy change can have an adverse impact on printed circuit assembly (PCA) manufacturing because of the higher melting point of some alloys. In particular, the change to low-silver ball alloys may require a change to PCA manufacturing processes. Thus it is important for the manufacturer (at a minimum) to know the alloy being used on all BGA/CSP parts and to be informed if a change in alloys is made. Results of the BGA/CSP component supplier survey described in Section 5.2.2 demonstrate the diversity of approaches currently taken by BGA/CSP component suppliers. Figure 5.19 shows that about half of the surveyed suppliers did not change part numbers when changing a ball alloy; another 42% provided a part number change when only the alloy changed or when it was coincident with other component changes.

The iNEMI Alloy Alternatives project team presented to the JEDEC JC-14 committee for Quality and Reliability the concerns about part numbers and customer notification when BGA/CSP suppliers changed ball alloys. These concerns were shared by the EMS Forum (EMSF), a consortium of EMS suppliers. The key point in the iNEMI and EMSF position was that a change in ball alloy was a change in form, fit, or function, and so necessitated the need for a part number change in order to alert the manufacturer of the potential impact on the reflow profile [51]. The iNEMI group requested that the committee consider mandating, or at least recommending, that new part numbers be issued when a BGA/CSP supplier changes the solder ball alloy such that a manufacturing process change may be needed. Based on this request, a new JEDEC JC-14 task group was formed to consider this issue and the iNEMI recommendation for standardization by JEDEC.

Another critical standard is J-STD-006, "Requirements for Electronic Grade Solder Alloys and Fluxed and Non-Fluxed Solid Solders for Electronic Soldering Applications." The committee responsible for this standard has been discussing how to update IPC J-STD-006 standard to account for new alloys, particularly those with micro-alloy additions. One major concern is that some micro-alloying elements are present in amounts that would normally be considered an impurity. Yet intellectual property concerns may limit the degree to which solder suppliers are at liberty to fully disclose alloy compositions. The IPC Solder Products Value Council (SPVC), iNEMI, and others are working to address these concerns together with the IPC standard committee responsible for this standard.

5.5.4 Industry Activities

Extensive study of new alloys continues to take place at individual companies, including solder suppliers, component manufacturers, EMS providers, and OEMs. The interest in this topic is evident by the number of papers presented at recent industry conferences. In addition a number of formal and informal consortia are investigating various aspects of new lead-free alloys. As mentioned earlier in this chapter, in 2007 iNEMI established an "Alloy Alternatives" project team, composed of representatives from 16 companies, including solder suppliers, BGA/CSP component producers, EMS providers, and OEMs. The activities and results of this group's work to date have been published elsewhere [19,27,52]. In summary, this group has focused on identifying knowledge gaps needing closure by the industry and on trying to help the industry manage alloy choice. One of the major knowledge gaps identified concerns about thermal fatigue resistance of low-silver and micro-alloyed solders. Listed below are a number of industry studies identified by iNEMI that may help address this gap. The results from most of these studies are expected to become public eventually.

- **HDPUG** A multiple alloy (10 alloys) screening study is in progress. Thermal cycling is being performed at 0°C to 100°C with 30-minute dwells. Data will be published shortly [53].
- **Industry Working Group (Flextronics, HP, Cisco, SUN, Xilinx, Motorola)** This group tested Sn3.5Ag, SAC105, and SAC305 using a Xilinx 676 PBGA component. ATC testing was completed using a 0°C to 100°C temperature cycle, with 10-minute ramps and dwells. Results were published in the proceedings of the 2009 IPC APEX conference [43].
- **Jabil Working Group (Jabil, Cookson, HP, Amkor, Cisco)** This group is testing SACX (Sn0.3Ag0.7Cu + bismuth), LF35 (Sn1.2Ag0.5Cu + nickel), SAC105, SAC205, SAC305, and Sn37Pb. Four Amkor organic package types and sizes are being tested. The program has two phases: (1) manufacturing impact and (2) reliability, including mechanical shock and thermal cycling. ATC testing will be performed with two thermal cycle profiles: 0°C to 100°C and –40°C to 125°C. The work is in progress [54–56] and the first set of ATC results were published in the proceedings of the IPC-APEX conference, 2010 [57].

- **Alcatel-Lucent Working Group (Alcatel-Lucent, LSI, Celestica)** This group is testing SAC405, SAC305, SAC105, and a SnPb baseline using a LSI 680 PBGA component. The program includes SMT and rework with thermal cycling at 0°C to 100°C with 10-, 30-, and 60-minute dwells. The work is in progress and the group is expected to publish results at the SMTAi conference in 2010.
- **Unovis** Tests are in progress to evaluate multiple alloys using commercial and in-house test vehicles. Data initially will be limited to consortium members but are likely to become available to the public at some point.

5.6 FUTURE WORK

The mission of the iNEMI Alloy Alternatives project has been to establish an industry state of knowledge regarding new lead-free solder alloys. Table 5.4 summarizes the major areas where industry understanding is relatively complete, or at least adequate, as determined by this group.

Clearly, there is much to be learned about new lead-free alloys and work to be done in successfully managing the transition to them. In addition to the efforts related to thermal fatigue listed earlier, Table 5.5 shows some other areas requiring work. EMS providers, solder and BGA/CSP component suppliers, and OEMs are encouraged to continue to study new alloys and to report on their findings.

The iNEMI Alloy Alternatives team has launched phase 2 of its efforts to help address both technical gaps and management issues. Two major areas are being addressed. First, the lack of information on the thermal fatigue performance for many new lead-free alloys has motivated the Alloy Alternatives team to plan accelerated thermal cycle experiments. The project team has considered many possible sets of experiments to answer a variety of questions. In the end the team decided on the following course of action:

- Validate the impact of silver concentration in the range of 0% to 4% on thermal fatigue resistance.
- Evaluate the impact of commercially common dopants, such as nickel, on thermal fatigue performance.
- Assess how alloy composition affects the acceleration behavior.

TABLE 5.4 Areas where knowledge about new lead-free alloys is complete or adequate

Sufficient Knowledge
Low-Ag alloys improve drop/shock resistance
Micro-alloy additions significantly improve drop/shock performance on Cu surfaces but not on Ni surfaces
Decreasing Ag content decreases elastic modulus, yield, and tensile strength of SAC
Decreasing Ag content decreases creep strength of SAC
Alloy additions can increase the creep strength of low-Ag SAC alloys
SAC alloys are not inherently brittle (needs to be better communicated, however)

TABLE 5.5 Key knowledge gaps regarding the performance and impact of new lead-free alloys

Gap or Concern
High priority
Advantages and disadvantages of specific alloys
Composition limits for microalloy additions; ranges of effectiveness
Standard method to assess new alloys; standard data requirements
Consistency of testing methods (test vehicles and assembly, test parameters, etc.)
Establish the microstructural characteristics of specific alloys
Long-term reliability data for new alloys, particularly low-Ag and micro-alloyed
Lack of thermal cycle data for evaluating new alloys; benchmark to Sn-Pb and SAC 305/405
Medium priority
Assessment of new alloys for use in "mission critical, long-life" products
Impact of rework on microstructure and properties
Mixed Sn-Pb/Pb-free assembly, including rework
Impact of alloy composition on work hardening rates and other flow properties; effect of strain rate and temperature
Impact of alloy composition on bend/flex limits (moderate strain rate; ICT, handing, card insertion, etc.)
Thermal fatigue accelerations factors (not yet fully established for SAC 305/405)
Impact of aging on microstructure and mechanical properties
Low priority
Solder process margins required for new alloys used in various product classifications
Mixing of different BGA ball alloys and paste alloys for various component and board designs

- Provide basic thermal fatigue data for several of the most common alternate alloys on the market today, benchmarking them against eutectic SnPb and SAC305.

Currently this team is in the process of finalizing details of the accelerated thermal cycle test plan, including which alloys to test and which thermal cycle profiles to investigate such that meaningful acceleration factor data will result. If successful, the data could be of major benefit to the industry, especially since such large studies would be nearly impossible for a single company to undertake.

In addition the iNEMI team is helping to develop standard test requirements and methods for new alloys. Work has begun with a review of the approach being developed by Hewlett-Packard, as described earlier. The iNEMI team is discussing the merits of this approach and possible modifications that may be necessary in order to meet the needs of the broader electronics industry. A dialogue also is taking place between the iNEMI Alloy Alternatives team and the IPC Solder Products Value Council (SPVC), with the ultimate goal of providing a formal starting point for the development of an IPC standard, or set of standards, addressing testing of new lead-free alloys.

Besides filling these technical knowledge gaps, the industry must continue to create or update standards that will help manage the increasing number of alloys choices, such as those discussed in Section 5.5.3. The goal is to drive standards that will help companies take advantage of the benefits of new alloys without suffering unintended negative consequences.

5.7 SUMMARY AND CONCLUSIONS

The drivers, benefits, and concerns associated with the deployment of second-generation, "alternative" lead-free solder alloys for use with area array packages have been presented. These new alloys represent both an opportunity to solve issues arising from the transition to lead-free solders and a reliability risk if not managed properly. While the number of alloy choices has risen dramatically in the past few years, it remains unclear if this situation will continue long term or if convergence to a handful of select alloys will occur. Generally, the new alloys employ reduced silver concentration and/or the addition of "micro-alloying" elements in concentrations of typically 0.1% or less. Based on a review of available literature and discussions with many who are active in the field, the following conclusions can be made at this time.

1. The number of lead-free alloy choices has expanded as various approaches are developed to address the limitations of near-eutectic SnAgCu alloys. The wide range of alloy choices is both an opportunity and a risk for the electronics industry. Clearly, the new alloys perform better in some respects than near-eutectic SnAgCu. However, many of these alloys have not yet been fully characterized over the full range of behavioral factors relevant to the industry as a whole. Furthermore having such a wide range of solder alloys in use adds complexity and risk that has not existed within the supply chain up to now.
2. Based on the survey results of mid 2007, low-silver SnAgCu ball alloys, specifically those based on Sn1Ag0.5Cu (SAC105), Sn1.2Ag0.5Cu (SAC125), and Sn1Ag0.1Cu (SAC101) were used by 77% of surveyed BGA/CSP component suppliers for at least some of their components. Typically these low-silver alloys contained micro-alloy additions. Many suppliers provided multiple alloys, dependent on package type and application.
3. Studies consistently showed that low-silver alloys performed better in mechanical shock than high-silver alloys. Many studies showed that micro-alloying additions improved the mechanical shock performance of SnAgCu alloys on copper surfaces. Reports of improved mechanical shock resistance are consistent with observations of changes in failure mode.
4. One complicating factor when assessing the impact of alloy composition on mechanical shock resistance was the impact of surface finish, both on the package and the PCB sides of the joint. Care would need to be taken when assessing data or making alloy selections to account for the materials used on both soldering surfaces.
5. A number of theories have been proposed to account for the improved performance of low-silver and micro-alloyed SnAgCu solders in mechanical shock. Some focused on the impacts of alloy composition on the mechanical properties of the bulk solder, while others focused on impacts to the composition,

crystal structure, growth rate, and mechanical properties of the IMC layers at pad interfaces. It was likely that the composition of the new alloys had multiple effects, so a number of theories could help account for the overall improvement in mechanical shock resistance.

6. OEMs and EMS providers are concerned about potential impacts of composition and melting point on proper solder joint formation during reflow and rework. The high melting point of low-silver alloys can impact assembly yields, or worse yet, create unreliable solder joints because the assembly was soldered at too low a temperature. Furthermore, when a low-silver ball is mixed with SnPb eutectic solder paste, the resulting alloy has a very wide "pasty" range, which can lead to incomplete solder joint formation.
7. Understanding of thermal fatigue performance is a major gap in knowledge for the new alloys. Thermal cycling studies for low silver content and micro-alloyed SnAgCu solders are extremely limited and sometimes conflicting. The performance of low-silver alloys relative to eutectic SnPb is not clear. The impact of significant alloy changes on the acceleration factor that relates field life to accelerated test life is unknown. Overall, the impact of ball alloy composition on thermal fatigue life in the field is difficult to judge at this time. Fortunately, a number of studies are underway to address this gap in knowledge.
8. Management of alloy choice presents challenges for OEMs. One situation that creates uncertainty in the industry regarding new alloys, and that may slow the adoption of improved materials, is the lack of defined information requirements for alloy acceptance. Hewlett-Packard recently published the outline of a test protocol designed to provide a balanced assessment of potential alloys [16]. If widely adopted and standardized, such a test protocol could increase the use of improved alloys while decreasing the risk of accidental misuse.
9. There are a number of key industry standards that require modification to address the new lead-free alloys. Updates to these standards could help manage alloy change and choice more smoothly, and reduce confusion throughout the industry. Three standards discussed in this chapter address labeling of components and PCAs, change notification, and defining solder alloy test requirements. Progress is being made to update these standards.
10. Extensive study of new alloys continues to take place at individual companies, including solder suppliers, BGA/CSP manufacturers, EMS providers, and OEMs. A number of formal and informal consortia also are investigating various aspects of new lead-free alloys. Thus we can expect further progress in understanding and managing lead-free alloys over the next few years.

ACKNOWLEDGMENTS

Many people contributed to the development and assessment of the information provided in this chapter. First, the author would like to thank his colleagues at Hewlett-Packard who have helped him tremendously over the past two years and have helped lead the industry in establishing alloy test methods: Helen Holder, Kris Troxel, Aileen Allen, Elizabeth Benedetto, Jian Miremadi, Guillermo Oviedo and Michael Roesch. Next, the author owes much to the iNEMI Alloy Alternatives Team, in particular the following: Steve Tisdale, Keith Sweatman, Keith Howell, Robert Healey, Ranjit Pandher, Richard Coyle, Thilo Sack, Polina Snugovsky, Fay Hua, Jim McElroy, and Jim Arnold. The author is also grateful for many excellent discussions and the efforts of his collaborators in the "Industry Working Group," especially Jasbir Bath, Keith Newman, Sundar Sethuraman, and M. J. Lee. Likewise the author would like to thank his colleagues in the "Jabil Working Team:" Quyen Chu, Chris Shea, Ahmer Syed, Ken Hubbard, and Girish Wable.

REFERENCES

1. NCMS Lead-Free Solder Project Final Report, NCMS, National Center for Manufacturing Sciences, 3025 Boardwalk, Ann Arbor, MI 48108-3266, Report 0401RE96, August 1997, and CD-ROM database of complete data set, including micrographs and raw data, August 1999. Information on how to order these can be obtained from http://www/ncsm.org/.
2. I. Artaki, D. Noctor, C. Desantis, W. Desaulnier, L. Felton, M. Palmer, J. Felty, J. Greaves, C. Handwerker, J. Mather, S. Schroeder, D. Napp, Y.-Y. Pan, J. Rosser, P. Vianco, G. Whitten, and Y. Zhu, "Lead-Free, High-Temperature, Fatigue-Resist and Solder," Final Report of NCMS Project No. 170503-96034, National Center for Manufacturing Sciences, 2001.
3. *Lead-Free Electronics: iNEMI Projects Lead to Successful Manufacturing*, E. Bradley, C. A. Handwerker, J. Bath, R. D. Parker, and R. W. Gedney, eds., IEEE Press Piscataway, New Jersey 2007.
4. J. Arnold and K. Sweatman "Reliability Testing of Ni-Modified SnCu and SAC305—Accelerated Thermal Cycling,", *Proc. SMTAI,* Surface Mount Technology Association, Edina, Minnesota, p. 187, 2008.
5. T. Gregorich, P. Holmes, J. Lee, and C. Lee, "SnNi and SnNiCu Intermetallic Compounds Found When Using SnAgCu Solders," IPC/Soldertec Global 2nd Int. Conf. on Lead Free Electronics, June 23, 2004.
6. L. Garner, S. Sane, D. Suh, T. Byrne, A. Dani, T. Martin, M. Mello, M. Patel, and R. Williams, "Finding Solutions to the Challenges in Package Interconnect Reliability," *Intel J.*, vol. 9, p. 297, 2005.
7. Y-S. Lai, P. Yang, and C. Yeh, "Experimental Studies of Board-Level Reliability of CSPs Subjected to JEDEC Drop Test Condition," *Microelectr. Reliab*, vol. 46, pp. 645–650, 2006.
8. R. Pandher, B. G. Lewis, R. Vangaveti, and B. Singh, "Drop Shock Reliability of Lead-Free Alloys—Effect of Micro-Additives," *Proc. ECTC*, IEEE, New York, New York, p. 669, 2007.
9. H. Kim, M. Zhang, C. M. Kumar, D. Suh, P. Liu, D. Kim, M. Xie, and Z. Wang, "Improved Drop Reliability Performance with Lead-Free Solders of Low Ag Content and Their Failure Modes," *Proc. ECTC*, IEEE, New York, New York, p. 962, 2007.
10. A. Syed, T.-S. Kim, S.-W. Cha, J. Scanlon, and C.-G. Ryu, "Effect of Pb free Alloy Composition on Drop/Impact Reliability of 0.4, 0.5 and 0.8 mm Pitch Chip Scale Packages with NiAu Pad Finish," *Proc. ECTC*, IEEE, New York, New York, p. 951, 2007.
11. D. Kim, D. Suh, T. Millard, H. Kim, C. Kumar, M. Zhu, and Y. Xu, "Evaluation of High Compliant Low Ag Solder Alloys on OSP as a Drop Solution for the 2nd Level Pb-Free Interconnection," *Proc. ECTC*, IEEE, New York, New York, p. 1614, 2007.
12. T. Kobayashi, Y. Kariya, T. Sasaki, M. Tanaka, and K. Tatsumi, "Effect of Ni Addition on Bending Properties of Sn-Ag-Cu Lead-Free Solder Joints," *Proc. ECTC*, IEEE, New York, New York, p. 684, 2007.

13. D. Cavasin, M. Anani, G. Rittman, and J. Casto, "Board Level Characterization of Pb-free Sn-3.5Ag versus Eutectic Pb-Sn Solder Joint Reliability," *Proc. ECTC*, IEEE, New York, New York, p. 129, 2007.
14. B. Huang, H.-S. Hwang, and N.-C. Lee, "A Compliant and Creep Resistant SAC-Al(Ni) Alloy," *Proc. ECTC*, IEEE, New York, New York, p. 184, 2007.
15. M. Pei and J. Qu, "Effect of Rare Earth Elements on Lead-Free Solder Microstructure Evolution," *Proc. ECTC*, IEEE, New York, New York, p. 198, 2007.
16. H. Holder, G. Henshall, A. Maloney, E. Benedetto, K. Troxel, G. Oviedo, J. Miremadi, and M Roesch, "Test Data Requirements for Assessment of Alternative Pb-Free Solder Alloys," *Proc. SMTAI*, Surface Mount Technology Association, Edina, Minnesota, p. 150, 2008.
17. J. Lau, J. Gleason, V. Schroeder, G. Henshall, W. Dauksher, and B. Sullivan, "Design, Materials, and Assembly Process of High-Density Packages with a Low-Temperature Lead-Free Solder (SnBiAg)," *Soldering Surf. Mount Technol.*, vol. 20, no. 2, p. 11, 2008.
18. US7179417, *Sn-Zn Lead-Free solder Alloy, Its Mixture, and Soldered Bond*, Yoshikawa, Nippon Metal, 2007.
19. G. Henshall, R. Healey, R. Pandher, K. Sweatman, K. Howell, R. Coyle, T. Sack, P. Snugovsky, S. Tisdale, and F. Hua, "iNEMI Pb-Free Alloy Alternatives Project Report: State of the Industry," *Proc. SMTAI*, Surface Mount Technology Association, Edina, Minnesota, p. 109, 2008.
20. M. Tanaka, T. Sasaki, T. Kobayshi, and K. Tatsumi, "Improvement in Drop Shock Reliability of Sn–1.2Ag–0.5Cu BGA Interconnects by Ni Addition," *Proc. ECTC*, IEEE, New York, New York, p. 78, 2006.
21. A. Syed, T. Kim, and S. Cha, "Alternate Solder Balls for Improving Drop/Shock Reliability," *Proc. SMTAI*, Surface Mount Technology Association, Edina, Minnesota, p. 390, 2007.
22. W. Liu and N.-C. Lee, "Novel SACX Solders with Superior Drop Test Performance," *Proc. SMTAI*, Surface Mount Technology Association, Edina, Minnesota, p. 134, 2006.
23. F. Che, J. Luan, and X. Baraton, "Effect of Silver Content and Nickel Dopant on Mechanical Properties of Sn-Ag-based Solders," *Proc. ECTC*, IEEE, New York, New York, p. 485, 2008.
24. F. Song, J. Lo, J. Lam, T. Jiang, and S. Lee, "A Comprehensive Parallel Study on the Board Level Reliability of SAC, SACX and SCN Solders," *Proc. ECTC*, IEEE, New York, New York, p. 146, 2008.
25. W. Liu, P. Bachorik, and N.-C. Lee, "The Superior Drop Test Performance of SAC-Ti Solders and Its Mechanism," *Proc. ECTC*, IEEE, New York, New York, p. 452, 2008.
26. K. Nogita and T. Nishimura, "Nickel-Stabilized Hexagonal $(Cu, Ni)_6Sn_5$ in Sn-Cu-Ni Lead-Free Solder Alloys," *Scripta Materialia*, vol. 59, no. 2 pp. 191–194, 2008.
27. G. Henshall, R. Healey, R. Pandher, K. Sweatman, K. Howell, R. Coyle, T. Sack, P. Snugovsky, S. Tisdale, F. Hua, and H. Fu, "Addressing Industry Knowledge Gaps Regarding New Pb-Free Solder Alloy Alternatives," *Proceedings. IEMT*, IEEE, New York, New York, p. B4.2, 2008.
28. J. Gleason, C. Reynolds, J. Bath, Q. Chu, M. Kelly K. Lyjak, and P. Roubaud, "Pb-Free Assembly, Rework, and Reliability Analysis of IPC Class 2 Assemblies," *Proceedings ECTC*, IEEE, New York, New York, p. 959, 2005.
29. C. A. Handwerker, U. Kattner, K. Moon, J. Bath, E. Bradley, and P. Snugovsky, "Alloy Selection," in *Lead-Free Electronics*, E. Bradley, C. A. Handwerker, J. Bath, R. D. Parker, R. W. Gedney, Eds., Wiley & Sons, Hoboken, New Jersey, p. 9, 2007.
30. National Institute of Standards and Technology, http://www.metallurgy.nist.gov/phase/solder/agcusn.html.

31. G. Henshall, M. Roesch, K. Troxel, H. Holder, J. Miremadi "Manufacturability and Reliability Impacts of Pb-Free BGA Ball Alloys," Unpublished research, Hewlett-Packard, 2007.
32. IPC/JEDEC J-STD-020 Standard, *Moisture/Reflow Sensitivity Classification for Nonhermetic Solid State Surface Mount Devices*, Rev. D, 2007.
33. F. Hua, R. Aspandiar, C. Anderson, G. Clemons, C.-K. Chung, M. Faizul, "Solder Joint Reliability Assessment of Sn-Ag-Cu BGA Components Attached with Eutectic Sn-Pb Solder," *Proc. SMTA International*, Surface Mount Technology Association, Edina, Minnesota, pp. 246–252, 2003.
34. B. Nandagopal, Z. Mei, and S. Teng, "Microstructure and Thermal Fatigue Life of BGAs with Eutectic Sn-Ag-Cu Balls Assembled at 210°C with Eutectic Sn-Pb Solder Paste," *Proc. ECTC*, IEEE, New York, New York, pp. 875–883, 2006.
35. M. Cole, M. Kelly, M. Interrante, G. Martin, C. Bergeron, M. Farooq, M. Hoffmeyer, S. Bagheri, P. Snugovsky, Z. Bagheri, and M. Romansky, "Reliability Study and Solder Joint Microstructure of Various SnAgCu Ceramic Ball Grid Array (CBGA) Geometries and Alloys," *Proc. SMTAI*, Surface Mount Technology Association, Edina, Minnesota, pp. 283–292, 2006.
36. H. McCormick, P. Snugovsky, and Z. Bagher, "Mixing Metallurgy: Reliability of SAC Balled Area Array Packages Assembled Using SnPb Solder," *Proc. SMTAI*, Surface Mount Technology Association, Edina, Minnesota, pp. 425–432, 2006.
37. R. Coyle, P. Read, and S. Kummerl "A Comprehensive Solder Joint Reliability Study of SnPb and Pb Free Plastic Ball Grid Arrays (PBGA) Using Backward and Forward Compatible Assembly Processes," *SMT J.*, vol. 21, no. 4, pp. 33–47, 2008.
38. M. Logterman and L. Gopalakrishanan, "A Product Feasibility Study of Assembling Pb-free BGAs in a Eutectic Sn/Pb Process," *Proc. ECTC*, IEEE, New York, New York, p. 742, 2009.
39. P. Snugovsky, H. McCormick, Z. Bagheri, S. Bagheri, C. Hamilton, and M. Romansky, "Microstructure, Defects, and Reliability of Mixed Pb Free/SnPb Assemblies", *Journal of Electronic Materials*, Vol 38, No. 2, pp. 292–302, 2008.
40. B. Smith, P. Snugovsky, M. Brizoux, and A. Grivon, "Industrial Backward Solution for Lead Free Exempted AHP Electronic Products: Process Technology Fundamentals and Failure Analysis," *Proc. IPC APEX*, IPC, Bannockburn, Illinois, p. S15–01, 2008.
41. S. Kang, P. Lauro, D.-Y. Shih, D. Henderson, T. Gosselin, J. Bartelo, S. Cain, C. Goldsmith, K. Puttlitz, and T.-K. Hwang, "Evaluation of Thermal Fatigue Life and Failure Mechanisms of Sn-Ag-Cu Solder Joints with Reduced Ag Contents," *Proc. ECTC*, IEEE, New York, New York p. 661, 2004.
42. S. Terashima, Y. Kariya, T. Hosoi, and M. Tanaka, "Effect of Silver Content on Thermal Fatigue Life of Sn-xAg-0.5Cu Flip-Chip Interconnects," *J. Electr. Mater.*, vol 32, no. 12, pp. 1527–1533, 2003.
43. G. Henshall, J. Bath, S. Sethuraman, D. Geiger, A. Syed, M. Lee, K. Newman, L. Hu, D. Kim, W. Xie, W. Eagar, and J. Waldvogel "Comparison of Thermal Fatigue Performance of SAC105, Sn-3.5Ag, and SAC305 BGA Components with SAC305 Solder Paste," *Proc. APEX*, pp. S05–03, 2009.
44. R. S. Pandher and Robert Healey, "Reliability of Pb-Free Solder Alloys in Demanding BGA and CSP Applications," *Proc. 58th Electronic Components and Packaging Technology*. (ECTC), IEEE, New York, New York, pp. 2018–2023, 2008
45. R. Darveaux and C. Reichman, "Mechanical Properties of Lead-Free Solders," *Proc. ECTC*, IEEE, New York, New York, p. 695, 2007.
46. M. Date, T. Shoji, M. Fujiyoshi, and K. Sato, "Pb-free Solder Ball with Higher Impact Reliability," Intel Pb-free Technology Forum, 2005.

47. B. Huang, A. Dasgupta, and N. C. Lee, "Effect of SAC Composition on Soldering Performance," *Proceedings 29th IEEE, SEMI Int. Electron. Manuf. Technol. Symp,* IEEE, New York, New York, pp. 45–55, 2004.
48. K. Nogita, J. Read, T. Nishimura, K. Sweatman, S. Suenaga, A. Dahle, "Solidification and Microstructure of Sn-0.7mass%Cu Alloys," Meeting of JIM, vol. 137, p. 295, 2005.
49. A. Allen, "HP Sn-Ag-Cu Solder Alloy Material Requirements—BGA/CSP Solder Ball Alloys," Hewlett-Packard specification EL-MF862-10, 2009.
50. G. Henshall, A. Allen, E. Benedetto, H. Holder, J. Miremadi, and K. Troxel, "Progress in Developing Industry Standard Test Requirements for Pb-Free Solder Alloys," *Proceedings of IPC APEX, IPC, Bannockburn, Illinois*, p. S35-01, 2010.
51. iNEMI, "iNEMI Members Call for Unique Part Numbers to Differentiate Ball Metallurgies on Pb-Free BGA Components," http://www.inemi.org/cms/newsroom/PR/2007/PR050707.html, 2007.
52. G. Henshall, R. Healey, R. Pandher, K. Sweatman, K. Howell, R. Coyle, T. Sack, P. Snugovsky, S. Tisdale, and F. Hua, "iNEMI Pb-Free Alloy Alternatives Project Reports: State of the Industry," *J. Surf. Mount Technol.*, vol. 21, no. 4, p. 11, 2008.
53. J. Smetana, R. Coyle, P. Read, T. Koshmeider, D. Love, M. Kolenik, and J. Nguyen, "Thermal Cycling Reliability Screening of Multiple Pb-free Solder Ball Alloys," *Proceedings of IPC APEX,* IPC, Bannockburn, Illinois, p. S23-01, 2010.
54. C. Shea, R. Pandher, K. Hubbard, G. Ramakrishna, A. Syed, G. Henshall, Q. Chu, N. Tokotch, L. Escuro, M. Lapitan, G. Ta, A. Babasa, and G. Wable, "Low-Silver BGA Assembly Phase I—Reflow Considerations and Joint Homogeneity Initial Report," *Proc. IPC APEX,* IPC, Bannockburn, Illinois, S26-02, 2008.
55. C. Shea, R. Pandher, K. Hubbard, G. Ramakrishna, A. Syed, G. Henshall, Q. Chu, N. Tokotch, L. Escuro, M. Lapitan, G. Ta, A. Babasa, and G. Wable, "Low-Silver BGA Assembly Phase I—Reflow Considerations and Joint Homogeneity Second Report: SAC105 Spheres with Tin-Lead Paste," *Proc. SMTAI*, Surface Mount Technology Association, Edina, Minnesota, p. 424, Aug. 2008.
56. C. Shea, R. Pandher, K. Hubbard, G. Ramakrishna, A. Syed, G. Henshall, Q. Chu, G. Wable, A. Babasa, E. Doxtad, M. Lapitan, M Santos, and J. Solon, "Low-Silver BGA Assembly Phase I—Reflow Considerations and Joint Homogeneity Third Report: Comparison of Four Low-Silver Sphere Alloys and Assembly Process Sensitivities," *Proc. APEX*, IPC, Bannockburn, Illinois, p. S05–01, 2009.
57. G. Henshall M. Fehrenbach, C. Shea, Q. Chu, G. Wable, R. Pandher, K. Hubbard, G. Ramakrishna, A. Syed, "Low-Silver BGA Assembly Phase II—Reliability Assessment Fifth Report: Preliminary Thermal Cycling Results," *Proceedings of IPC APEX,* IPC, Bannockburn, Illinois, p. S23–03, 2010.
58. Y. Zhang, Z. Cai, J. Suhling, P. Lall, and M. Bozack,"The Effects of Aging Temperature on SAC Solder Joint Material Behavior and Reliability," *Proc. ECTC*, IEEE, New York, New York, p. 99, 2008.
59. D. Rooney, D. Geiger, D. Shangguan, and J. Lau, "Metallurgical Analysis and Hot Storage Testing of Lead-Free Solder Interconnects: SAC versus SACC," *Proc. ECTC*, IEEE, New York, New York, p. 89, 2008.
60. K. Nogita, J. Read, T. Nishimura, K. Sweatman, S. Suenaga and A. K. Dahle, "Microstructure control in Sn-0.7mass%Cu alloys," *Mater. Trans.*, vol. 46, no. 11, pp. 2419–2425, 2005.
61. I. De Sousa, D. Henderson, L. Patry, and R. Martel, "Implementation of Increased Cu Levels (1%) in SAC Alloeys for PBGA Applications," *Proceedings SMTAi,* Surface Mount Technology Association, Edina, Minnesota, p. 435, 2008.

6

GROWTH MECHANISMS AND MITIGATION STRATEGIES OF TIN WHISKER GROWTH

Peng Su (Cisco Systems)

6.1 INTRODUCTION

Whisker growth is defined as the spontaneous formation and growth of filament-shaped grains on metal surfaces. This phenomenon has been observed on metal films and finishes, including cadmium, tin, zinc, and many others [1]. The growth rate of such whiskers can be high, and in some cases they can grow to hundreds of microns in a short period of time. Clearly, for electronic components and systems, such growth can be a serious reliability concern. For lead frame based components, shorting of adjacent leads could occur if excessive growth happens on both leads. Additionally, if these whiskers fall off or break from their original growth sites, they could be transported to other locations within the electronic systems and cause further issues.

Prior to the RoHS legislation, some research work had been performed on the driving force and growth process on whiskers. Recently such work has been reviewed [1], and some of the results will be discussed later in this chapter. The amount of effort to understand whisker growth increased sharply in the last few years due to the conversion to lead-free plating and solders for electronic components. Through the research and development work at component suppliers and research institutions, much new insight has been gained on the growth process and mechanisms as well as some effective mitigation strategies to reduce the risk of component reliability due to whisker growth.

Lead-Free Solder Process Development, Edited by Gregory Henshall, Jasbir Bath, and Carol A. Handwerker

The growth rates of whiskers vary greatly and are hastened by a wide range of factors including composition of the finish, plating chemistry, plating process, substrate (e.g., lead frame), and the application environment. Several types of lead-free finishes have been evaluated by the industry and matte tin is most frequently selected because it offers the optimal balance among manufacturability, reliability, solderability, and cost. In this chapter we will focus our discussion on pure tin, and more specifically, matte tin. The term "matte" at this time is defined by using grain size and carbon concentration in the deposit [2]. However, there has not been strong experimental evidence to correlate these parameters with the growth rate of whiskers. For the discussion of this chapter, matte tin will be used to denote today's most commonly used high-speed plating electrolytes for semiconductor devices.

At ambient conditions (e.g., office environment), whisker growth on matte tin is normally slow due to a combination of long incubation time and low growth rate of the whisker grains. To accelerate the growth rate, a set of standard acceleration tests has been developed [2]. During these tests, particularly the air-to-air thermal cycling (AATC) test, the whisker growth is highly accelerated by thermally induced stresses. Figure 6.1 gives an image of whisker growth on tin finish after the AATC test (–55°C to +85°C, 1000 cycles, alloy 42 lead frame). Experimental observations and analytical results from these accelerations tests have provided much new information about the growth process of whiskers.

In this chapter we will first review the current understanding and theory on the driving forces of whisker growth. Then we will discuss the dominant stress generation mechanism observed from the acceleration tests. And finally we will summarize some of the mitigation techniques to reduce the risk of whisker growth for electronic systems.

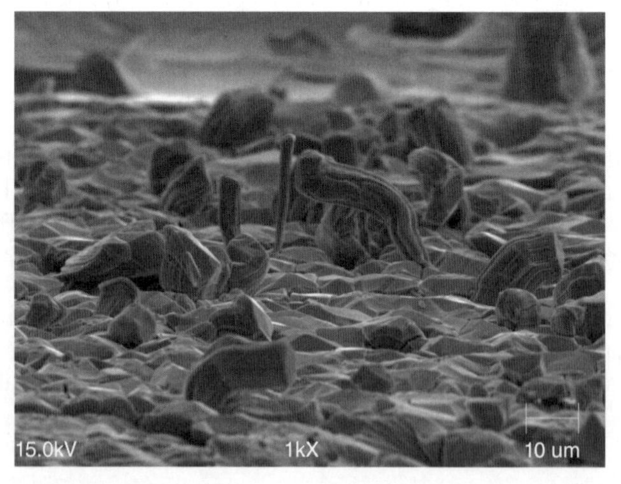

Figure 6.1 Whisker growth on pure tin finish after 1000 cycles of AATC between the temperatures of –55°C and 85°C

6.2 ROLE OF STRESS IN WHISKER GROWTH

The current consensus among researchers is that whisker growth is primarily a stress-induced process. While many sources and processes can contribute to the stress increase in Sn finishes, it is generally correct that the higher the stress levels, the greater is the whisker growth rate. In one of the early studies [3], whisker growth was induced by using a ring clamp to apply pressure on the tin-plated samples. Growth rate was observed to be mostly a function of clamp pressure, and increased clamp pressure correlated to increased whisker growth. Another example of the role of stress is the whisker growth induced by the AATC tests. Thermal stresses are induced in the tin finishes during such a test due to the CTE (coefficient of thermal expansion) mismatch between tin finish and the substrate material (e.g., lead frame). If the mismatch is high, then whisker growth rate and density are also high. Tin has an average CTE of approximately 22×10^{-6}/K. When the CTE value of the lead frame is low, such as for the commonly used alloy 42 lead frame, which has a CTE of approximately 5×10^{-6}/K, whisker growth is significantly higher than for lead frame materials with CTE values closer to tin, such as on copper alloy lead frames, which typically have CTE in the range of 15×10^{-6}/K.

Figure 6.2 shows the images of whisker growth on the same tin finish plated on a Cu lead frame (Figure 6.2*a*) and an alloy 42 lead frame (Figure 6.2*b*) after 500 cycles of AATC. There is a clear difference in growth rate, and the alloy 42 lead frame has generated much higher whisker density and whisker length [4].

It should be noted, however, that while the general statement that whisker growth is a stress-release phenomenon is widely accepted, it is very much debated as to which specific stress is the most important and what some of the major sources of the stresses are. Some work has been done in measuring the long-range (i.e., global) stress state in tin finishes using cantilever-type samples [5]. A stress plot from the work of Boettinger et al. is shown in Figure 6.3. (Note that the finishes used were bright tin and not the

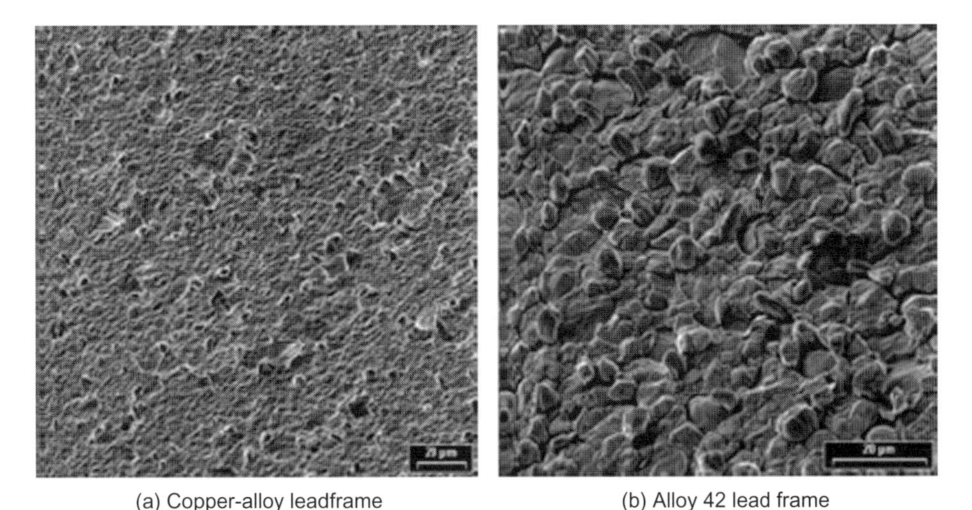

(a) Copper-alloy leadframe (b) Alloy 42 lead frame

Figure 6.2 Whisker growth after 500 cycle AATC on Copper and alloy 42 leadframes [4]

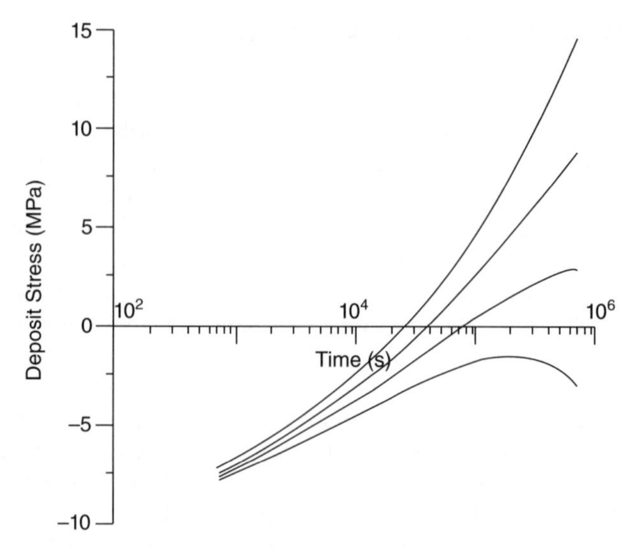

Figure 6.3 Stress in a 7-μm thick tin deposit as a function of time. Stress is calculated using measured cantilever beam deflection data for four possible values of the intermetallic stress-free transformation strain. The top curve is uncorrected stress (zero transformation strain) and lower curves show predictions for increasing values with the bottom curve being for a realistic value of transformation strain [5]

typically used matte tin chemistry.) In this particular study, deflection versus time curves of plated cantilever beams were measured. Because the deflection is due to the sum of the stress thickness products of the tin and the growing intermetallic, the stress in the tin must be deconvolved/separated from the measurements by a calculation procedure. With time, the initial deposit stress may decrease in magnitude through processes such as Coble creep, but depending on the growth rates and mechanism of the intermetallic compound at the deposit/substrate interface, a net increase in compressive stress magnitude may occur with time depending on the composition and thus microstructure of the finishes.

Furthermore, global stress cannot explain some of the details of the whisker growth process, such as why whiskers only grow at some but not all grain boundary intersections. Tin has a body-centered tetragonal (BCT) lattice structure (Figure 6.4) [6] with the lattice constants of $a = 5.82$ Å and $c = 3.17$ Å. This results in a highly anisotropic structure that has very different mechanical properties (e.g., Young's modulus and CTE values) on different planes. For electrolytically deposited tin finishes, the tin grains typically have a wide range of grain orientations, and even neighboring grains can have drastically different orientations. Figure 6.5*a* gives an optical image showing the tin grains on top of the copper substrate [7]. Note that the shade of gray is different for each of the tin grains. As reflectivity is affected by the Miller indices of the crystal planes, the difference in the shade of color suggests that these grains have different crystallographic orientations. Figure 6.5*b* gives a SEM image of the top-down view of

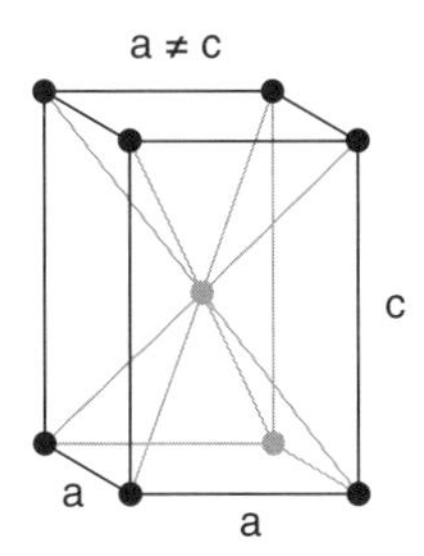

Figure 6.4 Lattice unit cell of Sn

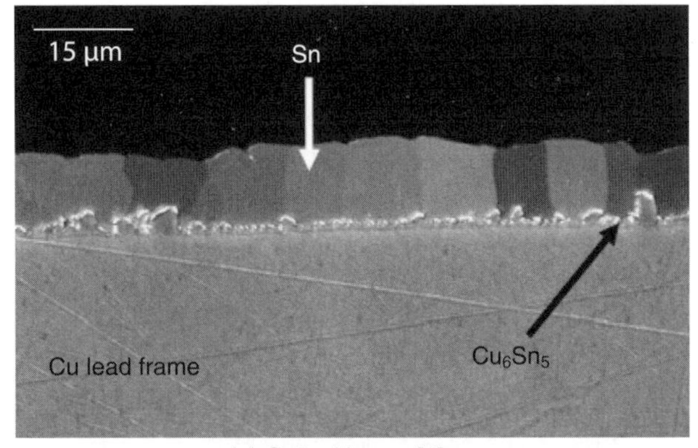

(a) Optical image [6]

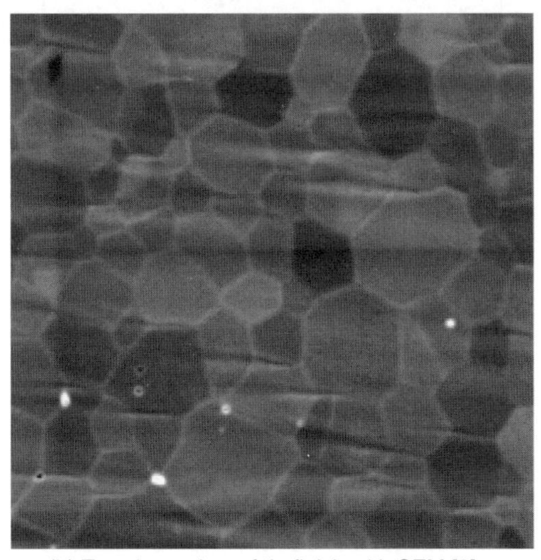

(b) Top-down view of tin finish with SEM [8]

Figure 6.5 Optical cross section and SEM image showing the different orientation of the plated tin grains

a matte tin film after being cleaned with FIB (focused Ion Beam) [8]. The difference in shading, indicating variations in grain orientation, is also clearly shown.

Grain boundary diffusion has been considered the main path for tin atoms to migrate to the growth site of the whiskers [9]. Recent research work [10] has shown that in extreme situations, such as during AATC tests, localized surface diffusion can also be an important path for tin diffusion. Shown in Figure 6.6 are a series of images

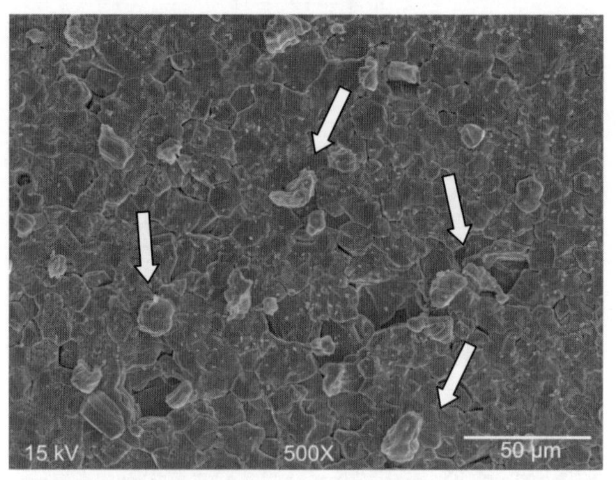

(a) Overview of the whisker growth on the surface of tin after AATC

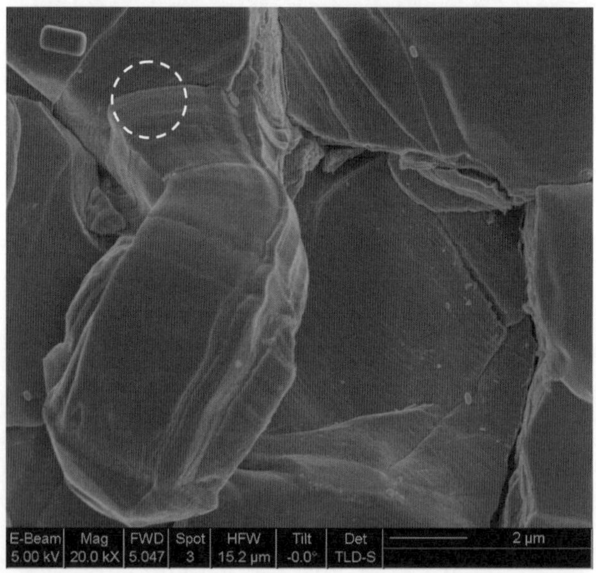

(b) Magnified view of a single whisker. Notice the smooth surface of the neighboring grains. A magnified view of the circled area is shown below in part (c).

Figure 6.6 Indications of whisker growth as assisted by surface diffusion [10]

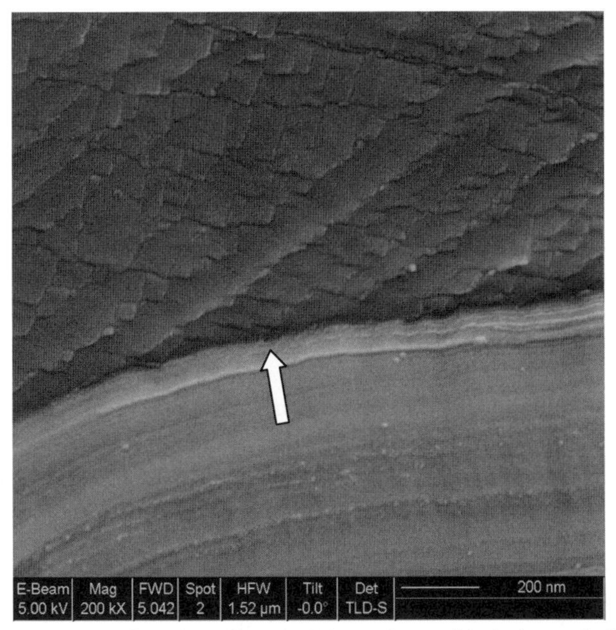

(c) A 200,000x magnified view of the circled area in part (b). At the top half are the "crumbled" surface planes of the neighboring grain, and at the bottom half the bottom portion of the whisker. The arrow indicates the bottom of the whisker.

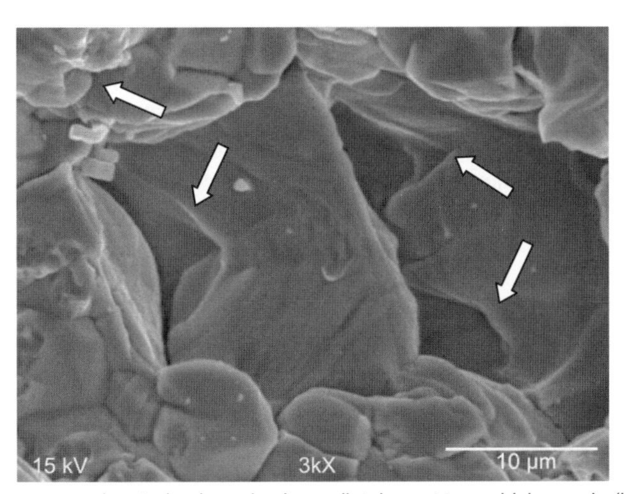

(d) Surface recess and cratering in grains immediately next to a whisker grain (indicated by the arrow in the top left of the image). The remaining arrows show the location of deep cratering at grain boundaries and intersections, suggesting localized stress increases and faster out-diffusion of the tin atoms.

Figure 6.6 *Continued*

taken on the surface of a tin finish after AATC testing. Figure 6.6*a* is a top-down overview of the surface. The extrusions, some indicated with arrows, are whiskers. During the AATC test, thermal stresses cause many of the grain boundaries to either shift or crack, providing a clear view of the geometry of the grains as illustrated by the visible grain boundaries. In Figure 6.6*a* notice that whiskers only grow at some of the grain intersections, and more importantly, grains immediately adjacent to the whisker grains often appear to have recessed from the original surface and the top surface of these grains appears to be smooth and different from the originally plated finish.

Figure 6.6*b* gives a magnified view of a single whisker growing at a grain intersection. Notice that the top surfaces of the grains next to the root of the whisker are extremely smooth, as if the top portions of the original grains have been cleaved along a single lattice plane. The final details of these top surfaces are revealed in Figure 6.6*c*. In this image the bottom half is the whisker, and the top half is the surface of the grain adjacent to the root of the whisker. It is apparent from this image that the thermal stress during the AATC test is so high for this particular grain that it caused the top planes to "crumble." The fractured segments from the surface then migrate along the "cleaved" planes toward the root of the whisker, providing the material needed for growth.

Figure 6.6*d* is another example of localized surface damage and diffusion. The top left corner of the image shows the bottom half of a whisker. In the two neighboring grains, the top surface has recessed caused by the out-migration of tin atoms along these planes. More important, note the even deeper indention within the grains (as indicated by the arrows) at grain boundaries. These are likely formed by even faster out-migration of tin at these locations, suggesting localized stress concentrations at these grain boundaries during the AATC test.

These experimental observations suggest that the nucleation and growth processes of whisker grains are more favored at grain intersections where the configuration of neighboring grains meet certain criteria. These criteria could be for grain crystallographic orientation, geometrical characteristics, or a combination of both. With simple finite element modeling (FEM) techniques, these effects can be visually demonstrated. In one study [8], tin grains with experimentally determined orientations were implemented in a finite element model. When thermal strains were applied on these grains, the resulting strain energy density varied greatly from grain to grain (Figure 6.7) and very noticeably at the areas near some of the grain boundaries and grain intersections. The contours of the strain energy density levels are nearly identical in shape and locations to the observed deep recession as shown in Figure 6.6.

These results suggest that the crystallographic orientation mix of tin grains is a critical factor in determining the overall whisker growth propensity of the tin finish. In the following discussion we will discuss some of the key factors that determine this property of the tin finish.

Another aspect of the whisker growth process—namely the detailed steps of whisker nucleation and growth—is still being investigated. In some of the published work, it has been proposed [11,12] that whiskers grow from dislocations located at their bases, and so operate through a diffusion-limited mechanism. One such mechanism [12] involves a rotating edge dislocation that is pinned to a screw dislocation, essentially at right angles to the surface. The rotating edge dislocation stays in the same plane after

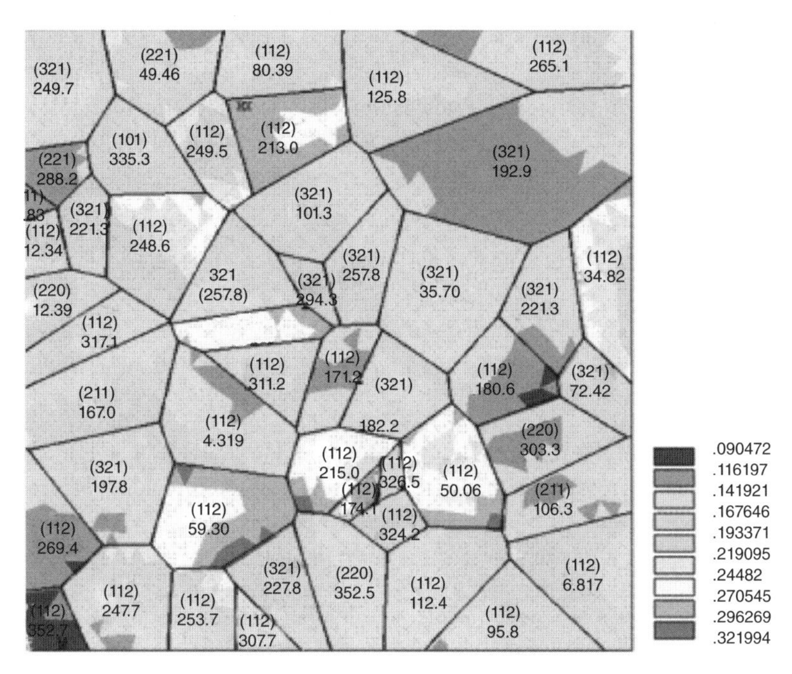

Figure 6.7 Strain energy density in tin finish due to thermal stress [8]

each revolution, and each revolution deposits an additional layer of tin atoms at the base of the whisker. Others have shown [13] that growth directions of whiskers are of small crystallographic indexes that are also the indexes for the glide planes. These are supporting arguments for the various dislocation models that utilize the concept of glide formation as an integral part of the whisker growth process. Some recent TEM work [14] has shown increased dislocation densities in grains adjacent to the roots of whiskers, which suggests that whiskers have nucleated from a plastically deformed grain due to slip or twinning within these grains. Other work [15] recorded the growth process continuously with images and has shown that certain whiskers rotated repeatedly as they grew, suggesting the growth to be propelled by screw dislocations at the base of these whiskers.

It should be mentioned that many other factors can contribute to the stress levels and diffusion processes in tin finishes and thus can affect the growth of whiskers. The electrolytic deposition process can introduce high residual stress in the tin finishes. As we will discuss in more detail later in this chapter, the makeup of the electrolyte and process parameters are both critical in determining the whisker growth propensity of tin finishes. Organic residue coming from the plating chemistry, such as carbon, is also believed to affect the growth process, although direct correlation between the two has not been experimentally established.

Many environmental factors have been known to accelerate the nucleation and growth of whiskers. Corrosion of the finish has also been positively correlated with accelerated nucleation and growth of whiskers. Intermetallic growth at the interface of finish and substrate, such as the Cu/Sn interface, has been connected to whisker growth in environments where there are no other apparent accelerating factors. We will also discuss the implication of these processes on the overall mitigation strategies for electronic systems.

6.3 UNDERSTANDING STANDARD ACCELERATION TESTS

At typical “ambient” conditions such as an office environment, matte tin finishes plated with today’s commercial electrolytes and processes do not typically have high rates of whisker growth. However, for electronic systems manufacturers, the long-term reliability of tin-plated components must be assessed and understood. In the past few years industry consortia, corporations, and research institutions have evaluated the effectiveness of a large variety of test methods and conditions. Based on the results from these evaluations, a set of standard tests methods have been accepted by the industry and published in multiple test standards [2,16,17]. In the following we will refer primarily to the test methods as recommended by the JEDEC standard JESD201.

Three types of tests are recommended by this JEDEC standard. The details of the testing requirements are listed in Table 6.1. Further details on variations in test durations, as well as recommended pass/fail criteria based on maximum observed whisker length, can be found in the standard. In this section we will discuss primarily the purpose of these specific tests methods, and the implication on mitigation strategies based on results collected on as-plated tin finishes. The implication on board-mounted components will be discussed in Section 6.5. The JEDEC standard has similar tin whisker test conditions to the JEITA [16] and IEC [17] test standards. For example, the IEC test standard mentions low-temperature storage conditions of 30°C and 60%RH or 25°C and 50%RH for up to 4000 hours. For high-temperature storage conditions, it mentions 55°C and 85%RH for 2,000 hours. For thermal cycling, it mentions a minimum temperature of –55°C or –40°C to a maximum temperature of +85°C or +125°C for

TABLE 6.1 Overview of the test methods as recommended by JEDECJESD201 standard [2]

Stress Type	Test Conditions	Test Durations
Air-to-air temperature cycling (AATC)	–55 + 0/–10°C to 85 + 10/–0°C, ~3 cycles/hr (10 min soak) –40 + 0/–10°C to 85 + 10/–0°C, ~3 cycles/hr (10 min soak)	Up to 1500 cycles
Low-temperature / humidity storage	30 ± 2°C and 60 ± 3% RH	Up to 4000 hours
High-temperature / humidity storage	55 ± 3°C and 85 ± 3% RH	Up to 4000 hours

1000 or 2000 cycles. The IEC standard also gives a typical acceptance criteria of less than or equal to 50 μm for whisker length during the testing.

6.3.1 Air-to-Air Thermal Cycling (AATC)

The first acceleration test is an AATC test intended to address thermal stress-induced whisker growth. For electronic systems that will experience wide environmental temperature variations, thermal stresses can be generated in the tin finishes and thus increase the risk for whisker growth. Testing data from many sources for these tests has shown that the AATC test generates the most consistent and reliable results among the three test methods. For the same material set and plating chemistry and process, whisker growth rate and whisker density can generally be repeated.

For tin finishes with similar thicknesseses, whisker growth from the AATC test appears to be strongly affected by the crystallographic orientations of the tin grains. From our discussion above, for the same AATC temperature ranges (and thus the same thermal strain), thermal stresses in the tin finish can be higher for finishes with certain orientation mixes. Advanced analytical techniques such as EBSD (electron backscattered diffraction) will be the most accurate and comprehensive techniques in collecting such data. However, for commercial plating process development and monitoring, simple techniques such as X-ray diffraction (XRD) can provide useful orientation information. Some successes have been achieved by using this technique [18]. By adjusting the makeup of the plating electrolyte and the plating parameters such as current density, the crystallographic texture of the tin finish can be improved which helps reduce whisker growth. We cannot yet definitively provide a set of universal guidelines on the best grain orientation mixes. Nevertheless, for a select electroplating electrolyte, once a plating process has been established and proved to pass the current standard test methods, XRD can be used as an effective monitoring tool to prevent significant process deviation.

The level of thermal stresses generated with current standard test conditions of −40°C or −55°C to +85°C is large enough to induce visible grain boundary cracking, particularly for lead frame materials such as alloy 42. This is somewhat problematic as stresses are at least partially released after the cracking occurs, resulting in lower stress levels in the tin finish in latter part of the tests. This in turn causes the whisker growth rate to decelerate and manifests as saturation of whisker length. Some work has been done to investigate the effects of temperature ranges [19,20]. In most cases it appears that smaller temperature ranges indeed reduce the growth rate of whiskers. Results from these works will be helpful in deriving a practical acceleration model for thermally induced whisker growth. Further investigation is certainly needed in this area, and additional test parameters such as dwell time at end temperatures also need to be investigated.

6.3.2 Low-Temperature/Low-Humidity Storage

The second standard test, the low-temperature/low-humidity storage test at the condition of 30°C/60%TH, is not necessarily an "accelerated" test as it is similar to the most common office environment. At this condition the dominant driving force for whisker growth is most likely the intermetallic growth at the tin finish–substrate interface. For some substrate (lead frame) materials such as copper alloys, the growth of intermetallic compound is very rapid, and significant growth can be observed at the interface within a matter of hours [21]. Shown in Figure 6.8 are a series of images of the exposed IMC (intermetallic compound) at tin grain boundaries after etching away the originally plated tin finish. The rapid growth of IMC grains is clearly evident.

Figure 6.9 gives a cross-sectional view of the IMC growth and whisker growth after the 30C/60%RH test. The copper substrate is at the bottom of the image, the two

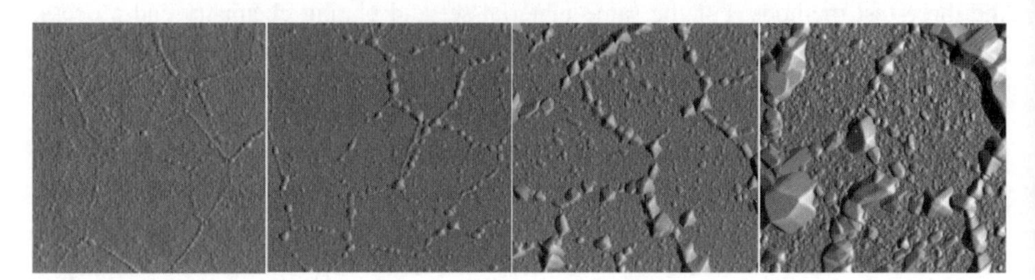

Figure 6.8 Rapid Cu/Sn intermetallic compound growth at the leadframe/tin interface, particularly at tin grain boundaries [21]. Images from left to right show the IMC growth at 0.5, 2, 15, and 120 hours, respectively

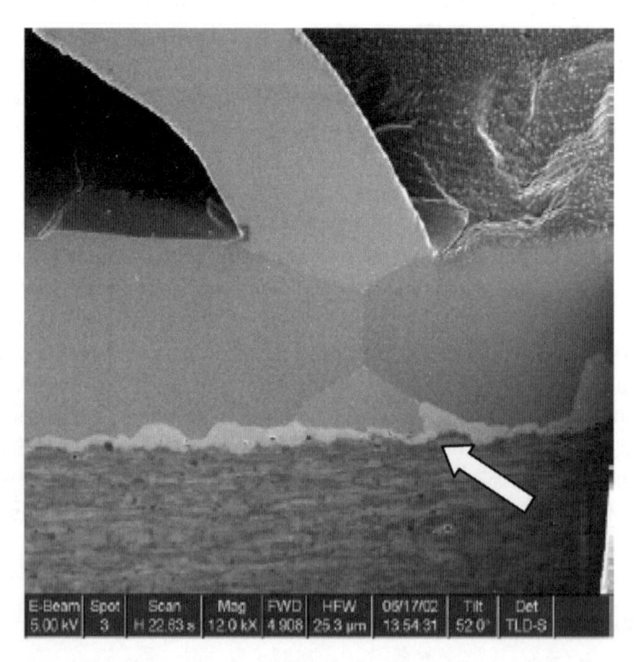

Figure 6.9 Cross section of a whisker by FIB. Cu/Sn IMC growth can be observed especially at grain boundaries

large grains are the originally plated tin, and the light-colored layer in between is the IMC layer, as marked by the arrow.

The exact role of the IMC grains on whisker nucleation and growth is still being investigated. It is likely the growth of the IMC grains introduces compressive stress into the tin finish. Additionally these IMC grains, sometimes nearly as thick as the tin finish itself, can act as pinning points that prevent any grain rotation or sliding, which in turn further increases the stress levels seen in the tin grains. The effects of IMC are also supported by results from situations where there is a lack of IMC growth. For lead frame materials such as alloy 42 (FeNi), or when a nickel barrier layer is plated before tin, where the IMC growth under the tin is greatly reduced compared with copper substrates, whisker growth during isothermal storage tests generally has been observed to be much reduced.

6.3.3 High-Temperature/High-Humidity Storage

The last test, the high-temperature/high-humidity test at the condition of 55°C/85%RH, has generated much debate on the whisker growth mechanism and growth behavior. In the majority of studies, the appearance of whiskers is preceded with the corrosion of the tin finish. Figure 6.10*a* gives a schematic view of a lead on a PQFP (plastic quad flat package) component. The lead-trimming process after plating leaves two exposed areas of copper, as indicated in the image. After extended exposure to the 55°C/85%RH conditions, the typical image of the tip of the leads is shown in Figure 6.10*b*. In this backscatter SEM image the darkened area shows corrosion has occurred in the tin finish. The extrusions on top of the corrosion are tin whiskers. Similar corrosion and whisker growth is also frequently observed at the top of the lead where the copper lead frame is also exposed.

The fact that corrosion is the most severe near the copper lead frame suggests that it is driven by the galvanic potential differences between tin and copper alloys. Being the anode in the Sn-Cu pair, the corrosion of tin is accelerated in the high-temperature/humidity testing environment. What is more important, however, is that the growth of whiskers is the most prominent within the boundary of the corroded areas. This phenomenon has been repeatedly observed [22,23], and many efforts have been invested on the investigation of this type of whisker growth. Recent TEM and electron diffraction work has shown that the corrosion product is crystalline SnO_2, (Figure 6.11) [24]. The exact mechanism, meaning how pure tin whiskers can continuously grow within an area of SnO_2, is still not understood, and further investigation is needed.

The data presented in this section showed that the current standard tests methods can effectively isolate and address the various mechanisms that can either increase the stress levels in the tin finish or assist the nucleation and growth process. While these test methods are by no means perfect in every situation, they provide an effective common platform for electronic system and component manufacturers to assess the whisker growth performance of tin-plated components. In the following sections, we will discuss some of the common techniques that can be used to develop a comprehensive mitigation strategy for whisker growth.

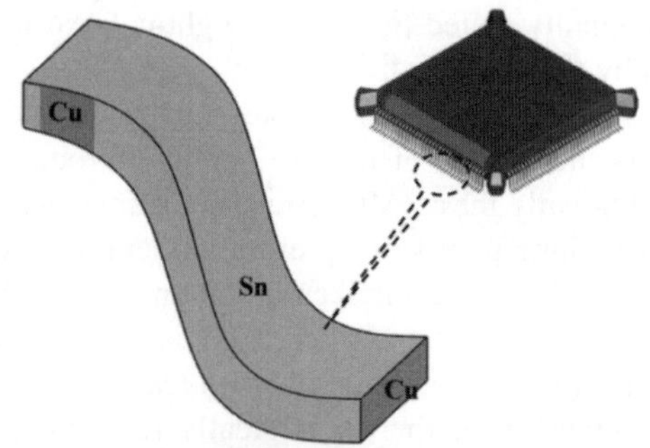

(a) Schematic of a QFP package and the exposed copper lead frame

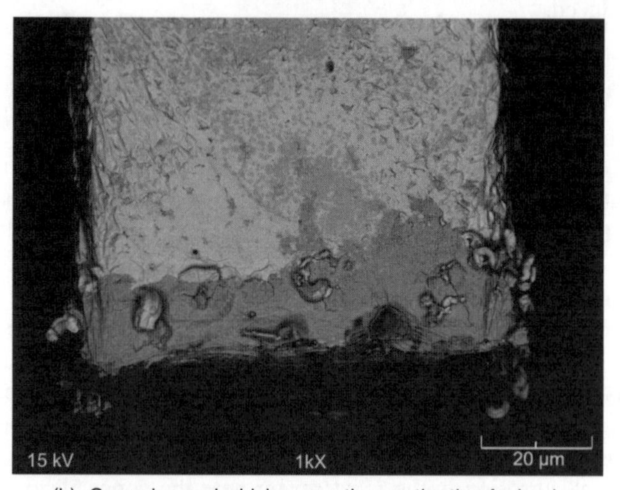

(b) Corrosion and whisker growth near the tip of a lead

Figure 6.10 Schematic of package lead and corrosion

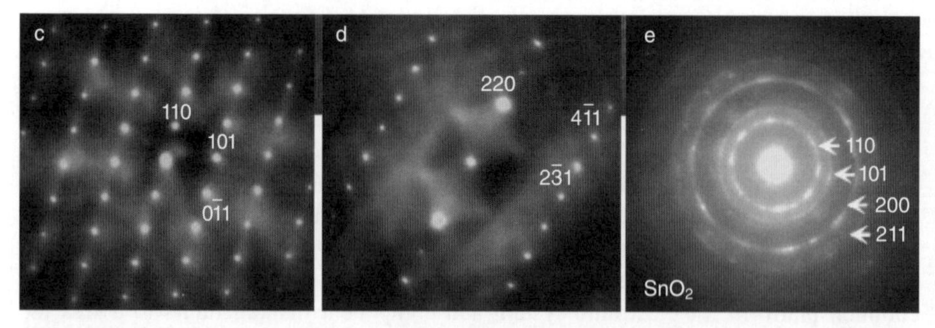

Figure 6.11 Electron diffraction patterns showing that the corrosion product is crystalline SnO_2 [24]

6.4 PLATING PROCESS OPTIMIZATION AND OTHER MITIGATION STRATEGIES

To mitigate the overall whisker growth risk for electronic systems, a systematic approach must be adopted to address all of the known acceleration factors. Recently the learning from research works has been reviewed, and a set of general guidelines has been addressed in a JEDEC/IPC publication JP002 [25]. In this section we focus on two of the steps important for component manufacturers, the plating process and the postplating treatment.

6.4.1 Electrolyte and Plating Process

The importance of plating electrolyte selection and plating process setup cannot be overstressed. The critical properties of the tin finishes, including grain size, geometry, and crystallographic orientation, are determined by the plating process and will have major impact on stress evolution and whisker growth. These factors have been investigated in some of the early studies [26], and the effects of electrolysis conditions during plating such as current density were well documented.

In one study [18], two levels of current densities were used for the evaluation and distinctively different XRD patterns were generated. In Figure 6.12*a* the top spectrum is from the low current density cell, and the bottom spectrum is from the high current density cell. The differences in peak intensity and location are obvious. Whisker growth in this study was assessed by using whisker density observed on the formed component lead (Figure 6.12*b*) after AATC testing. Within the windows outlined by the white rectangle, whisker density was collected by counting the number of whiskers on a large quantity of leads. The density of whiskers on samples from the two processes is compared in Figure 6.12*c*. Clearly, there is a significant difference in whisker growth between the two different processes.

The makeup of the plating electrolyte is equally critical. Currently the limited brands of commercial electrolytes have all gone through several rounds of improvements, and in most cases they can provide a stable plating process and thus maintain the quality of the finish. However, there are many chemical components in a plating chemistry, and changes in composition or concentration in any of these components could have a marked impact on whisker growth performance. In the same study discussed above, the type of antioxidant component was varied while other components were kept constant. The XRD patterns on the resulting tin finishes for type A and type B are showing in Figure 6.13*a*. The whisker density after AATC is plotted in Figure 6.13*b*. Again, a marked difference in whisker density is observed.

Results from these recent investigations have demonstrated that there are many opportunities on plating lines to fine-tune the many parameters for whisker growth performance. However, definitive guidelines on what microstructure characteristics to aim for cannot yet be provided because of the limited microstructure data available. Research work is continuing within industry consortia and universities, and data from these projects will provide further insight on this issue.

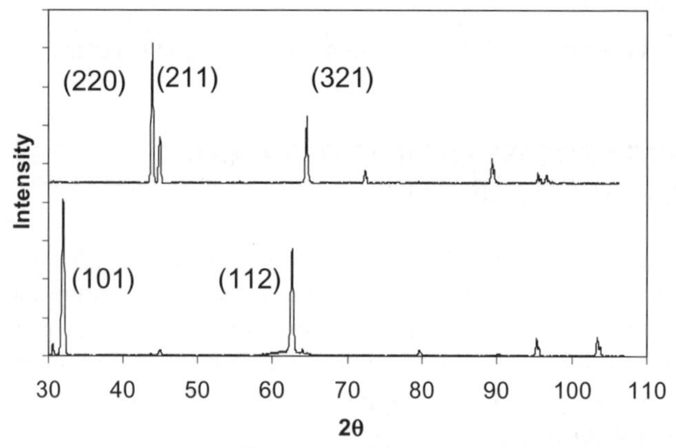

(a) XRD spectrum of tin finishes plated with two different current densities. The top spectrum is from the low current density and the bottom spectrum is from the high current density cell.

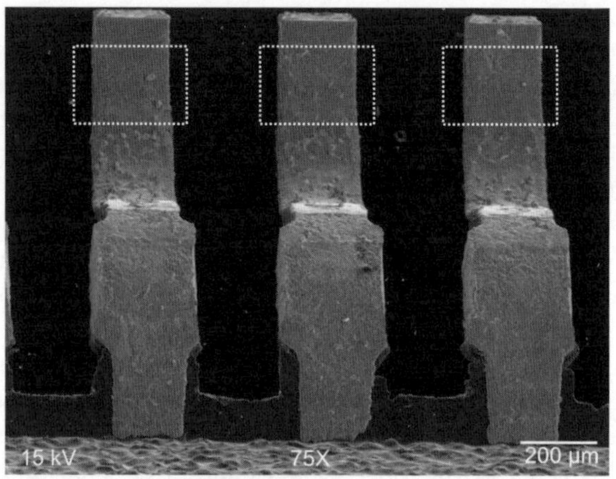

(b) Whisker density data is collected by counting the number of whiskers within the outlined rectangles

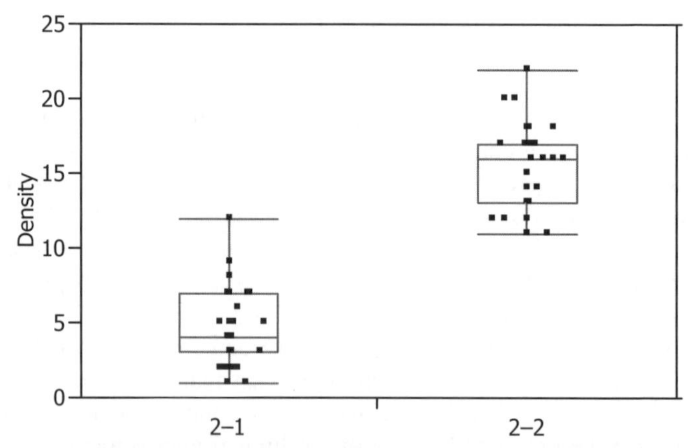

(c) Density of whisker growth observed on the two tin finish samples with low and high current densities

Figure 6.12 XRD spectrum of tin finishes plated with two current densities and the resultant whisker growth during AATC testing

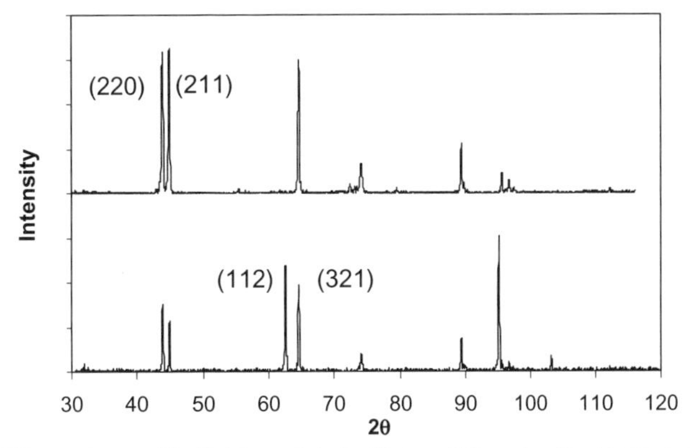

(a) XRD spectrum of tin finishes plated with two different antioxidants in the plating electrolyte. The top spectrum is from the electrolyte with antioxidant A, and the bottom is from electrolyte with antioxidant B.

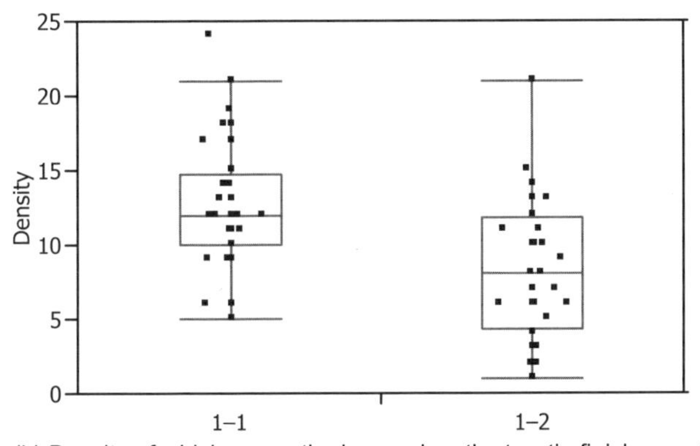

(b) Density of whisker growth observed on the two tin finish samples

Figure 6.13 XRD spectrum of tin finishes plated with two current antioxidants and the resultant whisker growth after AATC test

6.4.2 Underplate and Heat Treatment

An underplate (or barrier layer) is sometimes plated on the lead frame prior to the plating of tin finishes. This is used primarily for copper-alloy lead frames, and the barrier layer is usually nickel. Based on results from recent studies, the benefit of this process comes mostly from a reduction in IMC growth [27]. The IMC growth rate at the Sn–Ni interface is significantly slower than at the otherwise Sn–Cu interface, and thus likely reduces the stress buildup in the tin finish. This effect also correlates well with the fact that at ambient conditions whisker growth on alloy 42 lead frame is typically less compared with copper lead frames.

Based on the same principle, heat treatment is another effective method to reduce the impact of IMC growth for copper-based lead frames. The first study of the effect of heat treatment at temperatures from 100°C to 180°C dates back to the late 1980s [9],

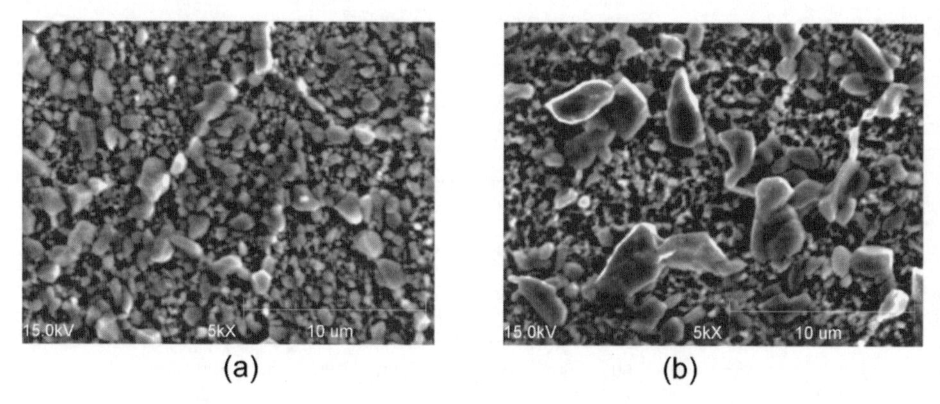

Figure 6.14 Morphology of IMC at the Cu/Sn interface after the sample has been stored at 30°C/60%RH for 3000 hours. The sample in (a) received a 1-hour 150°C postplating bake and the sample in (b) did not

which observed all heat treatment combinations to have a significant mitigation effect on the formation of tin whiskers. Currently in the industry a 1-hour bake at 150°C has become the standard process for all copper lead frame based components. The benefit of the heat treatment is primarily the uniform growth of IMC at the very beginning stage of the product life. As previously shown in Section 6.3, IMC growth is typically the most rapid at grain boundaries. At higher temperature, such as 150°C, the growth of IMC under the bulk of grains is increased relative to grain boundary growth, thus generating a more uniform layer of IMC.

Figure 6.14 shows the morphology of Cu–Sn IMC after 3000 hours of storage at the 30°C/60RH condition. The sample in Figure 6.14*a* had the 150°C/1-hour heat treatment, but the sample in Figure 6.14*b* did not. It can be seen that the IMC growth at grain boundaries is much greater in the latter case. Whisker growth rate on heat-treated tin finishes is generally reduced [27]. Particularly if the thickness of the tin finish is low (e.g., $<5\,\mu m$), the benefit of the postplating bake is even more significant.

It should be noted that the benefit of a barrier layer and the heat treatment is limited to a reduction in IMC growth, and thus is most effective for preventing whisker growth at low-temperature and low-humidity environments. For other application environments, additional mitigation measures must be jointly adopted.

6.5 WHISKER GROWTH ON BOARD-MOUNTED COMPONENTS

In the previous sections we discussed the microstructure characteristics and the impact of plating process on whisker growth for as-plated tin finishes. Using the standard test methods, we can generally obtain an initial assessment of the whisker growth propensity of these finishes. However, for semiconductor components, once they are mounted on circuit boards with solder pastes, significant changes occur to the microstructure of the tin component lead finish, and many new factors will have important impact on the growth rate of whiskers.

The first change comes from the coverage and alloying of the original finish. During the reflow process, molten solder paste wets and migrates along the leads of the components. For components with low component profiles, the molten paste can cover the entire surface of the original lead. One example is shown in Figure 6.15*a*, where the eutectic SnPb paste has covered the entire lead surface on a lead on a LQFP (low-profile quad flat pack) package. For packages with taller leads, solder paste can still cover a large portion of the lead. For the top portion of the lead, despite the visual appearance that solder coverage is less for thin packages, migration of the solder up the lead can still often be observed. Figure 6.15*b* shows the cross section of a lead on a PQFP package.

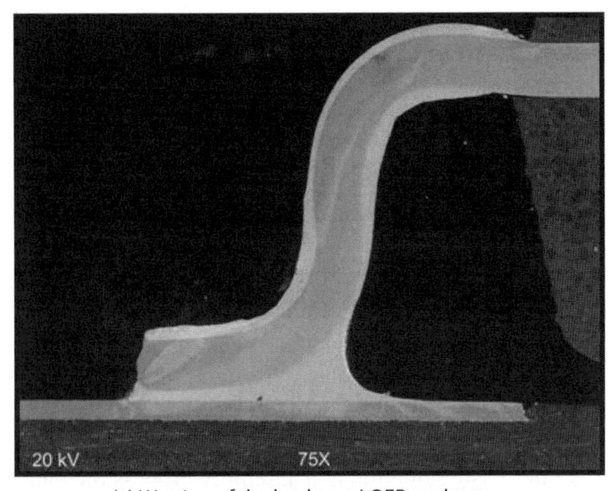

(a) Wetting of the lead on a LQFP package

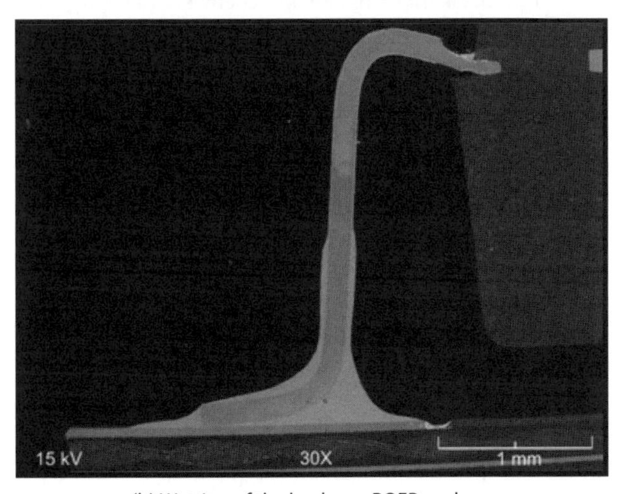

(b) Wetting of the lead on a PQFP package

Figure 6.15 Wetting of the component lead for package types with different dimensions (cross-sectional view)

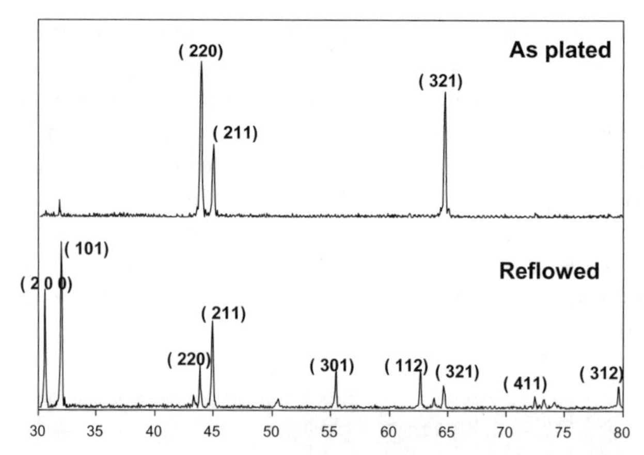

Figure 6.16 XRD spectra of as-plated and reflowed tin finish on lead frames

If the reflow temperature is above the melting temperature of 232°C, such as the case for the commonly used lead-free paste SnAgCu, melting of the original tin finish and mixing with the solder paste will occur. After solidification an entirely different microstructure is formed. Information on grain size and grain crystallographic orientation information on tin finishes after reflow is currently very limited. Based on XRD data collected on reflowed lead frames, reflow generated a very different orientation mix (Figure 6.16), and the average grain size also increased [28]. In some studies the grain sizes on the reflowed lead frames were analyzed with mechanical cross-sectioning and very high levels of non-uniformity observed.

The impact of solder paste selection and assembly process on solder paste coverage and whisker growth has been investigated in recent studies [20]. The experimental matrix is reproduced in Table 6.2. Eutectic tin-lead and lead-free solder paste with the composition of Sn3Ag0.5Cu (SAC305) were used to mount a component with the alloy 42 lead frame. Both pastes were no-clean types with low flux residue activity. For each of the solder pastes, 3 different peak temperatures and 6 different paste volumes were used (through variation of stencil opening and stencil thickness), resulting in a total of 18 temperature–paste volume combinations for each paste. Values of the peak temperatures and the solder paste volumes are listed in Table 6.2. For each assembly condition, 6 units were mounted. The assemblies were subject to the standard JEDEC AATC test condition and had a total test duration of 1500 cycles.

As discussed in previous sections, alloy 42 based components typically show high levels of whisker growth during the AATC test because of the CTE mismatch between tin and the alloy 42 lead frame. For the components used in this study, whisker growth was observed on 100% of the components and leads when loose components were tested. After just 500 cycles of AATC, the maximum observed whisker length had reached over 50 μm, and at 1500 cycles, the maximum whisker length was in the 80- to 90-μm range.

Once these components were mounted with solder paste, significantly reduced whisker growth was observed on the pure tin coated components, particularly when using the tin-lead solder paste. Table 6.3 lists the whisker growth rates for the tin-lead

TABLE 6.2 Experimental matrix for the board assembly study [20]

Solder Paste	Reflow Peak Temperature	Paste Volume versus Standard*	Atmosphere
Eutectic SnPb (no-clean)	205°C 215°C 225°C	–20% –10% 0% +10% +20% +30%	N_2
Sn3Ag0.5Cu (no-clean)	235°C 245°C 255°C	–20% –10% 0% +10% +20% +30%	N_2

*Standard solder paste volume = 5 mils stencil thickness with 90% stencil opening relative to board pad size.

TABLE 6.3 Whisker growth on components mounted with tin-lead solder paste for the various assembly process conditions

Peak Reflow Temperature (°C)	Paste Volume	Whisker Growth
205	–20%	4/6
	–10%	5/6
	0%	6/6
	10%	3/6
	20%	4/6
	30%	4/6
215	–20%	5/6
	–10%	0/6
	0%	0/6
	10%	0/6
	20%	0/6
	30%	0/6
225	–20%	0/6
	–10%	0/6
	0%	0/6
	10%	0/6
	20%	0/6
	30%	0/6

paste assembly cells. The level of whisker growth is indicated with the shading of the cell. Cells shaded with fine lines indicate the presence of whiskers of 45 μm or longer; cells shaded with solid grey indicate whiskers were observed but shorter than 45 μm; and cells with no shading indicate no whiskers were observed. Numbers in each cell indicate how many units of the total examined showed whisker growth; for example, 4/6 means that 4 out of the 6 total components tested showed whisker growth.

The peak reflow temperature showed a significant impact on the overall whisker growth propensity. The groups of components reflowed with the 215°C and 225°C peak temperatures showed almost no whisker growth with tin-lead paste. For the 205°C groups, even an overprint of 30% in paste volume did not effectively reduce whisker growth. These observations suggest that reflow temperature is the key parameter for driving the wetting behavior, and paste volume did not have significant effects on whisker mitigation.

Figure 6.17 is an example image of a component for the assembly condition with 20% paste underprint and 205°C peak temperature. Whisker growth is mostly observed near the body of the package, where wetting of the finish by the tin-lead paste is low. At the vertical section of the leadframe where the paste wetting is higher, there is nearly no whisker growth.

Whisker growth on components mounted with the Sn3Ag0.5Cu solder paste showed different behavior compared with what was observed on the tin-lead solder

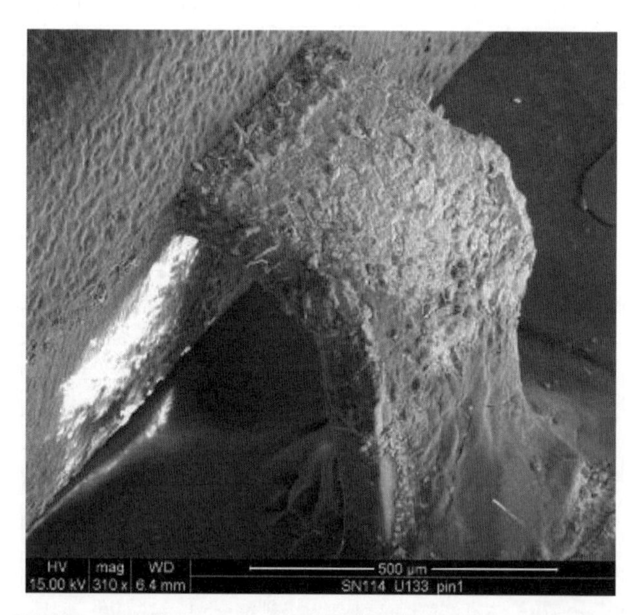

Figure 6.17 Whisker growth on a component mounted with tin-lead solder paste, 20% paste underprint, and 205°C peak temperature. Whiskers are observed at locations near the body of the package

TABLE 6.4 Whisker growth on pure tin-coated components mounted with Sn3Ag0.5Cu paste for the various assembly process conditions

Peak Reflow Temperature (°C)	Paste Volume	Whisker Growth
235	−20%	2/6
	−10%	2/6
	0%	0/6
	10%	0/6
	20%	1/6
	30%	3/6
245	−20%	0/6
	−10%	2/6
	0%	0/6
	10%	0/6
	20%	1/6
	30%	0/6
255	−20%	1/6
	−10%	4/6
	0%	0/6
	10%	0/6
	20%	4/6
	30%	2/6

paste soldered assemblies. Results from these cells are listed in Table 6.4. The shading of the cells has the same meaning as indicated in the previous section. Compared with what would be observed on loose components, reduction in whisker growth is still observed; however, the effects of solder paste volume and peak reflow temperatures are not clearly seen. An increase in paste volume and peak temperature did not result in a reduction in whisker growth. SEM inspection of the board-mounted components suggests that even for the lower end of the reflow temperature used in this study, a high level of wetting was achieved. So no additional benefit in wetting is achieved through increased reflow temperature or paste volume.

The results in Table 6.3 suggest that for application environments with high-temperature fluctuations, the tin-lead solder paste may offer sufficient mitigation in reducing thermally induced whisker growth if the assembly process is appropriately set up. The underlying mechanism is that if sufficient lead from the paste migrates and covers the original tin finish, growth of whiskers can be effectively curtailed.

For lead-free solder paste, the coverage by the solder paste offered some reduction in whisker growth compared to the growth rate seen on loose components. However,

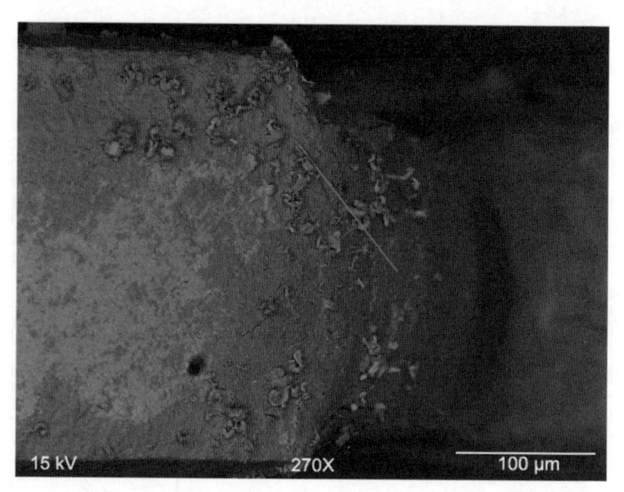

Figure 6.18 Corrosion-induced whisker growth on a lead mounted with SnAgCu paste

the process parameters did not seem to generate a consistent trend in whisker growth performance, maybe due to the fact that the solder paste too is high in tin content, and the copper and silver additions do not offer whisker growth mitigation benefits.

Corrosion-induced whisker growth can also occur for board-mounted components. For lead-free solder paste with high tin content, corrosion of the solder paste can be similarly accelerated by the galvanic potential differences with lead frames, as discussed in Section 6.3. Figure 6.18 shows the corrosion and whisker growth on a component mounted with Sn3Ag0.5Cu solder paste after 2000 hours of testing at the condition of 85°C/85%RH. Notice that the level of corrosion is higher near the edge of the lead, where the copper lead frame is exposed. Also, similar to what has been observed on loose components, whisker growth is mostly limited to the areas of corrosion.

For corrosion-driven whisker growth, experimental data suggest that as long as moisture is kept out of contact with the surface, corrosion will not occur and whisker growth can be essentially eliminated. In one study [22] components with pure tin finish and copper lead frames were first covered with an anticorrosive conformal coating and then stored in a 60°C/95%RH environment for over 7000 hours. None of the components showed any whisker growth at the end of testing, while in the same chamber, the components without coating showed high levels of growth after just 3000 hours. These results suggest that for electronic systems that may be exposed in harsh environments with high humidity levels, conformal coating can be an effective whisker mitigation measure.

6.6 SUMMARY

Much has been learned about the growth mechanisms of tin whiskers in the past few years from the large number of studies performed. Some mitigation techniques have been developed and are showing significant benefits in reducing the risk of whisker growth. The learning from these studies also suggests that a comprehensive and systematic approach must be adopted to address the known factors that accelerate whisker growth. Component and system manufacturers can both adopt process improvement measures to further reduce the risk of whisker growth.

At the fundamental level, more investigation of the details of the whisker growth process is needed. Only when we have a thorough understanding of the metallurgical processes for whisker nucleation and growth can we develop effective counter measures to further reduce or even eliminate the reliability risk from whisker growth.

REFERENCES

1. G. Gaylon, "Annotated Tin Whisker Bibliography and Anthology," *IEEE Trans. Electr. Pack. Manufact.*, vol. 28, no. 1, pp. 94–122, 2005.
2. JEDEC JESD201 Standard, *Environmental Acceptance Requirements for Tin Whisker Susceptibility of Tin and Tin Alloy Surface Finishes*, 2006.
3. R. M. Fisher, L. S. Darken, and K. G. Carroll, "Accelerated Growth of Tin Whiskers," *Acta Metal.*, vol, 2, pp. 368–372, 1954.
4. Y. Zhang, C. Fan, C. Xu, O. Khaselev, and J. A. Abys, "Tin Whisker Growth—Substrate Effect, Understanding CTE Mismatch and IMC Formation," Enthone and Cookson Electronic, http://www.enthone.com.
5. W. J. Boettinger, C. E. Johnson, L. A. Bendersky, K.-W. Moon, M. E. Williams, and G. R. Stafford, "Whisker and Hillock Formation on Sn, Sn–Cu and Sn–Pb Electrodeposits," *Acta Mate.*, vol. 53, pp. 5033–5050, 2005.
6. P. Oberndorff, M. Ditters, and L. Petit, "Intermetallic Formation in Relation to Tin Whisker," IPC/Soldertec, *Proceeding IPC/Soldertec International Conference on Lead Free Electronics—Towards Implementation of the ROHS Directive*, pp. 170–178, Brussels, 2003.
7. R. W. G. Wyckoff, *Crystal Structures*, 2nd ed., Interscience Publishers, New York, 1963.
8. J. Zhao, P. Su, M. Ding, S. Chopin, and P. S. Ho, "Micro-structure Based Stress Modeling of Tin Whisker Growth," *IEEE Trans. Electr. Pack. Manufact.*, vol, 29, no. 4, pp. 265–273, 2006.
9. B. D. Dunn, "A Laboratory Study of Tin Whisker Growth," European Space Agency (ESA) STR-223, pp. 1–50, 1987.
10. P. Su and M. Ding, "A Finite Element Study of Strain Energy Density Distribution near a Triple Grain Junction and Its Implication on Whisker Growth," Third iNEMI Sn Whisker Workshop, 2006.
11. M. O. Peach, "Mechanism of Growth of Whiskers on Cadmium," *J. Appl. Phys.*, vol. 23, pp. 1401–1403, 1953.
12. F. C. Frank, "On Tin Whiskers," *Phil. Mag.*, pp. 854–860, 1953.
13. W. C. Ellis, "Morphology of Whisker Crystals of Tin, Zinc, and Cadmium Grown Spontaneously from the Solid", *Trans. Met., Soc. AIME*, vol. 236, pp. 872–875, 1966.
14. Y. Mizuguchi, Y. Murakami, S. Tanaka, S. Tomiya, T. Asai, and T. Kiga, "Whisker and Nodule Formation on Lead-Free Tin Plating by External Stress," 2nd Int. Symp. Tin Whisker, Tokyo, 2008.

15. L. Panashchenko, S. Mathew, S. Han, M. Osterman, and M. Pecht, "Tin Whisker Growth Measurements and Observations," 2nd Int. Symp. Tin Whiskers, Tokyo, 2008.
16. JEITA Standard ET-7410, *Whisker Test Methods on Components for Use in Electrical and Electronic Equipment*, 2005.
17. IEC Standard, IEC 60068-2-82, *Whisker Test Methods for Electronic and Electric Components*, 2007.
18. P. Su, S. Bai, M. Ding, and S. Chopin, "The Effects of Plating Parameters on the Microstructure and Whisker Growth Propensity of Pure Sn Finish," *SMTA Int. Conf.*, 2006.
19. M. Dittes, P. Oberndorff, P. Crema, and V. Schroeder, "The Effect of Temperature Cycling on Tin Whisker Formation," IPC/JEDEC, *Proc. IPC-JEDEC Conf.*, pp. 105–111, Frankfurt, 2003.
20. P. Su, C. Lee, L. Li, J. Xue, B. Khan, R. Moazeni, and M. Hartranft, "Practical Assessment of Tin Whisker Growth Risk due to Environmental Temperature Variations," IEEC, *Proc. 59th ECTC*, pp. 736–741, San Diego, 2009.
21. W. Zhang, A. Egli, F. Schwager, and N. Brown, "Investigation of Sn–Cu Intermetallic Compounds by AFM: New Aspects of the Role of Intermetallic Compounds in Whisker Formation," *IEEE Trans. Electr. Pack. Manufact.*, vol. 28, no. 1, pp. 85–93, 2005.
22. P. Su, J. Howell, and S. Chopin, "A Statistical Study of Sn Whisker Population and Growth during Elevated Temperature and Humidity Tests," *IEEE Trans. Electr. Pack. Manufact.*, vol, 29, no. 4, pp. 246–251, 2006.
23. P. Oberndorff, M. Dittes, P. Crema, P. Su, and E. Yu, "Humidity Effects on Sn Whisker Formation," *IEEE Trans. Electr. Pack. Manufact.*, vol, 29, no. 4, pp. 239–245, 2006.
24. H. Sosiati, N. Hirokado, N. Kuwano, and Y. Ohno, "Transmission Electron Microscopy of Spontaneous Tin Whisker Growth under High Temperature/Humidity Storage," 10th Electronics Packaging Technology Conf., 2008.
25. JEDEC/IPC JP002 Publication, *Current Tin Whiskers Theory and Mitigation Practices Guideline*, 2006.
26. V. K. Glazunova and N. T. Kudryavtsev, "An Investigation of the Conditions of Spontaneous Growth of Filament Crystals on Electrolytic Coatings," transl. from *Zhurnal Prikladnoi Khimii*, vol, 36, pp. 543–550, 1963.
27. M. Dittes, P. Oberndorff, and L. Petit, "Tin Whisker Formation–Results, Test Methods and Countermeasures," IEEE, *Proc. 53rd ECTC Conf.*, pp. 822–826, New Orleans, 2003.
28. P. Su, M. Ding, and S. Chopin, "Effects of Reflow on the Microstructure and Whisker Growth Propensity of Sn Finish," IEEE, *Proc. 55th ECTC Conf.*, pp. 434–440, Orlando, 2005.

7

TESTABILITY OF LEAD-FREE PRINTED CIRCUIT ASSEMBLIES

Rosa D. Reinosa (Hewlett-Packard) and
Aileen M. Allen (Hewlett-Packard)

7.1 INTRODUCTION

This chapter focuses on testability of printed circuit assemblies (PCAs) and the challenges associated with the transition to lead-free solders. First, the results of several contact repeatability studies demonstrating the difficulties associated with probing lead-free solder pastes and fluxes are discussed. Second, the increased wear and contamination on in-circuit test (ICT) probes due to lead-free materials is described. The final section of this chapter discusses the impact of transient bend flexure modes during ICT on lead-free PCAs.

7.2 CONTACT REPEATABILITY OF LEAD-FREE BOARDS

To verify the electrical integrity of printed circuit assemblies, in-circuit testing and/or board functional testing are performed after board assembly. During these test stages a variety of fixture test probes are used to contact pad and via test targets of various geometries. It is important that fixture test probes make reliable electrical contact with test targets to avoid re-testing and to maintain a high first-pass test yield in production.

Lead-Free Solder Process Development, Edited by Gregory Henshall, Jasbir Bath, and Carol A. Handwerker

Fixture test probes are expected to function seamlessly after thousands of cycles and not become dull over time. Lead-free solders, however, have brought major challenges to board manufacturing and the electronic industry as a whole and to electrical test in particular. Because of the increased levels of flux residues (after processing lead-free fluxes and pastes compared with tin-lead fluxes and pastes) remaining on test targets, test probes have a difficult time penetrating and making electrical contact with test targets. In-circuit and functional re-test rates have drastically increased relative to those for tin-lead solders. In some cases up to 100% of the boards needs to be re-tested beyond six cycles. Test fixtures are being programmed to re-test boards for multiple cycles, which can lead to latent or permanent mechanical damage to the PCAs.

There are currently no industry test standards to qualify the probeability of solder pastes and fluxes. Some methodologies have been developed and published by original equipment manufacturers (OEMs) and by paste and probe vendors to assess performance of probes using lead-free pastes and fluxes. They normally require the development of a test vehicle with thousands of test targets of various geometries (Figure 7.1). This section describes the results of testing on one such test vehicle, and the conclusions that can be drawn regarding the impact of the lead-free transition on test probeability.

Flux residues are left on test pads and vias after processing boards using no-clean solder pastes or wave fluxes. These residues, especially for lead-free solder formulations, are tougher to penetrate with conventional probes. There is a tendency in the industry to increase the probe force used on test fixtures, but this can lead to damage caused by increased board strain levels, as discussed later in this chapter. During the transition to lead-free solders, the industry focused on solder process development, without recognizing the detrimental impact to probeability. Although new lead-free probes have been developed with sharper edges to better penetrate

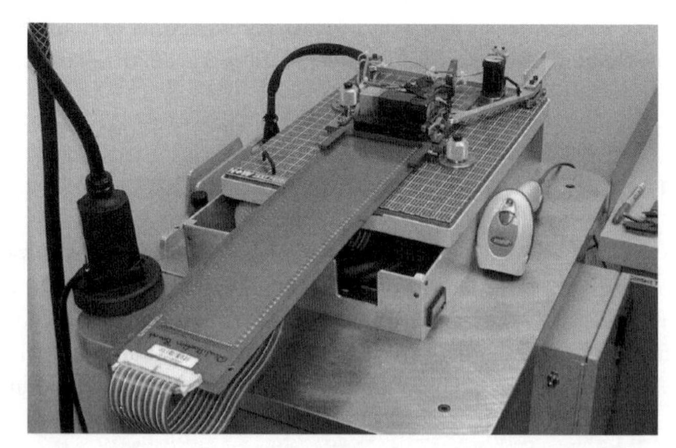

Figure 7.1 Example of a test vehicle for contact repeatability testing

flux residues, not all lead-free probes are 100% effective. There is a need to optimize the board manufacturing process, flux, and probe used in order to achieve the best results.

Examples of the flux residues left on test targets after assembly are shown in Figures 7.2, 7.3, 7.4, and 7.5. Conventional probes (with blunt edges) routinely fail to make electrical contact because they become trapped in the pool of flux, which is

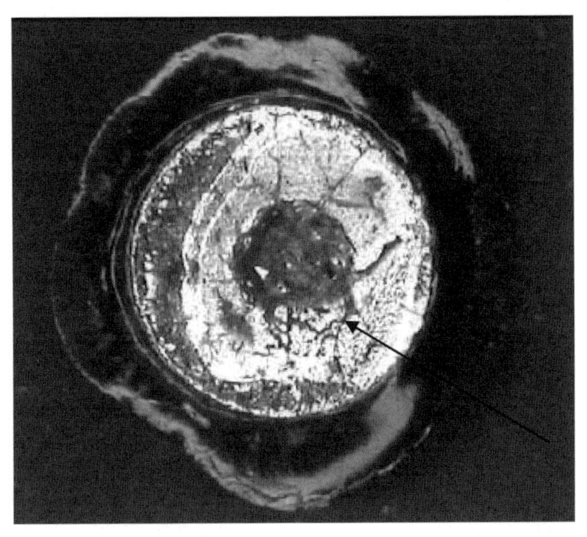

Figure 7.2 Enlarged view of test via showing solder flux pools in the cavity

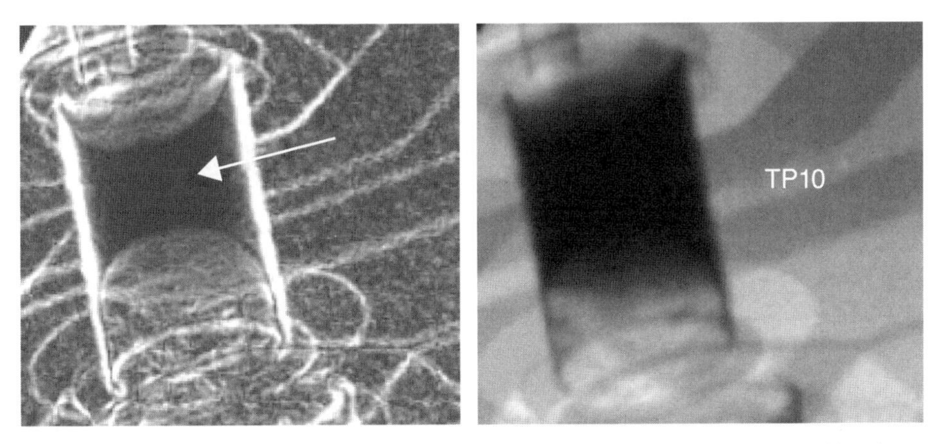

Figure 7.3 X-ray view of the same test via as in Figure 7.2, showing hourglass solder shape and flux collected at each end

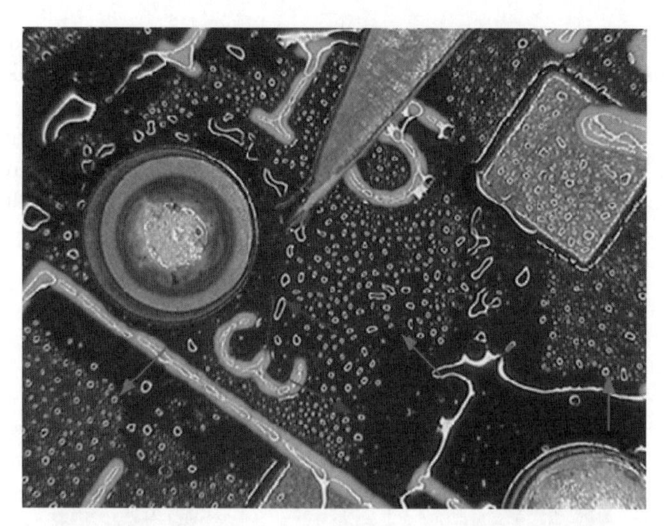

Figure 7.4 Selective wave soldering can negatively affect contact repeatability. There is unburned flux from the wave process as well as excessive flux from the wave pallet opening

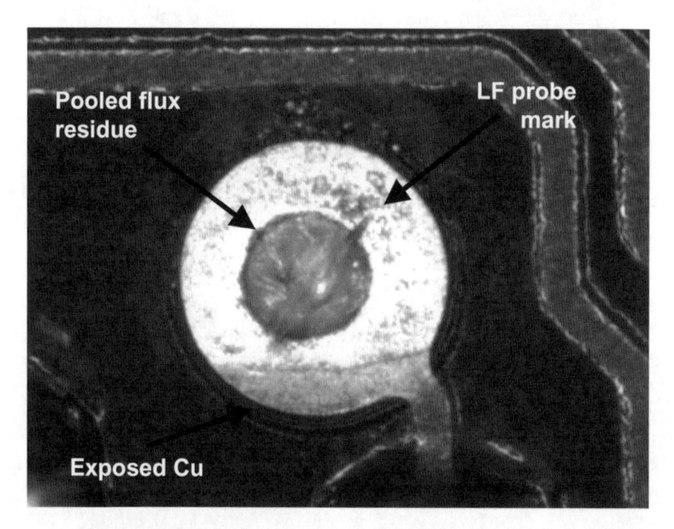

Figure 7.5 Pooled lead-free flux residue on test via

typically in the center of the test target (Figure 7.6). Even a small percentage of contact failure rates in ICT or functional test fixtures with high probe counts can translate into disastrous board re-test rates. Production throughput, PCA reliability, and yields are negatively affected by board re-test. Board re-test is particularly undesirable because it further exposes the assembly/solder joints to high levels of strain.

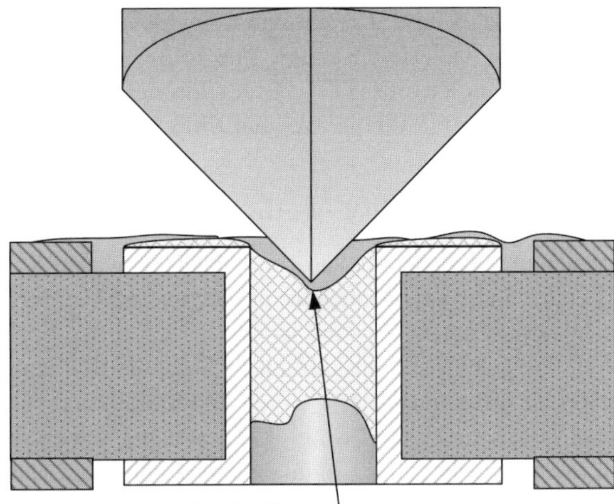

Figure 7.6 Schematic illustrating difficulty of probing test vias using conventional probe geometries when large pools of flux residue are present

TABLE 7.1 Effect of contact failures on board retest rates for THT (through-hole technology) targets

Contact Failure Rate (PPM)	Probe Cycles per Failure	Equivalent Contact % Fail	Equivalent Contact % Pass	Board Retest % Rate Based on Number of THT Targets in ICT Fixture			
				100 Probe Fixture	200 Probe Fixture	300 Probe Fixture	500 Probe Fixture
100	10,000	0.01	99.99	1	2	3	5
200	5,000	0.02	99.98	2	4	6	10
500	2,000	0.05	99.95	5	10	15	25
1,000	1,000	0.1	99.9	10	20	30	50
2,000	500	0.2	99.8	20	40	60	100

Note: "Equivalent contact % fail" is the failure rate in percent, "equivalent contact % pass" is 100—(equivalent contact % fail).

As the number of probes in an ICT fixture increases, the board re-test rate can magnify in value, and the test yields can be radically affected (Table 7.1). However, this effect is less pronounced for through-hole technology (THT) targets due to the fact that in a typical PCA there are usually smaller quantities of these targets as opposed to pads or vias. A typical maximum acceptable contact failure rate for test pads is 0.003% and 0.02% for THT test vias. The following summarizes the test results on surface mount technology (SMT) [1] and through-hole test targets (including through-hole reflowed/paste-in-hole, wave, and SMT with waving using a selective pallet [2]).

7.2.1 Probeability Methodology

A custom test system for probe contact verification was used that provides a high-speed *x-y* table and *z*-axis control for probe penetrability verification for various combinations of materials and ICT probe types.

The system was programmed so that the ICT probe pins on the probe-head sequentially hit thousands of test targets (pads/vias) and collected measurement data at each target. Resistance data were obtained using 4-wire measurements between the socket of the probe pin contacting a test target and a connector at the edge of the test vehicle. Given the fact that most measurements in ICT do not require a very tight tolerance, the threshold was set to 0.5 Ohms. A schematic of the measurement approach is provided in Figure 7.7. A summary of the SMT and THT test vehicles tested is provided in Table 7.2.

SMT Test Vehicle Description SMT test vehicles were solder paste printed and then reflowed once or twice (as specified by the bullets below), and probed on one

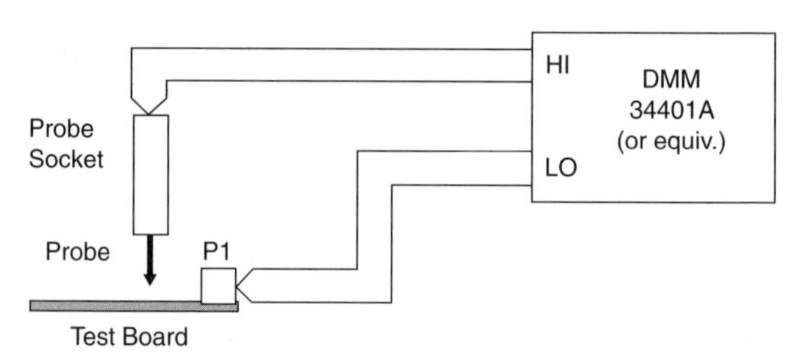

Figure 7.7 Schematic of probe repeatability test apparatus

TABLE 7.2 Summary of test targets for the SMT and THT (through-hole technology) test vehicles

Surface Mount Test Targets Reflow	Through-Hole Test Targets		
	Through-hole reflow	Through-hole wave	Selective wave
Pads Vias	Connectors	Pins Pads Vias	Pads Vias

side (the side that they were filled on). SMT test targets included both unpasted (for control) and pasted/reflowed pads and vias. Each SMT test vehicle consisted of 25,200 total test targets: 12,600 pads and 6300 vias that could be probed on either side. The variety of materials used on the SMT test vehicles included:

- Five different lead-free solder pastes in addition to a "no-paste" control.
- Multiple surface finishes
  - Single- and double-reflow OSP(organic solderability preservative) (high-temperature rated)
  - Double-reflowed immersion tin (Sn)
  - Double-reflowed immersion silver (Ag)
  - Double-reflowed electrolytic nickel-gold (Ni/Au)

THT Test Vehicle Description Through-hole technology (THT) targets were produced using three processes: through hole reflow (THR)/paste-in-hole), wave and surface mount technology (SMT), and selective wave. The THR or paste-in-hole printing process involved overprinting solder paste on the pads of through-hole components, inserting the component leads and then reflowing the board. For the waved process boards, all THT pins, pads, and vias had flux simultaneously applied and then waved. For SMT plus selective waved boards, test pads and test vias had solder paste applied and reflowed followed by the boards being run through a selective wave machine. A special selective wave pallet was used to provide various amounts of shielding to the test pads and vias.

The through-hole test vehicle consisted of 3150 pins, 1575 test pads, and 1650 test vias for a total of 6375 targets per board (Figure 7.8). Multiple material combinations were also used.

Figure 7.8 Through-hole test vehicle

- For through-hole reflow, two different lead-free solder pastes were tested on through-hole components, with the following surface finishes:

  OSP (high-temperature rated)

  Electrolytic nickel–gold
- For through-hole wave (of pins, pads, and vias), six different wave fluxes were tested, with the following surface finishes:

  OSP 106A HT(high-temperature rated)

  Immersion tin

  Immersion silver

  Electrolytic nickel-gold
- For SMT reflow followed by selective wave soldering onto test pads and vias, testing was done with nine combinations of six fluxes and six pastes and the following surface finishes:

  OSP (high-temperature rated)

  Immersion tin

  Immersion silver

  Electrolytic nickel-gold

Finally, multiple probe geometries with varying probe forces were used to determine how these parameters affected contact repeatability. Probe selection was done carefully to ensure acceptable contact rates without increasing the risk of damage, since higher force probes can cause excessive damage and reduce PCA reliability. All probe types tested in this evaluation are described in Table 7.3.

TABLE 7.3 Probes used on waved test targets

Target	Probe Type	Probe Style/Force	Picture
Pins	Crown	36*, 38, and 45 mil 5.5 and 6.5 oz	Ø.036 [0.91]
Pads	Chisel	15°, 10°, and 30°; 5.5, and 6.5 oz	Ø.035 [0.89] 15° (3) 28° (3)
	Blade	40° and 90°; 5.5 and 6.5 oz	Ø.036 [0.92] 40°
Vias	Chisel	15°, 10°, and 30°; 5.5 and 6.5 oz	Ø.035 [0.89] 15° (3) 28° (3)

Note: The 36-mil probe style with 5.5 oz probe force for crown probes and the 15-degree probe style with 5.5 oz probe force for chisel probes were the geometries and forces most often used in the experimental studies.

TABLE 7.4 Contact repeatability failure rates of pasted and nonpasted test pads by board surface finish

Test Target Type	Contact Repeatability Failure Rates
Nonpasted OSP single-reflow test pad	Up to 8.2% (82 k PPM)
Nonpasted OSP double-reflow test pad	Up to 5.6% (56 k PPM)
Nonpasted test pads with 1 mm Ag, 1 mm Sn and electrolytic Ni Au (not OSP)	0.005 to 0.08% (50 to 800 PPM)
Pasted test pads with OSP, electrolytic Ni Au and 1 mm Ag finishes.	0–0.018% (0–180 PPM)
Pasted test pads with 1 mm Sn	0.00% (0 PPM)

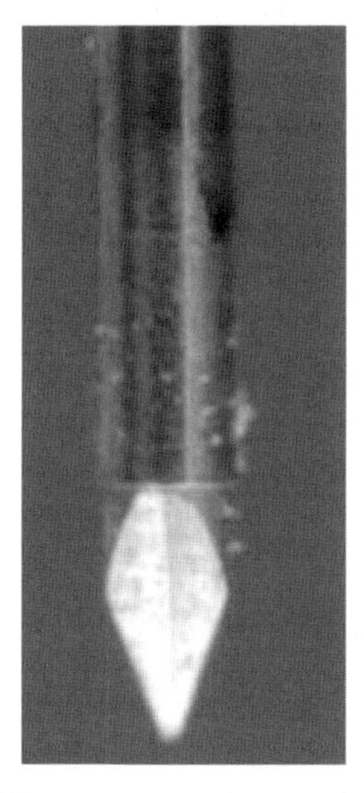

Figure 7.9 5.5-oz. 15-degree chisel probe

7.2.2 SMT Target Results

A low force probe, specifically 5.5-oz, 15-degree chisel (Figure 7.9), was found to provide a relatively good performance over a variety of test pads and surface finish combinations [1]. Better contact repeatability was achieved with solder paste printed test pads than with solder paste printed test vias (Figures 7.10 and 7.11). It was also noted that board surface finish had an impact on contact repeatability performance for both test pads and vias. In this study, all solder paste printed test pads performed acceptably across all surface finishes (Table 7.4). In this study, paste 3 with nickel-gold and immersion silver board surface finishes was not probed (Figures 7.10 and 7.11). The worst contact repeatability was observed with non–solder paste printed OSP test pads

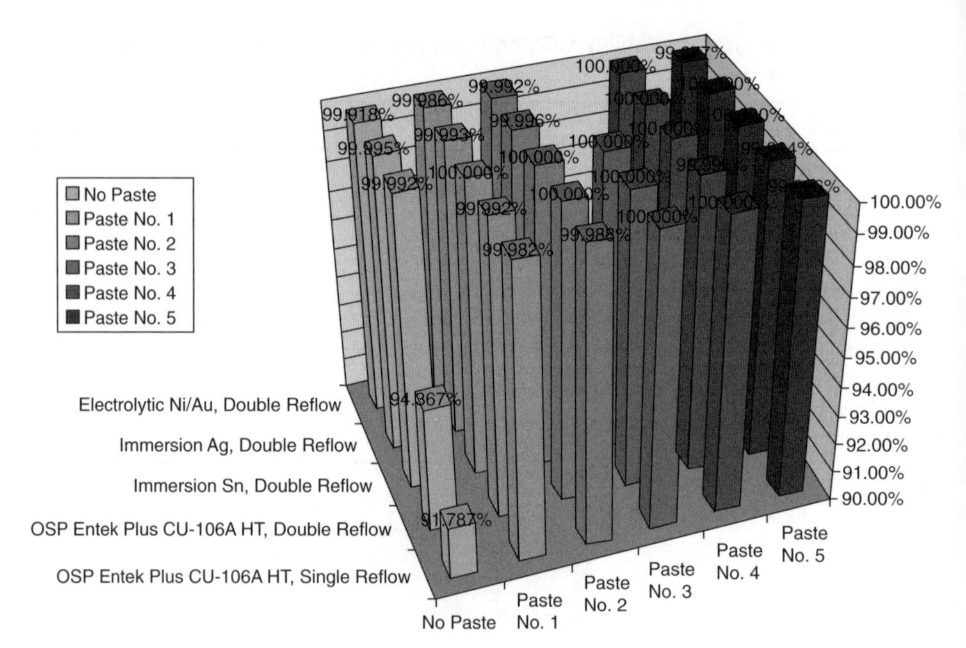

Figure 7.10 Contact yields on SMT reflowed test pads. Paste 3 with Ni/Au and Imm Ag board surface finishes was not probed

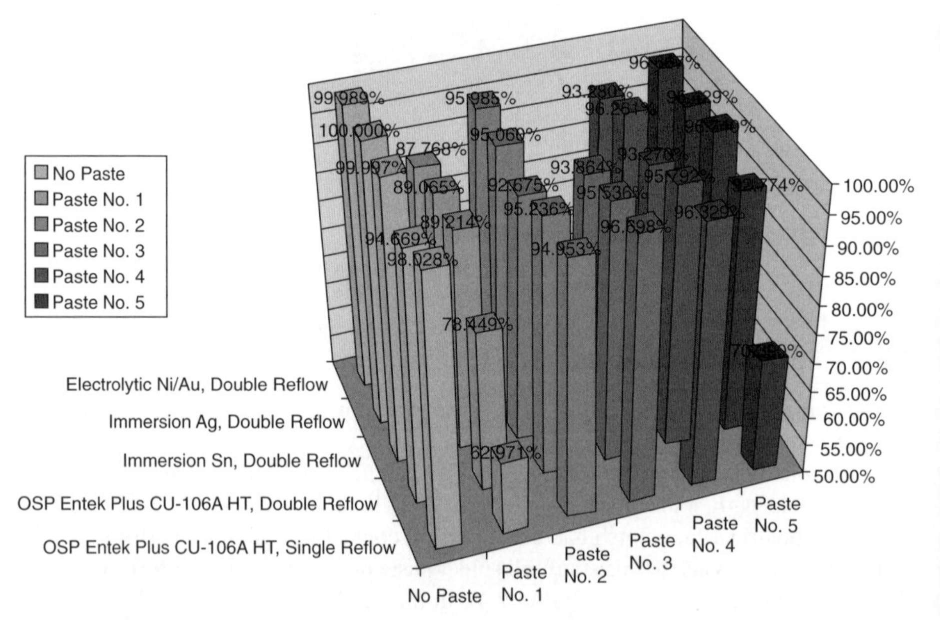

Figure 7.11 Contact yields on SMT reflowed test vias. Paste 3 with Ni/Au and Imm Ag board surface finishes was not probed

TABLE 7.5 Contact repeatability failure rates of pasted and non-pasted test vias by board surface finish

Test Target Type	Contact Repeatability Failure Rates
Nonpasted OSP single-reflow or double-reflow test vias	Up to 5.4% (54 k ppm)
Nonpasted test vias with 1 mm Ag and 1 mm Sn	0.00 and 0.003% (0–30 ppm)
Nonpasted test vias with electrolytic Ni Au	0.01% (100 ppm)
Pasted OSP single- or double-reflow test via	Up to 37% (370 k ppm)
Pasted test vias with other finishes (1 mm Ag, 1 mm Sn, electrolytic Ni Au)	Up to 12.3% (123 k ppm)

with up to an 8% contact failure rate In contrast, solder paste printed OSP test pads had as low as a 0% to 0.02% contact failure rate. Solder paste printed OSP test targets would be preferred over non–solder paste printed test targets. This would be because Cu-OSP coated surfaces oxidize with reflow, and thick oxides would inhibit electrical probing during ICT.

The best overall results on non–solder paste printed test vias were achieved with immersion tin and immersion silver followed by electrolytic nickel-gold surface finish (Table 7.5). Test vias on OSP finish boards are the most difficult type of targets to test because the probes typically hit the center of the via where the flux residues accumulate (Figure 7.6). In this case lead-free probes with sharp edges should be used to improve contact.

7.2.3 Through-Hole Test Target Results

Based on the results shown in Figure 7.12, THR (through-hole reflow) or paste-in-hole solder paste printed through-hole connectors were not probe-testable, regardless of probe type and applied force. Ni/Au board finish was better than OSP due to oxidation of Cu. To maximize improvements in future through-hole test studies, it is recommended that the paste-in-hole process be fine-tuned, with experiments on other probe geometries, in addition to an optimization of the flux application process.

Generally, wave-soldered pads (Figure 7.13) had worse contact repeatability than solder paste printed SMT through-hole reflow pads (Figure 7.10), although acceptable testability performance was achieved with the 5.5-oz, 15-degree chisel probe with the waved pads.

THR/ paste-in-hole and waved THT connectors (Figures 7.12 and 7.15) generally have an order of magnitude worse contact repeatability compared to waved test pads (Figure 7.13) and test vias (Figure 7.14) The best performance was obtained using a 36-mil crown probe. There was no consistency in testability performance across various combinations of lead-free fluxes and surface finishes.

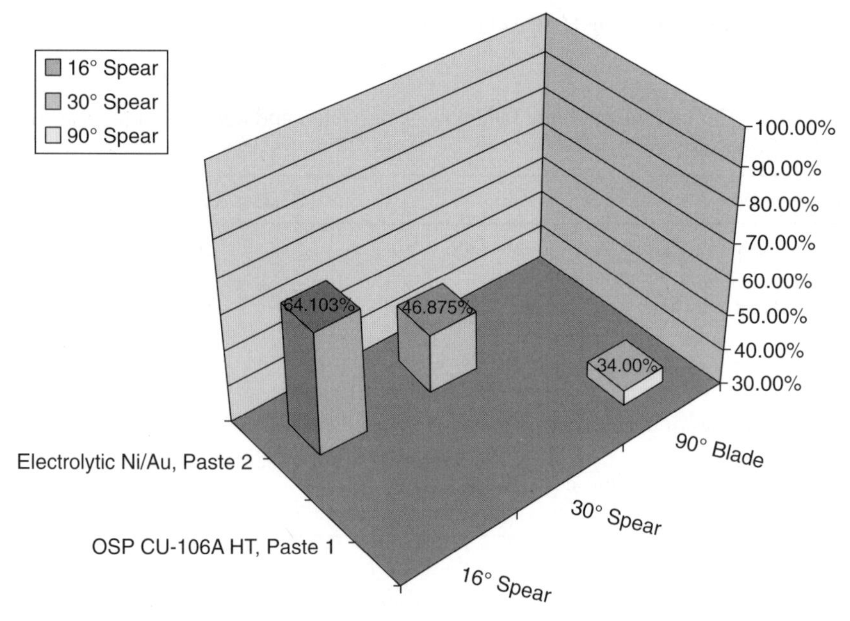

Figure 7.12 Contact yields on through-hole reflowed connectors

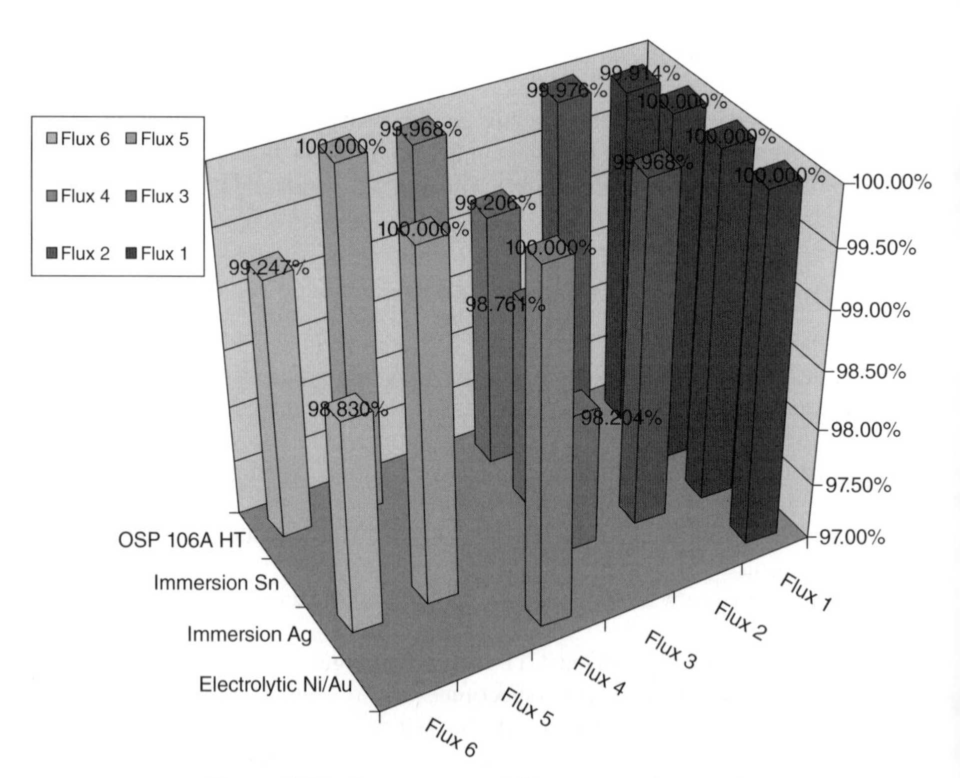

Figure 7.13 Contact repeatability on waved test pads

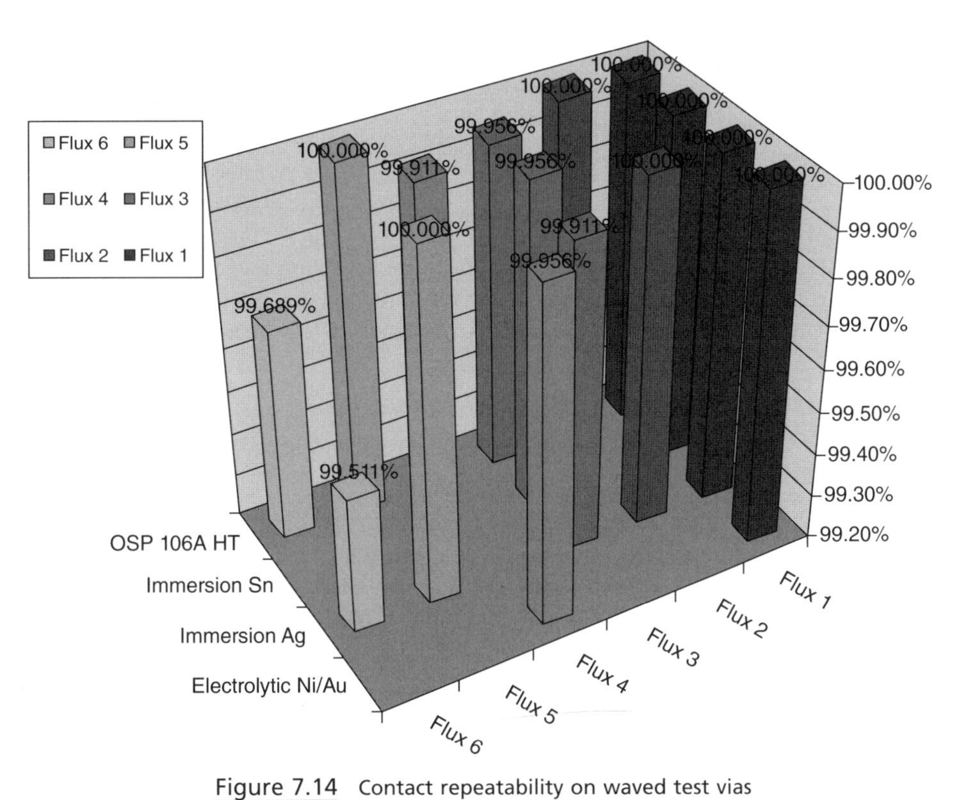

Figure 7.14 Contact repeatability on waved test vias

Figure 7.15 Contact repeatability on waved THT components

When applied correctly, the lead-free wave process has a positive impact on probeability of test vias. Contact repeatability performance for wave-soldered test vias was improved over SMT paste-printed test vias (Figures 7.14 and 7.11). Acceptable contact repeatability was observed when a 5.5-oz, 15-degree chisel probe was used.

During these experiments it was observed that SMT paste plus selective wave negatively impacted the contact repeatability of solder paste printed test pads and test

TABLE 7.6 Contact repeatability failure rates on pasted and selectively waved test pads by paste and flux type and distance from the pallet opening

Paste	Flux	Average Failure Rate (0 mils from Palette)	Average Failure Rate (15 mils from Palette)	Average Failure Rate (70 mils from Palette)
Paste 1	Flux 1	0.00%	0.00%	0.00%
Paste 2	Flux 1	0.11%	0.00%	0.00%
	Flux 4	0.33%	0.00%	0.00%
Paste 3	Flux 3	0.89%	0.26%	0.60%
Paste 5	Flux 1	0.11%	0.00%	0.04%
	Flux 2	0.56%	0.07%	0.04%
	Flux 4	0.22%	0.00%	0.00%
	Flux 5	0.00%	0.00%	0.00%
Paste 6	Flux 6	0.00%	0.04%	0.00%
Passed combinations		3/9	6/9	6/9

TABLE 7.7 Contact repeatability failure rates on pasted and selectively waved test vias, by paste and flux type and distance from the pallet opening

Paste	Flux	Average Failure Rate (0 mils from Palette)	Average Failure Rate (15 mils from Palette)	Average Failure Rate (70 mils from Palette)
Paste 1	Flux 1	0.00%	0.00%	0.00%
Paste 2	Flux 1	0.00%	0.07%	0.17%
	Flux 4	0.00%	0.00%	0.00%
Paste 3	Flux 3	0.00%	0.00%	0.00%
Paste 5	Flux 1	0.00%	0.00%	0.00%
	Flux 2	0.00%	0.00%	0.04%
	Flux 4	0.00%	0.00%	0.04%
	Flux 5	0.33%	0.18%	0.07%
Paste 6	Flux 6	0.19%	0.12%	0.20%
Passed combinations vs. distance		7/9	6/9	4/9

vias because of flux seepage on test targets in the neighborhood of the pallet opening (Tables 7.6 and 7.7). Closely controlling all aspect of the SMT plus selective wave process could help to avoid contact issues during the manufacturing test process. However, it was found that paste-flux combinations that worked well on test pads did not work well on test vias [2].

7.2.4 Trends and Solutions

The key issue when probing test vias was that traditional probes could make contact at the center of the vias but not to penetrate through the lead-free flux residues left on vias during a no-clean process by the solder paste after reflow or from the wave flux during selective wave/wave soldering. The conventional type of sharp-tipped probes would routinely fail to make contact because they would become lodged in the pooled flux at the center of vias. Improved probe tip geometries would enable better performance on test vias. Some of the lead-free probes developed recently have addressed these challenges, and can perform well when combined with a good quality manufacturing process. Other industry approaches have involved the elimination of test vias on boards, use of via plugging technologies (where applicable), use of domed vias (Figure 7.16), change to non-OSP board finish, and use of a water-clean process. Note that changes to non-OSP board finishes may not be an option due to other reliability concerns for these board finishes (brittle solder joint failures, corrosion, tin whiskering, etc.). Lead-free hot-air solder leveling (HASL) appeared to be emerging as a PCB surface finish. However, limited testability performance data was available at this time. Lead-free HASL possesses other reliability risks related to coplanarity issues, PCB delamination and is not suited for boards with gold fingers.

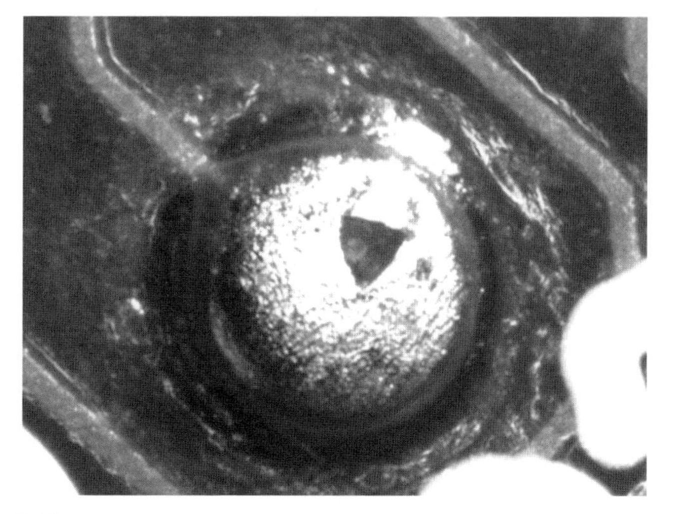

Figure 7.16 An example of a domed via probed with a 15-degree chisel probe

7.3 PROBE WEAR AND CONTAMINATION

In general, pin probes are designed to provide accurate readings for tens of thousands of contacts. However, probe contamination and tip blunting can limit contact repeatability long before ten thousand contacts.

More contamination is observed on solder paste printed test vias than on solder paste printed test pads (Figure 7.17). This is because of the tendency of more flux residue to get trapped in the test via holes than in the test pads. PCA surface finish and solder paste can also have an effect on probe contamination. Increased levels of contamination are observed on probes used to test OSP pasted via targets compared to other finishes. High solid content in a solder paste causes a large amount of flux residue to remain after reflow, making probing of the flux residue more difficult.

The higher yield strength and stiffness (Young's modulus E) of the lead-free SnAgCu solders ($E = 51.0$ GPa for Sn3Ag0.5Cu) compared to. Sn37Pb ($E = 40.2$ GPa [3]) produce faster wear on the tip of the probe. Higher levels of contaminants on probes (Figure 7.18) will also affect the wear of the shaft and the barrel. Probing non-solder paste printed OSP finish test pads would not be recommended because of the increased wear of the probe tip and the risk of oxidation. Any of these failure mechanisms will cause contact repeatability failures. Increased levels of probe contamination due to flux residues will drive more frequent maintenance of test fixtures. With lead-free solders, the life of test probes is significantly reduced due to the higher levels of contaminants, as well as the yield strength and stiffness of the SnAgCu solders (typically 20,000 cycles instead of 60,000 cycles for PCAs with tin-lead solder).

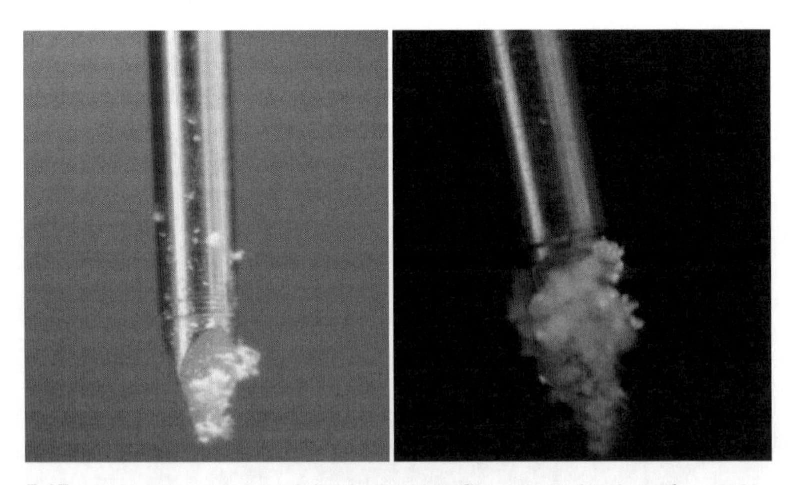

Figure 7.17 Probes on OSP board finish showing flux contamination after 4200 test pad contacts (left) and 200 test via contacts (right)

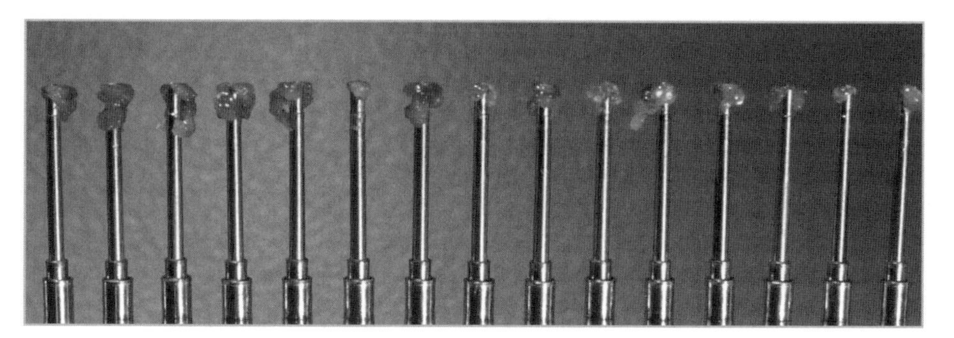

Figure 7.18 Lead-free solder flux residue contamination

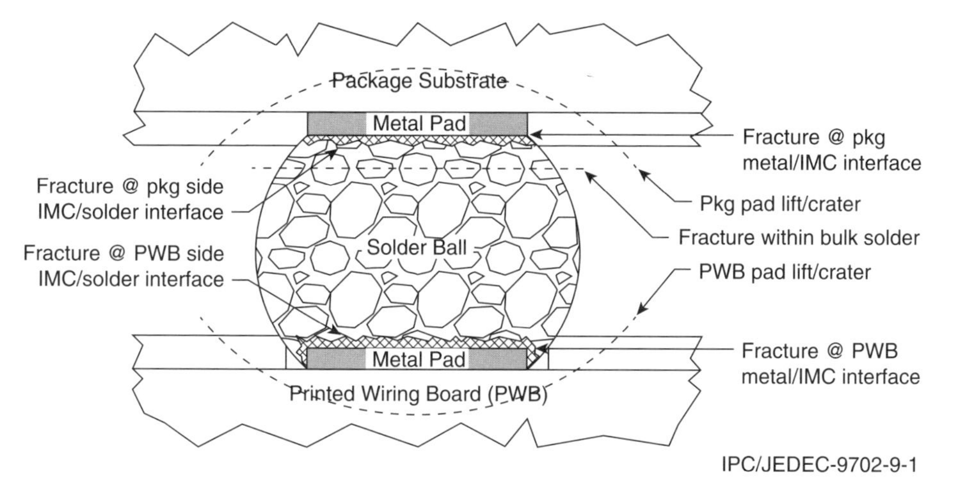

Figure 7.19 Potential solder joint failure modes in a typical BGA/PCB assembly. In a lead-free PCA bending/flexure, PWB pad cratering is typically the greatest risk [5]

7.4 BOARD FLEXURE

The transition to lead-free materials has increased testability concerns beyond the contact repeatability issues described earlier. Two other aspects of this transition have also negatively impacted the mechanical reliability of PCAs. First, when higher probe force levels are used to penetrate flux residues, the PCAs are subjected to greater transient bend levels, which in turn transfer larger stresses to the solder joints. Additionally SnAgCu solders are stiffer and stronger than SnPb solders, so the larger stresses on the solder joint are transferred more to the pad on the PCB side of the joint for lead-free SnAgCu solder. The PCB laminate is then the weakest link in the solder joint system, and as a result the failure mode for bending PCAs soldered with SnAgCu alloys is pad cratering and trace cracking (Figures 7.19, 7.20, 7.21, and 7.22). With the higher processing temperatures for lead-free solder, as well as the industry trend to move toward halogen-free laminates,

Figure 7.20 Example of pad lifting (left two arrows) and trace breaking (right arrow)

Figure 7.21 Example of pad cratering (circle in the figure)

PCBs may also become more prone to mechanical damage. So the risk of damage due to overflexure of boards is greater in the lead-free environment.

There is currently work being done in the iNEMI Board Flexure project to assess the risk of laminate damage due to flexure of lead-free PCAs from ICT, assembly, and handling [4]. There are multiple ways of evaluating bend resistance, including four-point monotonic bend described in the IPC/JEDEC 9702 standard [5]. An alternate technique, the spherical bend test method, has been identified as approximating the "worst-case" bend mode observed in ICT. In Figure 7.23 the PCB strain to electrical failure is lower in spherical bending than in four-point bending. Further the corner solder ball will reach a given load at lower PCB strains under spherical bend rather

Figure 7.22 Damage to the solder joint (under the pad on the PCB side) resulting from overflexure of the PCA

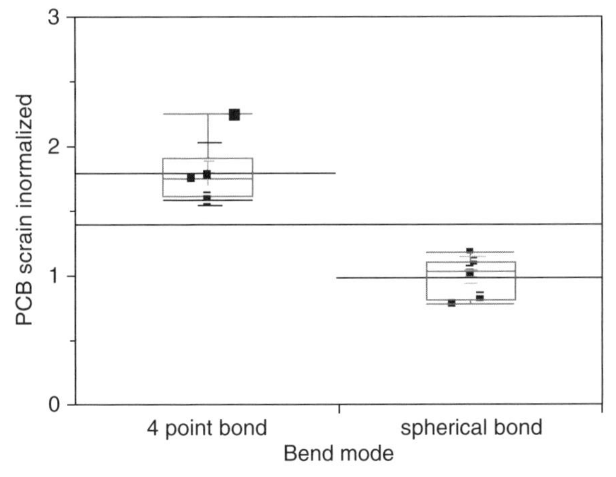

Figure 7.23 PCB strain to electrical failure is smaller in spherical bend testing than in four-point bend testing [6]

than under four-point or twist bending (Figure 7.24) [6]. Spherical bend concentrates the stresses on the corner solder joints in a BGA, which is typically the most vulnerable part on a PCA during bending.

A joint industry draft standard detailing the spherical bend test methodology (IPC/JEDEC 9707) [4] is being proposed by iNEMI through the IPC 6-10d and JEDEC JC14.1 standard committee groups. Figure 7.25 shows the test setup in which eight support pins provide the "spherical" bending when a single push pin is forced downward onto the back side of the PCB with the BGA component facing down. This bend test has been successfully used to set strain limits for PCAs and BGA packages. The PCA strain limit guidance is unique for each material set, solder alloy, layout, component package, and so forth. In general, SnAgCu solder alloys have a lower strain limit

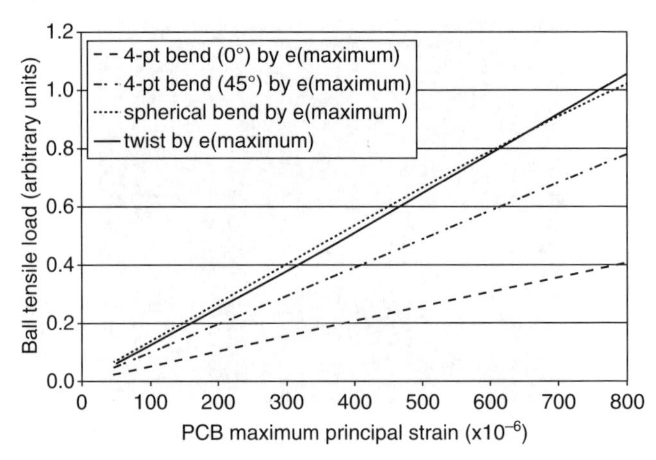

Figure 7.24 Tensile load versus PCB strains, comparing four-point bend test and spherical bend test [6]

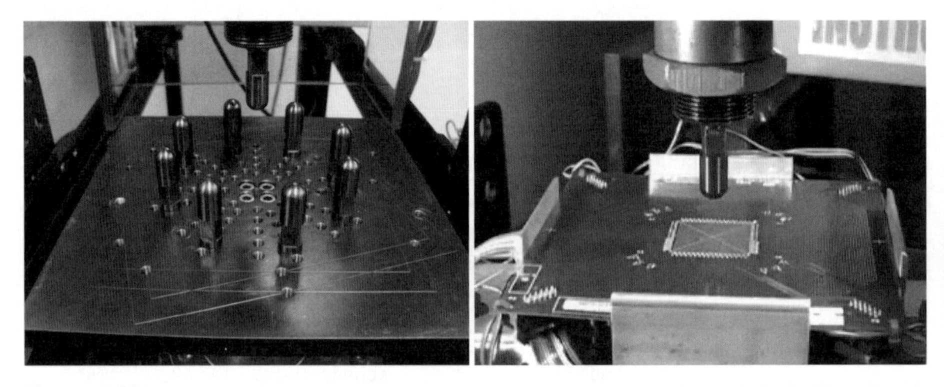

Figure 7.25 Spherical bend test setup, with support pins arranged in a circle (left) and with the load pin positioned to come down onto the board (right) the test board (right)

than SnPb, but strain limits for each product can be determined through spherical bend testing. Such guidance can then be used during manufacturing and assembly as a strain not to be exceeded. The primary concern is during ICT when some of the worst flexure conditions are induced on a board.

In sum, spherical bend testing can be used to aid in board and fixture designs such that no mechanical damage is induced. The use of high probe forces may damage the solder joint, package interconnect, board land, substrate, or the components. The ICT fixture design should be optimized to counterbalance probe and push down finger forces. Selection of the probe force must balance the need for a large enough force to maintain contact repeatability while reducing bending risk due to the larger probe forces. Optimization can be guided by spherical bend test data showing where the strain limits are for a given PCA.

7.5 CONCLUSIONS

OSP board surface finish is currently the one of the most popular surface finishes for lead-free boards because of its low cost and high reliability outside of the test process. OSP, however, cannot be reliably probed due to oxidized copper and the high yield strength of copper. For these reasons exposed copper is an issue during probe testing. OSP test points need to have solder applied for improved testability. The higher temperatures during lead-free reflow and lead-free wave soldering require the use of new paste and flux chemistries. Flux residues from the lead-free solder paste and wave fluxes make it harder for probes to penetrate and make contact with test points. As flux residue pools form in a solder dimple at via openings, the testability of vias is affected more than pads. Flux pools impede conventional probes from making contact with test targets. Potential solutions include the use of sharper lead-free probes, test pads wherever possible, and pad-like geometries on through-hole vias. Waved test points that are close to the pallet opening are generally at risk. Excessive wave flux floods the bottom of the board or pallet opening and may seep beyond the pallet opening edge. In addition temperature cycles result in pads covered with dried flux and vias that are difficult to probe.

With the transition to lead-free solders, the focus is on component damage caused by overflexure of printed circuit boards during manufacturing, test, and assembly processes. Because BGA package size and ball pitch have decreased, second-level interconnect failures due to overflexure have become an increasing concern, especially when combined with other PCA material changes (laminate type, solder alloy, pad geometry, etc.). Overflexure can also cause partial cracking of the second-level interconnect, which can only be found by cross-sectioning the BGA component. The industry standard IPC/JEDEC 9704 [7] describes the strain gauge methodology that can be used to measure a board's response to flexure, induced by using the transient bend test methodologies detailed in IPC/JEDEC standard 9702 [5] and the draft IPC/JEDEC standard 9707 [4]. OEMs and component suppliers should be able to use the IPC/JEDEC 9707 standard (when published) to determine a maximum strain level for components or PCAs and to serve as guidelines in manufacturing, test, and assembly.

Testability of PCAs has become more challenging in the lead-free environment, due to reduced contact repeatability, increased probe wear and contamination, and decreased robustness to board flexure, but solutions to overcome these challenges have and will continue to be worked out by the industry.

ACKNOWLEDGMENTS

The authors wish to acknowledge Alex Leon for his vision and technical contributions to the work described in this chapter and would also like to thank Teik Ju Choo (Hewlett-Packard Company) for his technical contributions and Contract Manufacturer support. The technical leadership and contributions of QA Technologies, as well as the management support of Michael Roesch (Hewlett-Packard Company), are also gratefully acknowledged.

REFERENCES

1. R. D. Reinosa, "Effect of lead free solders on in-circuit test process," *Proceedings, IEEE, International Test Conference, Paper 26.3*, New York, NY, Nov 2005.
2. R. D. Reinosa, "Lead-free Through-Hole Technology (THT) and Contact Repeatability in In-Circuit Test," *International Test Conference, Paper 5.3*, New York, NY, Oct 2006.
3. Dongwook Kim, Daewoong Suh, Thomas Millard, Hyunchul Kim, Chetan Kumar, Mark Zhu, Youren Xu, "Evaluation of High Compliant Low Ag Solder Alloys on OSP as a Drop Solution for the 2nd Level Pb-Free Interconnection," IEEE, *Proc. 57th Electronics Components and Technology Conf.*, New York, NY, May 2007, p. 1614.
4. IPC/JEDEC Draft Standard IPC/JEDEC-9707, *Spherical Bend Test Method: Mechanical Characterization of PCA Interconnects and Determination of Strain Limits*, 2010.
5. IPC/JEDEC Standard IPC/JEDEC-9702, *Monotonic Bend Characterization of Board-Level Interconnects*, June 2004.
6. G. Hsieh and A. Mcallister, "Flip Chip Ball Grid Array Component Testing under Board Flexure," IEEE, *Proc. 55th Electronic Components and Technology Conf.*, New York, NY, June 2005, pp. 937–944.
7. IPC/JEDEC Standard IPC/JEDEC-9704, *Printed Wiring Board Strain Gage Test Guideline*, June 2005.

8

BOARD-LEVEL SOLDER JOINT RELIABILITY OF HIGH-PERFORMANCE COMPUTERS UNDER MECHANICAL LOADING

Keith Newman (Sun Microsystems)

8.1 INTRODUCTION

A solder joint figuratively and physically defines the border between a surface mount (SMT) component and a printed wiring board (PWB). Consequently, solder joint failures of a SMT device during system assembly often result in finger pointing from both the device supplier and the board assembler—each blames the other! Indeed similar conflicts can occur between the PWB laminate supplier and the board assembler. But how does one gauge whether the solder joint fracture is an expected event for the given mechanical loading condition, or is likely due to a defective component, material, surface finish, PWB, or SMT assembly process? Which corrective action will result in the most timely, cost-effective solution?

Figure 8.1 shows the principal solder joint fracture locations of a typical BGA (ball grid array) package assembled on a PWB. Depending on the specific solder alloy, surface finish, PWB laminate, SMT component, SMT assembly process, thermal aging, and so forth, and the mechanical stress condition, solder joint brittle fracture may occur at any of the identified locations in Figure 8.1, with the typical exception of the bulk solder.

The discussion of board-level solder joint reliability can be broken down into the following major sections:

Lead-Free Solder Process Development, Edited by Gregory Henshall, Jasbir Bath, and Carol A. Handwerker

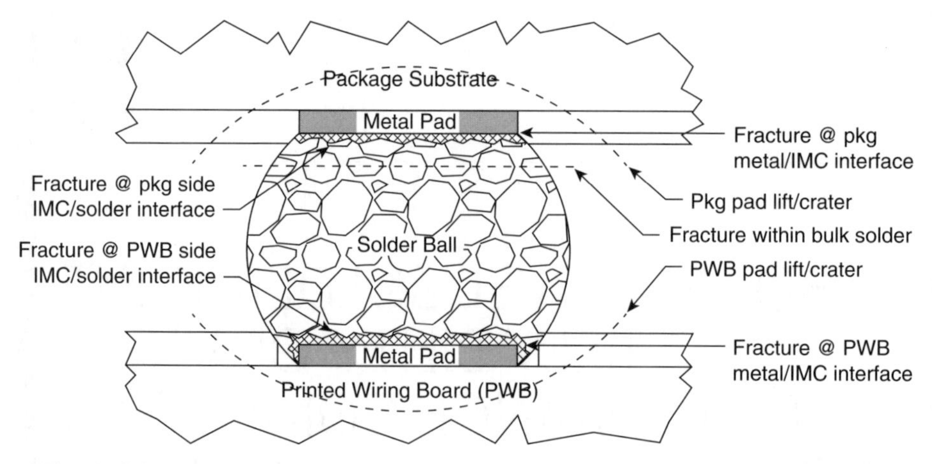

Figure 8.1 Various solder joint fracture locations for a typical BGA/PWB assembly [1]

- Establishing PWB strain limits for manufacturing
- SMT component fracture strength characterization
- PWB fracture strength characterization
- PWB strain characterization
- Solder joint fracture prediction
- Fracture strength optimization

These major categories encompass different aspects of the reliability of solder joints subjected to mechanical loading conditions, and must all be considered to derive a robust product with improved brittle fracture strength. Due to chapter page limitations, not all possible mechanical stress concerns are necessarily addressed within this single chapter. In particular, mechanical fatigue-related testing such as vibration and cyclic bend/twist are not included. This chapter reflects the author's approach to characterizing mechanical reliability of solder joints in high performance computers, and does not necessarily serve as an exhaustive reference of all known mechanical reliability testing/analysis techniques.

8.2 ESTABLISHING PWB STRAIN LIMITS FOR MANUFACTURING

Original equipment manufactureres (OEMs) and/or board assemblers may set limits on the amount of PWB strain allowed during board assembly and test operations to eliminate brittle fractures in production. Figure 8.2, based on work performed in 2003 by Sun Microsystems, illustrates the initial set of guidance for maximum allowable PWB strain based on production failure data and laboratory testing. This graph defined an "acceptable" region in which the PWB board strain (immediately adjacent to an SMT component), would not be expected to result in solder joint brittle fractures.

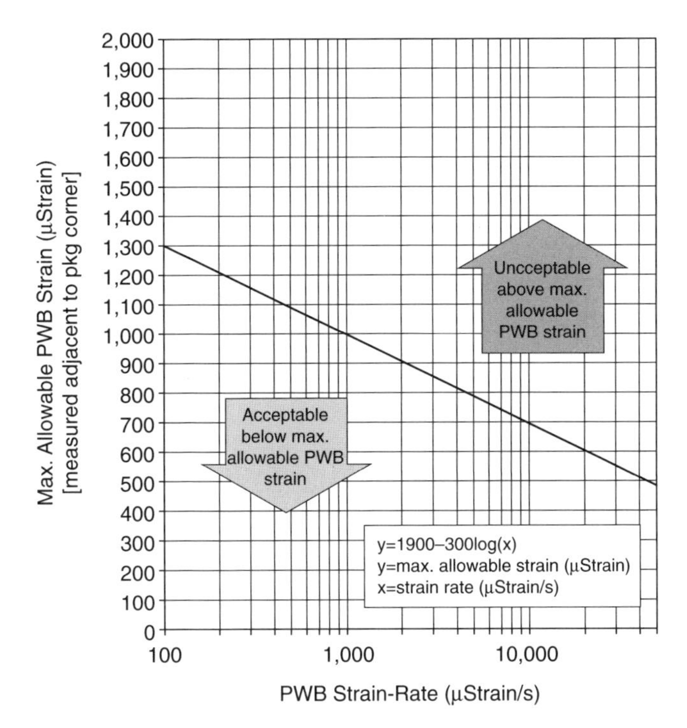

Figure 8.2 Maximum allowable PWB strain (original 2003 guidance)

Roughly speaking, this "line-in-the-sand" reflected an estimated 0.1% cumulative failure rate (1000 dpm). The graph was constructed with a limited empirical data set, but it does reflect the strain-rate dependent material properties of tin-based solder alloys. Significantly, the acceptance criteria defined maximum allowable strain as a function of strain-rate, rather than a fixed value as some other companies have established and continue to use.

By 2004, finite element analysis and additional empirical results were incorporated into the guidance, which was modified to accommodate varying board thicknesses (Figure 8.3). This guidance was published in Appendix A of the IPC/JEDEC-9704 standard [2]. A list of the numerous assumptions and simplifications of this guidance is provided in Appendix B of the IPC/JEDEC-9704 standard [2], with the appendix detailing limitations on failure mode, solder alloy, surface finish, package body size, PWB material properties, and the like, as they relate to applicability of the graph. Critical factors such as solder joint metallurgy, surface finish, and laminate impact the solder joint reliability and are discussed in later sections; these factors affect the failure mode and mechanisms and change the PWB strain rates needed for failure to occur.

By 2005, however, new mechanical shock test efforts and additional production data resulted in the guidance being extended to cover drop impact conditions (Figure 8.4). This updated graph was included in the *Printed Circuits Handbook*, 6th edition, published in 2007 [3].

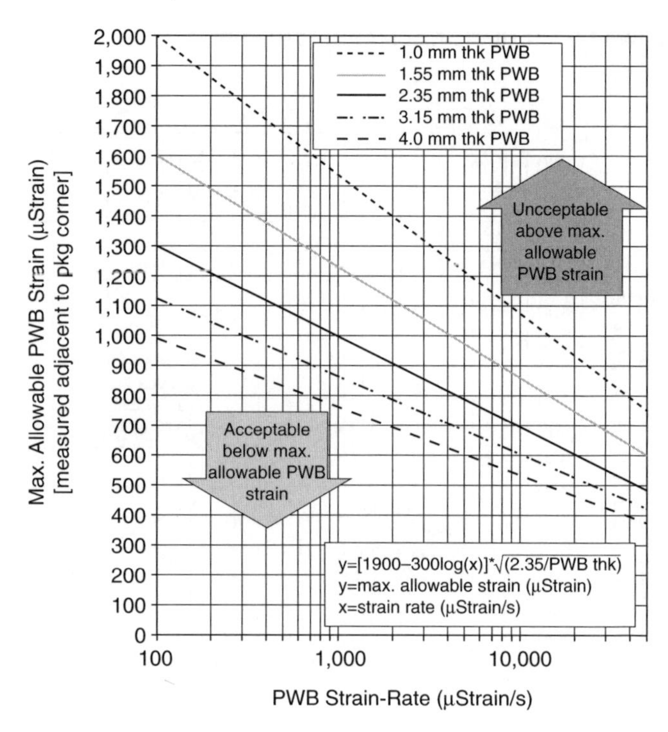

Figure 8.3 Maximum allowable PWB strain (2004 guidance, published in IPC/JEDEC-9704 Standard Appendix A) [2]

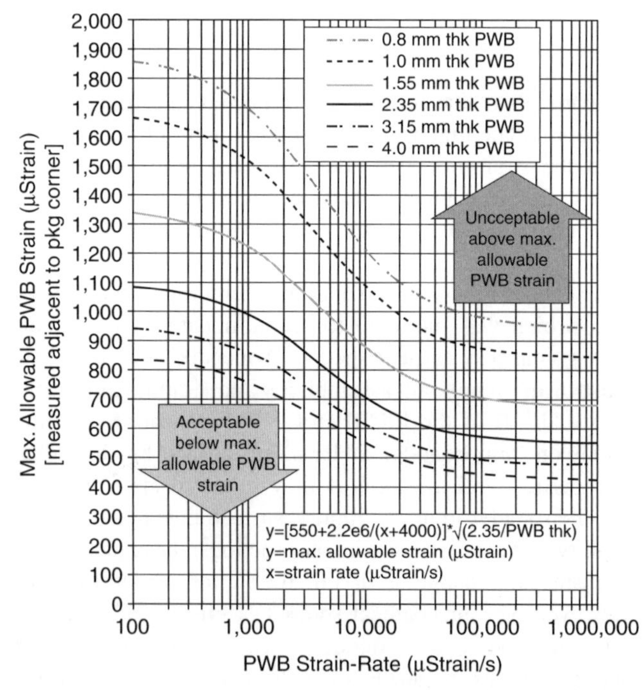

Figure 8.4 Maximum allowable PWB strain (2005 guidance) [3]

At the present time, the guidance shown in Figure 8.4 reflects the current available data, but it remains subject to future revision due to continued learning. Given the many simplifications associated with these acceptance criteria, it is surprising that they have proved to be a reasonable predictor of solder joint brittle fracture (~0.1% cumulative failure) across a wide range of components, PWB constructions, and solder alloys.

It is important to consider that the guidance reflects the tail-end of the fracture strength distribution (approx. $\bar{x} - 3\sigma$), where $\bar{x}$ is the average value and σ is the standard deviation), and not the average response ($\bar{x}$). Consequently a material/process combination with higher average fracture strength, but greater variation, may underperform a nominally weaker alternative material/process at the restrictive 0.1% cumulative failure rate.

Indisputably, however, the acceptance criteria are primarily derived from empirical data, and lack a fundamental, physics-based model; their "validity" is limited to the high performance computer systems comprising the solder joint failure data set. Further more empirically measured strain will always be limited to the adjacent PWB surface, and not the actual solder joint or solder interface, due to physical strain gage placement restrictions.

8.3 SMT COMPONENT FRACTURE STRENGTH CHARACTERIZATION

8.3.1 Bend

The maximum allowable PWB strain limits have been largely embraced by component and board assembly suppliers because they establish a quantitative metric for determining whether the solder joint brittle fractures are likely attributable to a component/PWB defect or a mechanical overstress. For initial cases where a SMT component was determined to be the primary root cause, the simple concept of improving the component's fracture strength proved difficult to implement.

A significant barrier to improvement was the lack of standardized test methods to define solder joint brittle fracture strength. As will be discussed later in this chapter, the industry standard solder ball shear test in use at the time did not prove to be a useful, validated measure of interfacial, brittle fracture strength.

An industry standard test method, IPC/JEDEC-9702 [1], was published in 2004 and established a board-level monotonic bend test procedure to characterize SMT component brittle fracture strength (Figure 8.5). The IPC/JEDEC-9702 standard [1] defines a highly reproducible, easily modeled, 4-point monotonic bend test, with a prescribed PWB global strain-rate of 5000 μStrain/s, minimum.

The monotonic bend test condition is within the typical strain rate range (1000–30,000 μStrain/s) measured during actual PWB board assembly, handling, and test conditions; however, it is 1 to 2 orders of magnitude lower than strain rates associated with a drop impact condition. Still the test method defined in the IPC/JEDEC-9702 standard has proved to reproduce the solder joint brittle fractures observed in actual board assembly, handling, and test operations. Consequently, this test method has provided an important brittle fracture characterization tool to component suppliers, though it is not sufficient to define component solder joint fracture strength across all use conditions, such as shipment and drop impact.

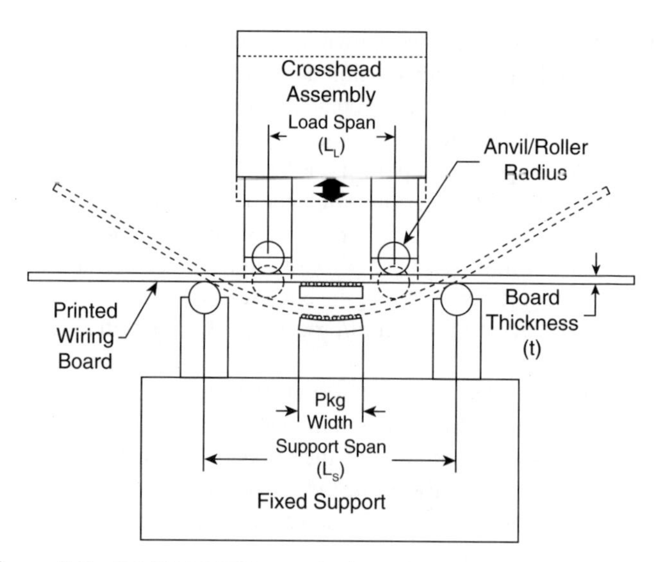

Figure 8.5 IPC/JEDEC-9702 standard 4-point monotonic bend test [1]

8.3.2 Drop

A board level drop test standard, JEDEC JESD22-B111 [4], was introduced in 2003. Although highly referenced in the literature, this useful solder joint fracture characterization test is limited to SMT devices smaller than 15 mm, and is intended for handheld applications. High-performance computers typically contain SMT devices larger than 15 mm, and certainly have a more benign operating condition than a handheld product, such as the cell phone.

Recognizing the importance of a drop test for larger components/assemblies, and nonportable applications, a more broadly applicable mechanical shock test standard was created, the IPC/JEDEC-9703 standard [5], and recently released.

Currently, some OEMs have instead relied on in-house specifications to define board-level drop test procedures for their component suppliers. The tests provide critical fracture strength characterization of components at strain rates of approximately 50,000 to 1,000,000 μStrain/s, consistent with measured values during unpackaged (unprotected) assembly drop, and packaged assembly shipment conditions. Figures 8.6 and 8.7 show some examples of drop tower equipment used to obtain mechanical shock test results.

Mechanical shock test results have generally shown good correlation with the brittle fracture failure modes (interfacial, pad cratering, etc.; Figure 8.1) observed during actual assembled board shipment and impact conditions. Given the variation in end-use conditions, a number of mechanical shock service conditions may be specified as defined in the JEDEC JESD22-B110A standard [6]. Depending on acceleration conditions, multiple machine types may be able to provide an appropriate half-sine pulse shock profile as defined by the JESD22-B110A standard [6].

Although the PWB strain guidance limits obviously assume that PWB strain relates to solder joint reliability under mechanical loading, there certainly exist many

Figure 8.6 Example of mechanical shock drop tower

Figure 8.7 Example of mechanical shock test board and fixture

circumstances of material, geometry, configuration, and the like, that may preclude such a simplified relationship. However, an internal company study showed that for their typical component and board constructions, an empirical relationship between PWB strain and solder joint fracture strength continued to be observed, as illustrated

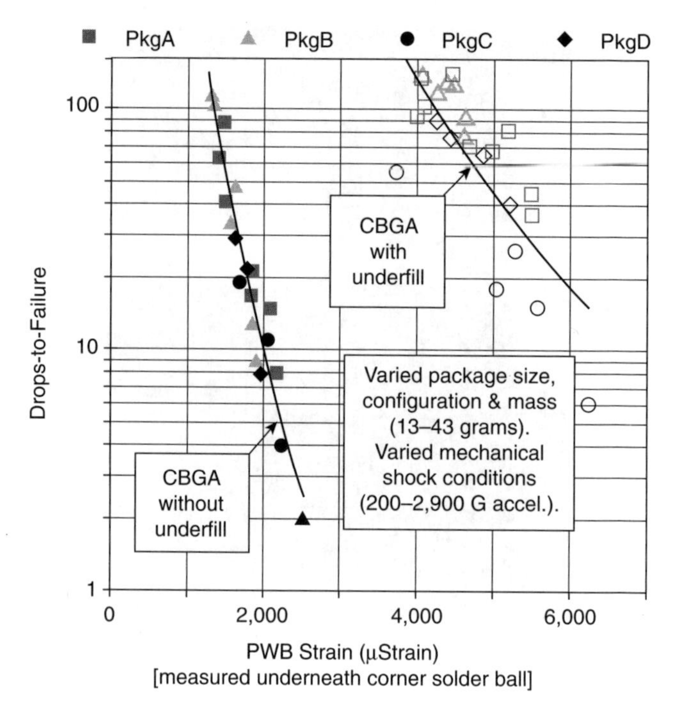

Figure 8.8 Example correlation of mechanical shock fracture strength and PWB strain from internal company studies [Sun Microsystems]

in Figure 8.8. Figure 8.8 references multiple configurations of CBGA devices, evaluated across the range of JESD22-B110A standard [6] service conditions. Similar correlations between drops-to-failure and PWB strain have been observed for other package constructions, such as PBGA devices.

8.3.3 High-Speed Solder Ball Shear/Pull

The drop and 4-point monotonic bend tests described above approximate actual conditions associated with solder joint brittle fracture, and can yield relevant assessments of brittle fracture strength. These characterization methods, however, are not a replacement for SMT component manufacturers to provide in-line solder joint integrity monitoring, and are associated with considerable test cost and complexity.

Historically SMT component manufacturers have used solder ball shear testing to monitor solder joint integrity. This test method, albeit destructive, requires no test board or SMT assembly, and can be readily conducted with specialized bond testing equipment. Unfortunately, over an investigation period of approximately eight years, semiconductor suppliers were unable to establish correlation of occasional solder joint brittle fractures observed during system assembly with conventionally practiced solder ball shear testing (using a shear tool speed of approx. 0.3–0.6 mm/s).

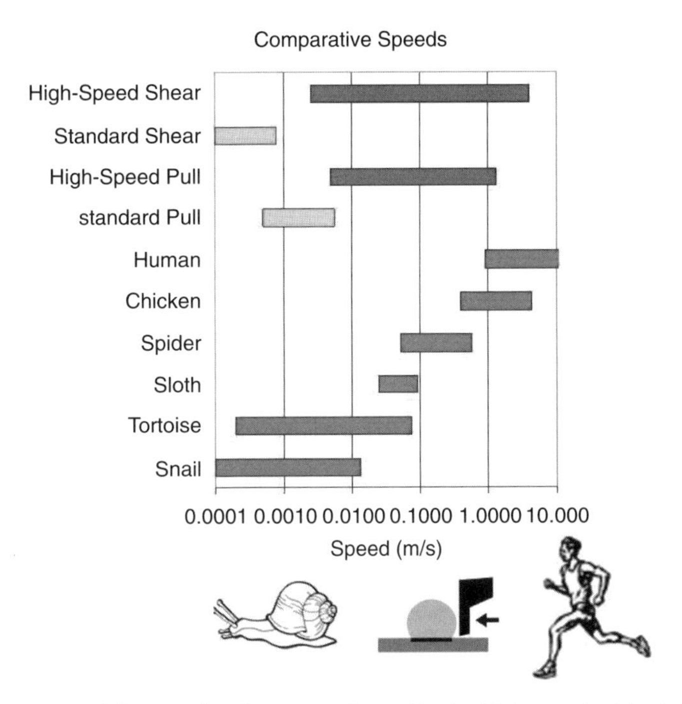

Figure 8.9 Range of shear and pull test speeds used in the high-speed solder ball shear/pull test study [7]

The lack of correlation was suspected to be due to the extremely low strain rates generated in shear testing relative to the strain rates during actual assembly, test, handling, shipment, and use conditions. Consequently high-speed solder ball shear/pull test equipment was developed to address this concern.

In conjunction with five electronic manufacturers, and with the support of a test equipment supplier, a prototype high-speed solder ball shear/pull tester was evaluated in 2004 for 27 unique package constructions, under a wide variety of test conditions. Shear speeds ranged from 0.1 mm/s to as high as 4000 mm/s, while pull testing ranged from speeds of 0.5 mm/s to 1300 mm/s. A summary report of the study was published [7]. Figure 8.9 provides an illustration of the wide range of test speeds evaluated in this study.

Results of the study indicated that high-speed solder ball shear and pull testing offered improved correlation to observed solder joint brittle failures in system manufacturing. A series of follow-on studies were conducted with results published in conference proceedings [8–12]. These results confirmed the observations of the original study, and established several correlations between failure mode, fracture energy, IMC (intermetallic compound), and thermal aging for drop, and solder ball shear/pull testing.

Figure 8.10 illustrates the agreement in failure mode that was established between mechanical shock testing and high-speed solder ball shear and pull testing [8,12]. In this example, the brittle fracture occurred at the interface between the electroless Ni-P plated surface on the package substrate and the solder IMC for each of the tests: drop,

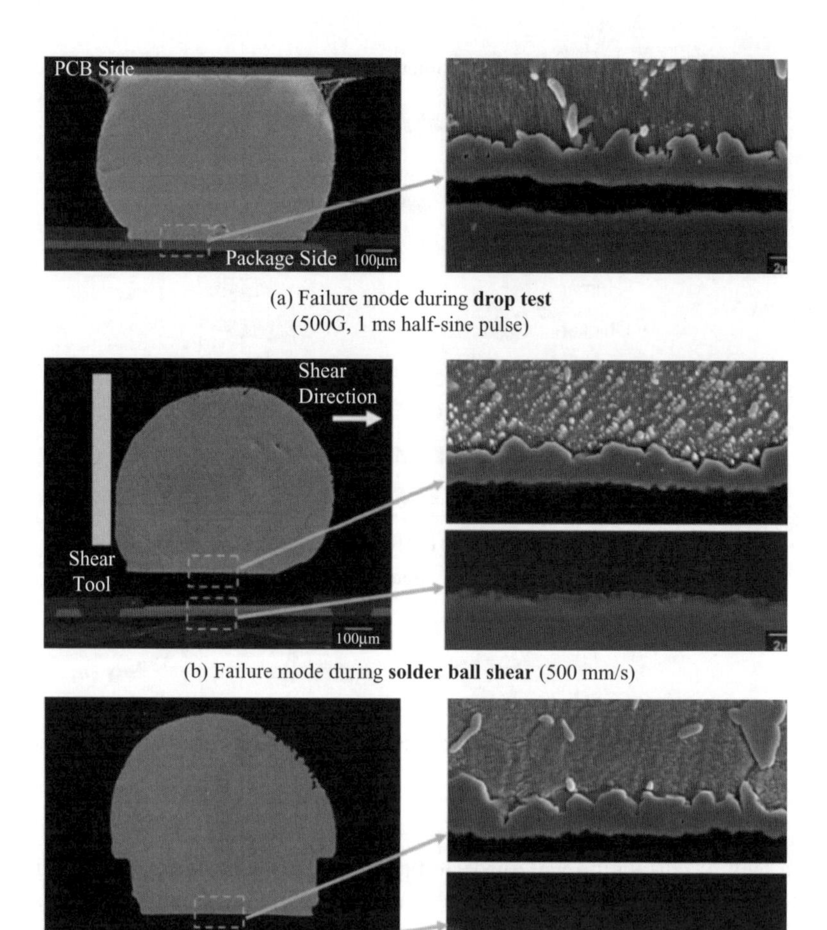

(a) Failure mode during **drop test** (500G, 1 ms half-sine pulse)

(b) Failure mode during **solder ball shear** (500 mm/s)

(c) Failure mode during **solder ball pull** (50 mm/s)

Figure 8.10 Cross-sectional comparison of fracture location during drop test, high-speed ball shear/pull tests (Sn4Ag0.5Cu + ENIG, 500 hours@125°C) [8,12]

shear and pull. The reference papers [8,12] detail the fracture interfaces, define the optimization process to select the appropriate test speeds, and describe the various package constructions, surface finishes and solder alloys that were evaluated.

These papers also established quantitative correlations for both high-speed shear and pull fracture strength to drop testing lifetimes of the evaluated devices (Figures 8.11 and 8.12). These example figures show that a reduction of solder ball shear or pull fracture energy of 20% from an initial, normal component-level condition each corresponds to an approximate 60% reduction in board-level drop lifetime.

Of course, this specific correlation represents but a small fraction of potential device constructions, materials, and solder alloys. It demonstrates the basic utility of the high-speed shear–pull testing as a SMT component brittle fracture characterization

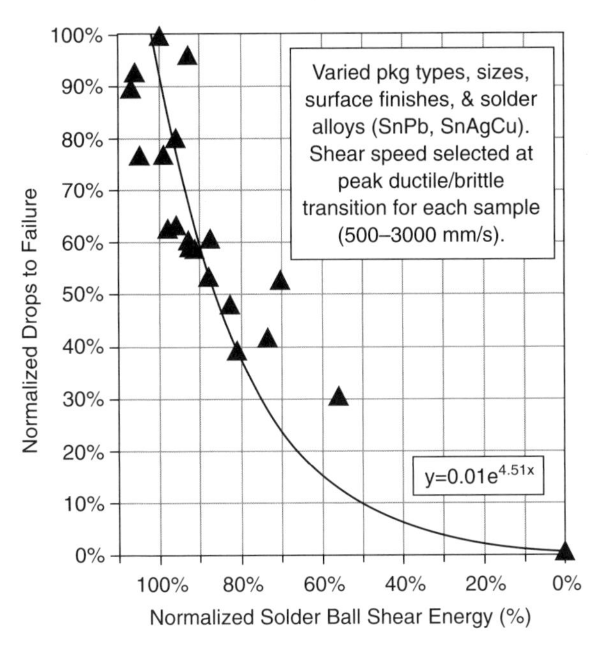

Figure 8.11 Correlation of normalized drop number and high-speed solder ball shear energy across multiple package types and materials [12]

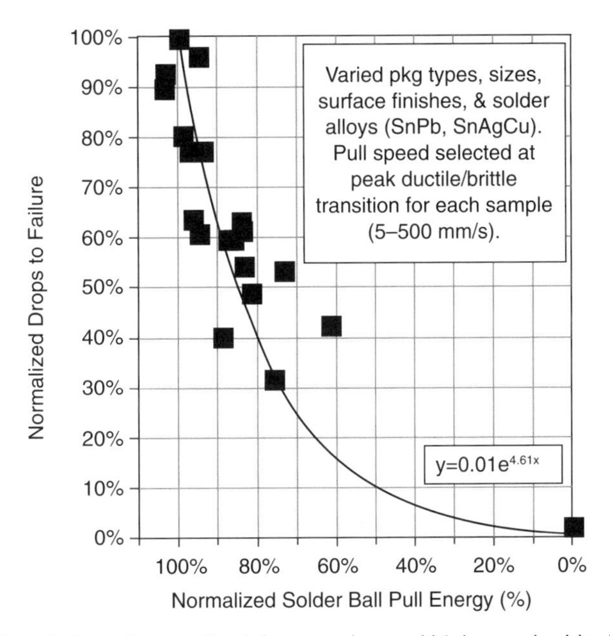

Figure 8.12 Correlation of normalized drop number and high-speed solder ball pull energy across multiple package types and materials [12]

tool. Additional research in this area could establish a more fundamental relationship between high-speed shear or pull results across all device types and material sets.

Although these studies have provided a significant data set demonstrating the usefulness of high-speed solder ball shear and pull testing to characterize component solder joint brittle fracture strength, the need for an industry standard test method has been evident from the first study conducted in 2004 [7]. A revised solder ball shear test procedure (JEDEC JESD22-B117A standard [13]) and a new solder ball pull test method (JEDEC JESD22-B115 standard [14]) were published that included high-speed testing. Figure 8.13 illustrates the failure mode categories of these JEDEC test methods.

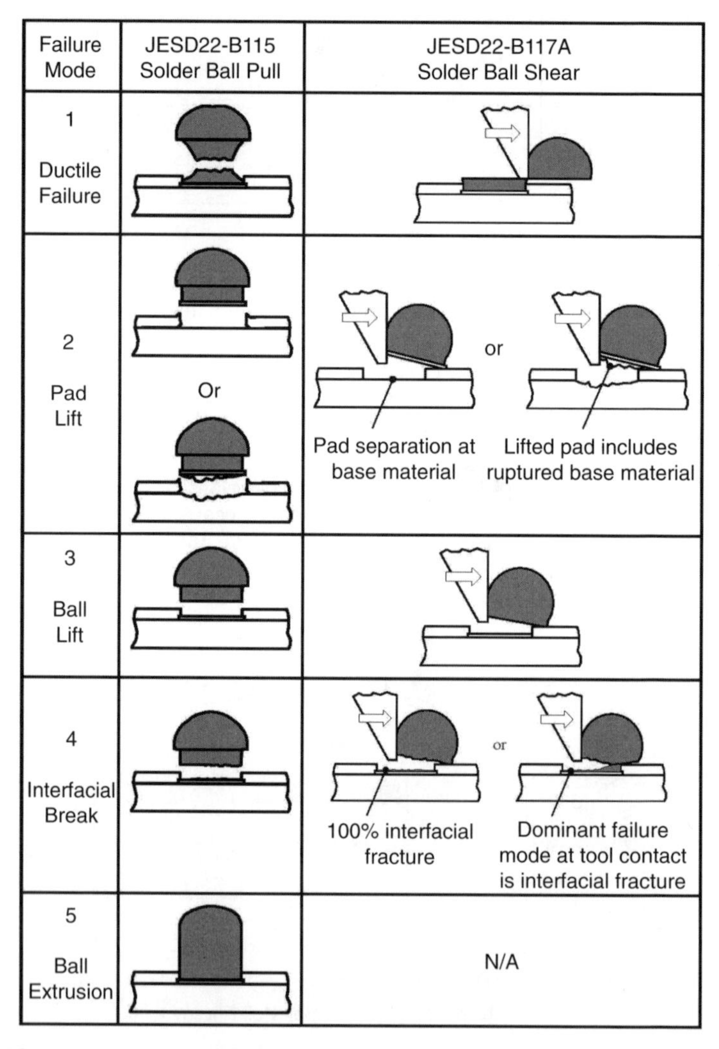

Figure 8.13 JEDEC JESD22-B115 and JEDEC JESD22-B117A standard solder ball failure mode categories [13,14]

8.3.4 Other Test Method Examples

The myriad component–PWB configurations and use-conditions demand multiple test methods to adequately characterize component solder joint brittle fracture strength. Figures 8.14 and 8.15 illustrate just two of the many examples of alternative solder joint reliability mechanical test methods described in industry publications [15,16].

Figure 8.14 Spherical bend test method (courtesy of Intel Corp.) [15]

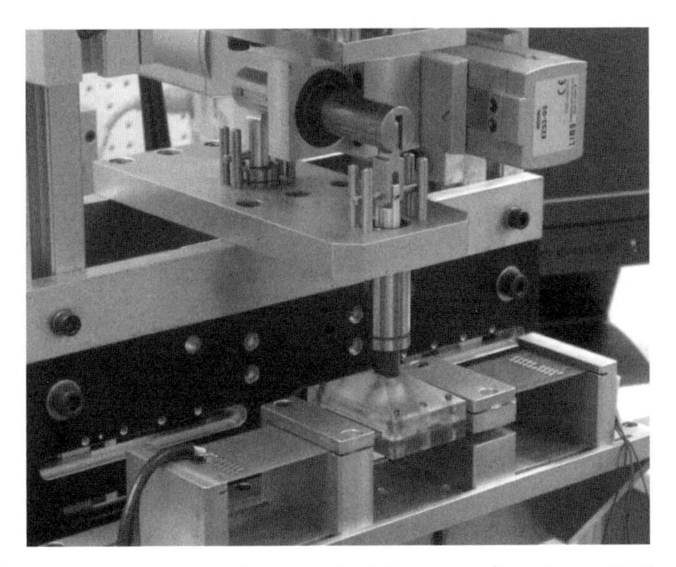

Figure 8.15 Four-point cyclic bend test method (courtesy of Institute of Microelectronics, Singapore) [16]

The test configuration shown in Figure 8.14 creates a spherical bend shape in the test board assembly; this may provide a more severe loading condition than a 4-point bend test, and better represent PWB strain in some actual system configurations. Further details of the spherical bend test method are given in Chapter 7 of this book. The high-speed, cyclic bend test apparatus of Figure 8.15 was developed to perform displacement-controlled bend tests at high flexing frequencies. Results from the high-speed bend tests could be used to construct constant amplitude power law fatigue curves.

8.4 PWB FRACTURE STRENGTH CHARACTERIZATION

Although the solder joint brittle fractures observed over the last several years typically related to failures at the solder–component or solder–PWB interface, the recent implementation of lead-free compatible solders and PWB laminates has resulted in a more frequently observed failure mode—PWB pad cratering (Figures 8.16 and 8.17) [17].

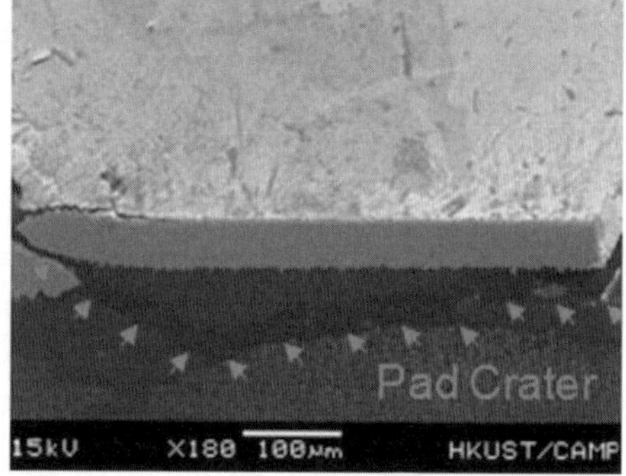

Figure 8.16 PWB pad cratering example (courtesy of Hong Kong University of Science and Technology [HKUST])

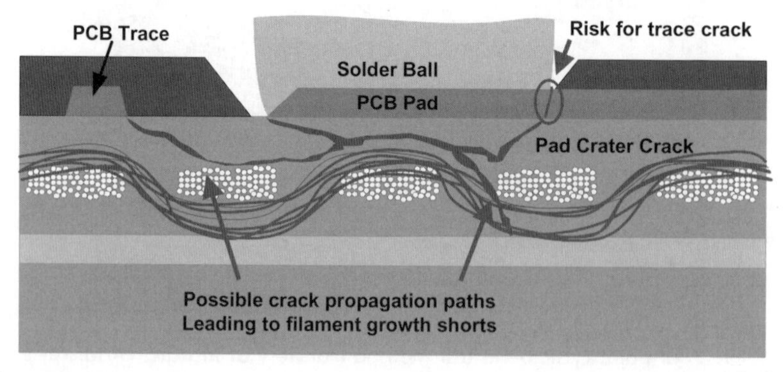

Figure 8.17 Potential PWB pad crater crack patterns (courtesy of Intel Corp.) [17]

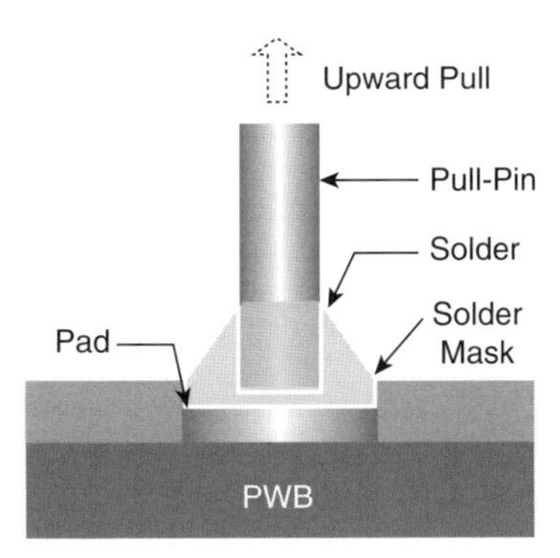

Figure 8.18 Hot pin pull test method (courtesy of Cisco Systems) [18,19]

Contributing factors likely include the less compliant nature of many lead-free solders, as well as the more brittle nature of the newest generation of PWB laminates designed to accommodate the higher SMT reflow conditions.

In recent years several industry working groups have investigated various test methods to characterize the pad rupture strength of PWB laminate. Efforts to date have mostly evaluated the options of hot pin pull test method (EIAJ ET-7407 standard) [18,19], shown in Figure 8.18, or room temperature solder ball shear (JESD22-B117A standard [13]) and pull (JESD22-B115 standard [14]) testing. These test methods require the application of either solder paste or a solder ball to the PWB pad to allow assessment of the pad rupture strength.

In addition to PWB pad brittle fracture strength characterization using shear or pull testing, an alternative method has been evaluated [20] that uses a 50-mm diameter steel ball dropped on the backside of the PWB, opposite the SMT component, from a height typically ranging from 10 to 100 mm. Resultant PWB pad craters from this ball drop test replicated the fracture signatures observed in full system drop tests, but yielded a more controlled and specific analysis of the laminate fracture characteristics.

As development of the various PWB pad crater test methods continue, for example, with the proposed IPC/JEDEC-9708 standard, "Test Methods for Characterization of PCB Pad Cratering," correlations will potentially be established with existing PWB material characterization tests: TGA (thermogravimetric analysis), TMA (thermomechanical analysis), and hardness.

8.5 PWB STRAIN CHARACTERIZATION

8.5.1 Board Assembly and Test

The introduction of the maximum allowable PWB strain guidelines in 2004 demanded a corresponding strain gage measurement procedure. The measurement procedure, however, varied in some regards dependent on the board assembler's customers. To establish consistency within the industry and incorporate best practices from a broad selection of electronics manufacturers, a joint industry standard, IPC/JEDEC-9704 standard, "Printed Wiring Board Strain Gage Test Guideline," was published in 2005 (Figures 8.19 and 8.20) [2].

Characterization of PWB strain during board assembly, test and handling operations not only defines the expected mechanical strain conditions, but also identifies the process operations associated with the greatest risk of creating solder joint brittle

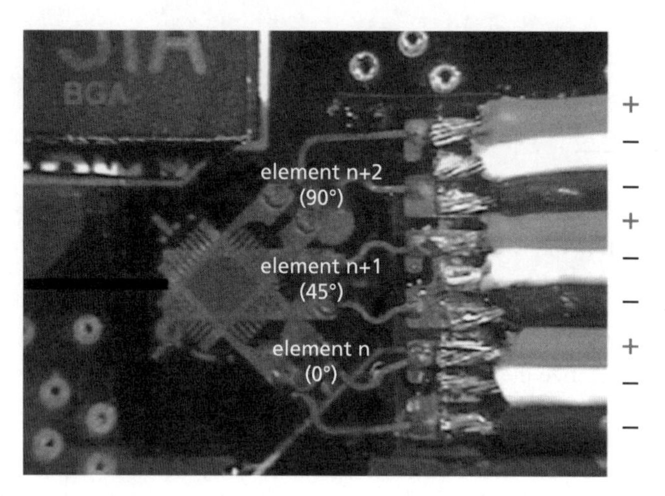

Figure 8.19 Example of stacked rosette strain gage described in IPC/JEDEC-9704 standard to determine maximum PWB principal strain [2]

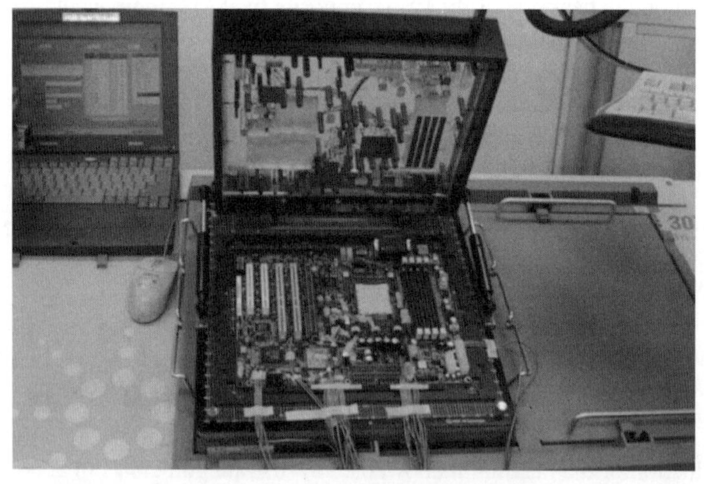

Figure 8.20 Example of typical PWB strain characterization described in IPC/JEDEC-9704 standard; in this case during ICT (in-circuit test) operation [2]

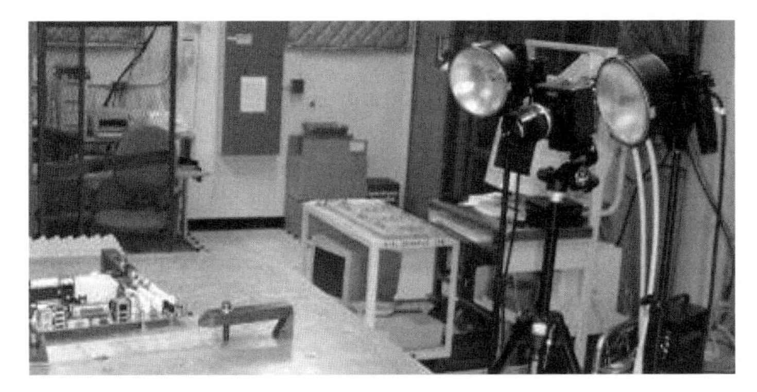

Figure 8.21 Example of high-speed camera used to determine PWB strain during mechanical shock testing (courtesy of Intel Corp.) [15]

fractures. Potential corrective measures to reduce these peak strain events can be readily identified and quantified, facilitating a more rapid assembly process improvement effort.

8.5.2 Optical Methods

The IPC/JEDEC-9704 standard [2] relies on strain gages to measure dynamic PWB strain, but optical, noncontact measurement procedures have also been developed using high-speed digital cameras (Figure 8.21) [15,21,22]. In some procedures an array of low-mass fiducial targets is attached to the PWB to provide reference markers, and traingulation methods are employed to transform the digital images into three-dimensional surface deflection data. Other procedures employ random, high-contrast paint speckles on the PWB, and DIC (digital image correlation) techniques are used with dual cameras to yield the dynamic surface deformation.

The collected deflection data from these techniques can be subsequently transformed to provide full-field PWB strain plots. The optical strain measurement methods provide complete, full-area, strain characterization of the PWB but are limited to dynamic load conditions where line-of-sight is possible.

8.5.3 Unpackaged and Packaged Drop

The 2005 maximum allowable PWB strain guidelines established acceptance criteria for drop/impact conditions that high-performance computer products might experience during anticipated shipment and handling conditions. For packaged and unpackaged (unprotected) drop testing, PWB strain measurements follow the IPC/JEDEC-9704 standard [2], but the drop/impact stress conditions for telecom/server products are typically defined by the Telecordia NEBS GR-63-CORE 5.3.2 standard [23] and the IEC 60068-2-31 standard [24,25].

PWB strain measurements are preferably conducted during both unpackaged handling conditions (with a typical drop height of 25–100 mm), and packaged drop (with a typical drop height of 150–750 mm). These drop tests may involve multiple system/package orientations (e.g., top, bottom, side, corner) and various drop configurations (e.g., topple, flat-drop, complex orientation). (Figures 8.22–8.24). Drop testing of actual systems measures the PWB strain during anticipated unpackaged handling, and packaged

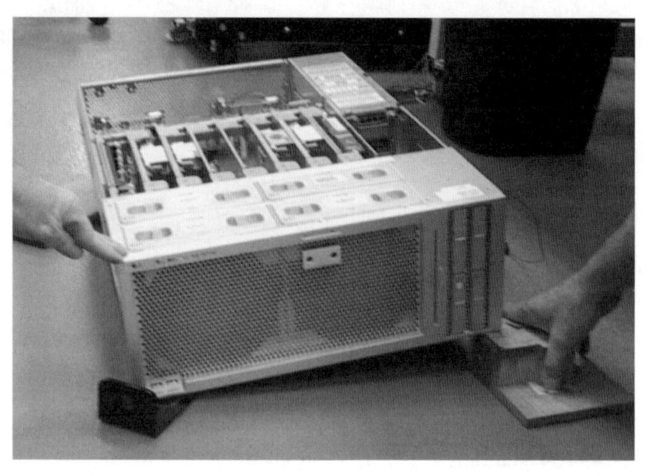

Figure 8.22 Unpackaged system drop example (10–20-mm supports and 60-mm corner drop height per IEC 60068-2-31 standards [24,25])

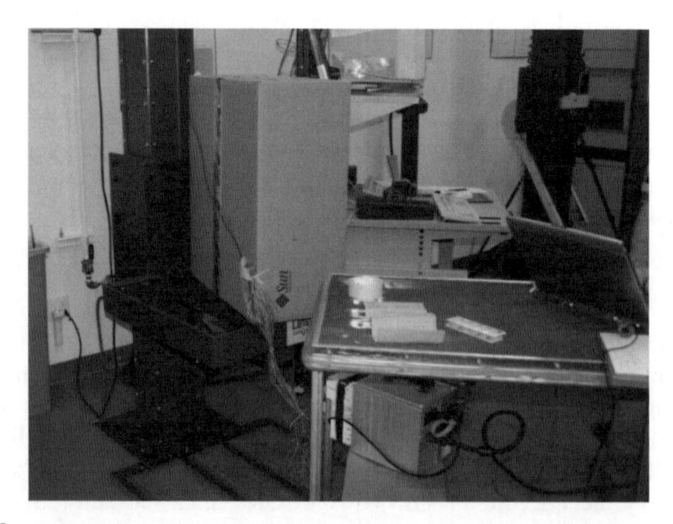

Figure 8.23 Packaged system drop example (600-mm end drop per Telecordia NEBS GR-63-CORE 5.3.2 standard [23])

Figure 8.24 Packaged system drop example (300-mm edge drop per Telecordia NEBS GR-63-CORE 5.3.2 standard [23])

shipment, allowing optimization of the shipment packaging and the system frame, chasis, enclosure or board assembly.

8.6 SOLDER JOINT FRACTURE PREDICTION—MODELING

Previous sections discussed various empirical characterization techniques to describe solder joint brittle fracture strength under mechanical stress conditions. Analytical and/or computational models to predict component brittle fracture, however, remain compelling and attractive alternatives. An accurate, validated dynamic model enables a broad evaluation of numerous material and geometry parameters, provides quantitative understanding of internal, interfacial and surface strain levels under multiple dynamic mechanical loading conditions, and allows for optimization of design, materials and construction.

Many obstacles lie in the path to a broadly validated dynamic model for the complex material systems typical of most lead-free electronic products. Variations in solder material properties as a function of sample geometry, sample formation, reflow parameters, thermal history, time after reflow, dynamic hardening, and so forth, create exceptional modeling challenges. Figures 8.25 and 8.26 provide examples of a dynamic finite element analysis (FEA) using cohesive elements [26,27]. These figures are derived from a collaborative effort to establish a validated material model for Sn1Ag0.5Cu BGA solder joints during simulated high-speed solder ball shear testing [27].

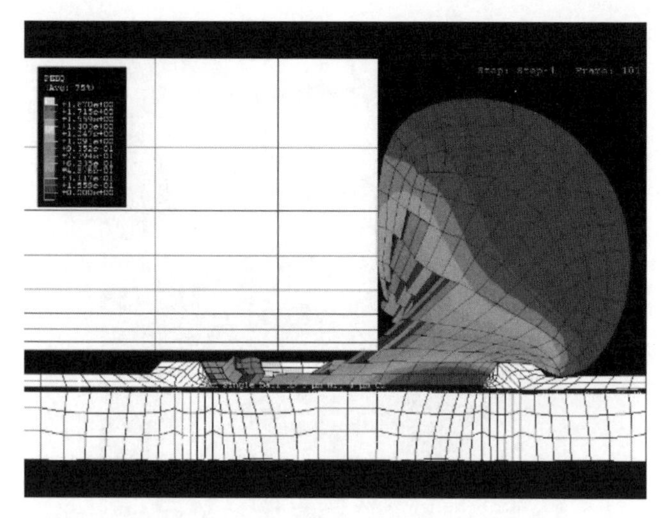

Figure 8.25 Mixed ductile–brittle solder ball shear failure mode at simulated 1 m/s (courtesy of Fraunhofer IZM) [27]

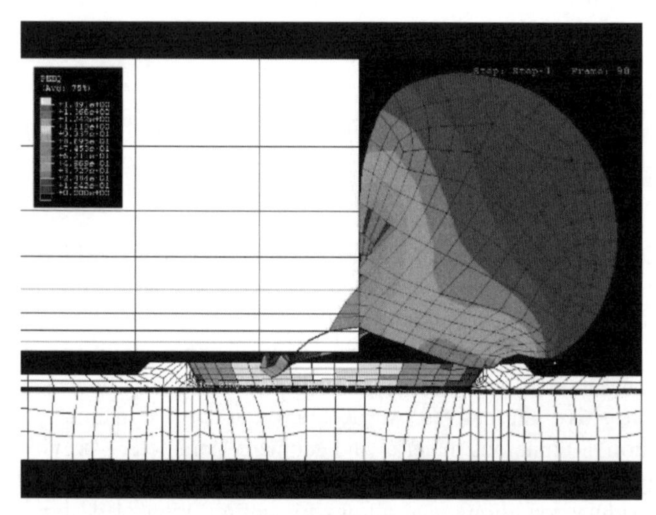

Figure 8.26 Brittle solder ball shear failure mode at simulated 2 m/s (courtesy of Fraunhofer IZM) [27]

Figure 8.27 presents an example to predict solder joint fracture during a dynamically simulated mechanical shock test using cohesive elements correlated to empirical studies. Figure 8.28 illustrates a solder joint modeling approach to incorporate two-node fastener elements to dramatically reduce the number of FEA nodes, a technique

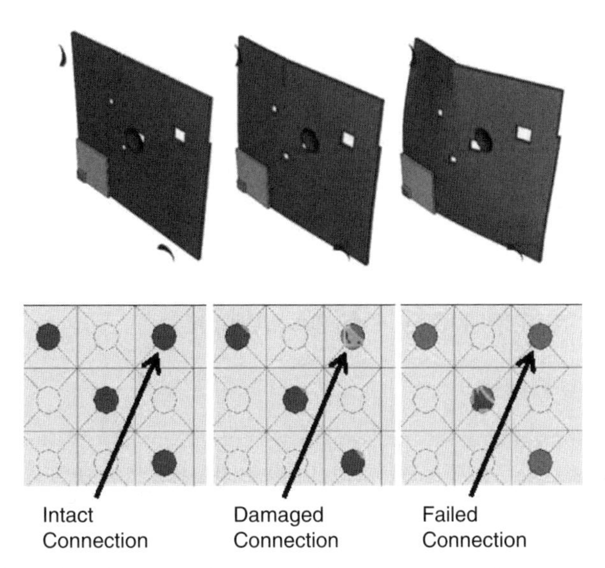

Figure 8.27 Example of simulated, progressive damage and fracture in a dynamic FEA model using cohesive elements, correlated to empirical studies (courtesy of Dassault Systèmes)

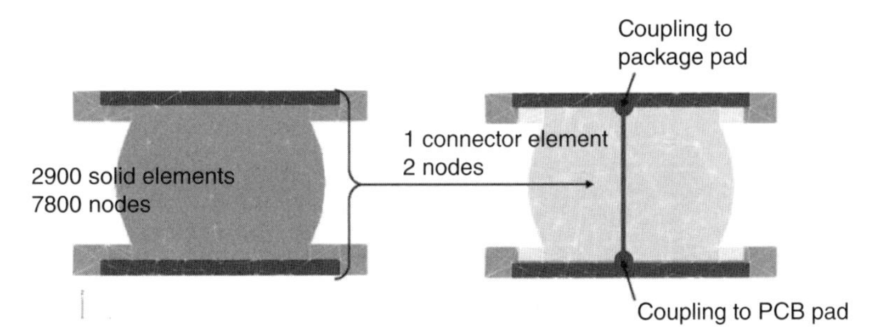

Figure 8.28 Example of reduction of detailed FEA solder joint model to a mesh-independent fastener element (courtesy of Dassault Systèmes)

particularily useful in reducing the computational burden involved in full system modeling of simulated dynamic events. This example method was used to evaluate the time-dependent solder joint fractures of a complete cellular radio during simulated system-level drop [28].

Other researchers [22] have used Timoshenko-beam models for solder joints, smeared property models, and explicit submodels to provide a computationally efficient dynamic simulation of board assemblies subjected to shock in various orientations. This study showed strong correlation to experimental results.

In addition to the complex, computational modeling examples noted above, others have pursued analytical approaches using spring-mass, end-support beams and edge-support plate models to describe the dynamic response of a PWB during drop impact testing, and provide a more fundamental understanding of the dynamic system [29].

The examples above demonstrate that accurate, validated dynamic models to predict solder joint brittle fracture have been developed for specific applications; however, computational and analytical modeling have not presently advanced to the level required for widespread replacement of mechanical solder joint reliability testing.

8.7 FRACTURE STRENGTH OPTIMIZATION

8.7.2 Surface Finish

Of course, once the solder joint fracture strength has been defined, and the PWB strain characterized, a critical corrective action to mitigate excessive product failures might very well prove to be enhancement in the fracture strength of the component and/or PWB. In such instances one parameter that might be considered for optimization is the surface finish associated with the fracture interface.

Figure 8.29 illustrates one example of a plating characteristic that was determined to be closely associated with solder joint brittle fracture events. The specific combination of Sn37Pb solder and ENIG (electroless Ni, immersion Au) surface plating typically yielded microvoids at the interface between the P-enriched Ni-P surface and

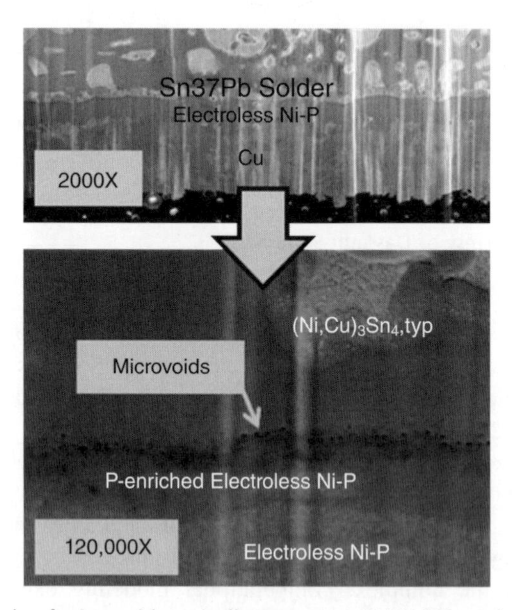

Figure 8.29 Example of microvoids typically present at Ni-P/IMC interface for Sn37Pb solder over ENIG-plated surface (courtesy of LSI Corp.)

the solder IMC. This characteristic of ENIG plating after SnPb solder attach was found to be most specifically associated with the ENIG plating chemistry and individual processing equipment, but it has always been found to exist at some void density in multiple internal company evaluations.

Unfortunately, the necessity of ion beam milling to prevent smearing of the submicron voids, and SEM magnifications typically in excess of 60,000×, make sample preparation and examination of this type the domain of relatively few analytical labs in the industry.

Nonetheless, with this advanced analytical tool in hand, many component suppliers have found that alternative surface finishes such as electrolytic nickel-gold, or "unplated" copper—organic solderability preservative (OSP), solder-on-pad, hot-air solder level (HASL), and immersion tin thief among these—did not have evidence of interfacial microvoids after Sn37Pb solder joint formation (Figures 8.30 and 8.31). Indeed, multiple brittle fracture characterization tests with Sn37Pb solder joints verified an improved strength for these alternative surface finishes.

These examples are not meant to suggest that replacement of ENIG surface plating will enhance solder joint reliability (ENIG may in fact even be preferred over some surface finishes for various lead-free solder applications) but to highlight that the surface finish of the SMT component and/or PWB may prove an important solder joint brittle fracture strength optimization parameter. It is important to understand that although the micrographs in the figures above describe Sn37Pb solder joints, the

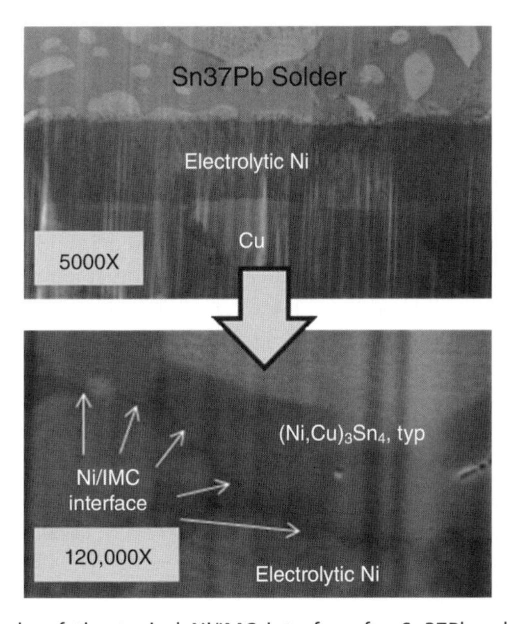

Figure 8.30 Example of the typical Ni/IMC interface for Sn37Pb solder over electrolytic Ni-plated surface (courtesy of LSI Corp.)

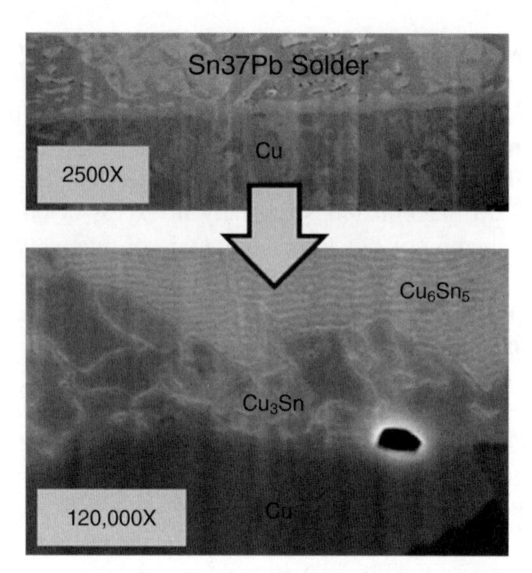

Figure 8.31 Example of typical copper–Cu_3Sn interface for Sn37Pb solder over unplated copper surface (courtesy of LSI Corp.)

criticality of surface finish illustrated in these figures is also applicable to lead-free solder alloys. Indeed, Chapter 5 of this book describes some of the drop/shock lifetime dependencies between Cu-OSP and nickel surface finishes and various lead-free solder alloys.

8.7.2 Solder Alloy

A multitude of technical papers, articles, and presentations have covered the solder joint reliability implications associated with the transition to lead-free solder materials. Typically the most common SnAgCu solder alloys used in electronic products have been associated with a reduction in solder joint brittle fracture strength relative to Sn37Pb solder (Figures 8.32 and 8.33) [16,30,31]. This observation has led toward the development of a number of alternative lead-free solder alloys, optimized for improved mechanical fracture strength but often requiring higher reflow temperatures. Chapter 5 of this book discusses this topic in greater detail.

Unfortunately, the proliferation of lead-free alloys has many assembly and reliability repercussions for high-performance computer manufacturers; the higher degree of component mix on large system boards, coupled with varying solder alloys (and associated liquidus temperatures) present on a single PWB, creates significant potential for improperly formed solder joints across typical lead-free SMT processing windows. Again, the reader is encouraged to refer to Chapter 5 of this book for a detailed discussion of this topic.

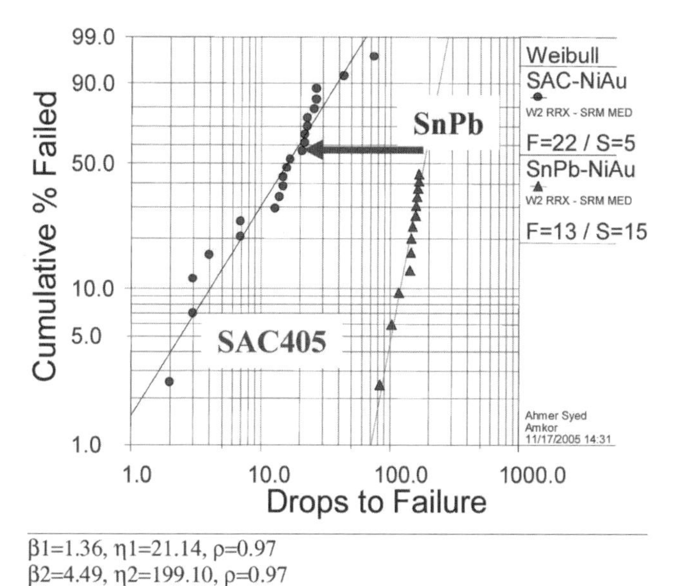

Figure 8.32 Example of relative drop test performance between Sn37Pb and a common lead-free solder (courtesy of Amkor Technology) [30]

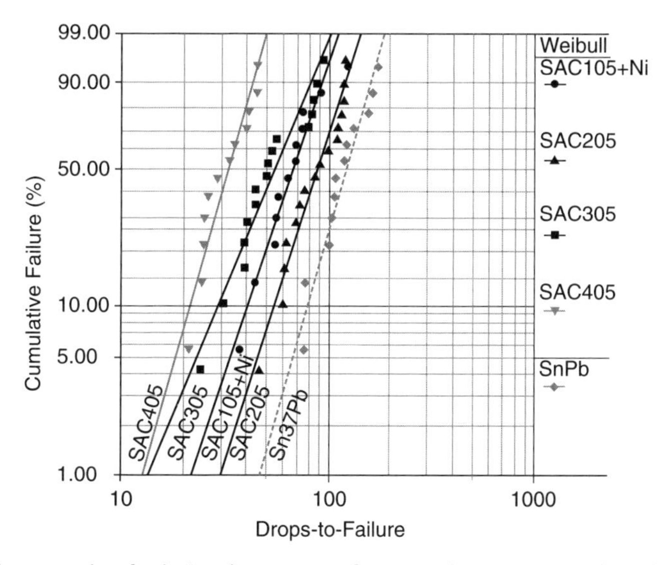

Figure 8.33 Example of relative drop test performance between Sn37Pb and various lead-free solders (data courtesy of Unovis Solutions) [31]

8.7.3 PWB Laminate

As noted previously, PWB pad rupture has been more frequently observed with lead-free compatible laminate materials than traditional Sn37Pb compatible circuit boards. Lead-free PWB materials have been developed to withstand the higher reflow temperatures of lead-free solder, and to maintain high via reliability and laminate stability during projected end-use power/thermal cycling. Unfortunately, optimization for thermal cycle lifetime, high-temperature exposure, and the like, typically yields a more brittle material. With improved PWB pad crater characterization techniques, laminate suppliers would be better able to quantify the various thermal and mechanical reliability trade-offs, and develop a more robust product in all expected assembly, transport, and end-use conditions.

8.7.4 Thermal Exposure

Previous sections have described fracture strength optimization parameters that were related to material selection and mechanical assembly/test/handling operations; however, thermal aging and SMT reflow conditions can also significantly affect solder joint brittle fracture strength. A recent study was done to investigate the relationship between IMC and solder joint fracture strength during mechanical shock and high-speed solder ball shear/pull testing (Figures 8.34–8.36) [11]. From the figures it can be seen that multiple solder reflow passes or extended high-temperature exposure can result in a significant reduction in brittle fracture strength, depending on specific surface finish and solder alloy.

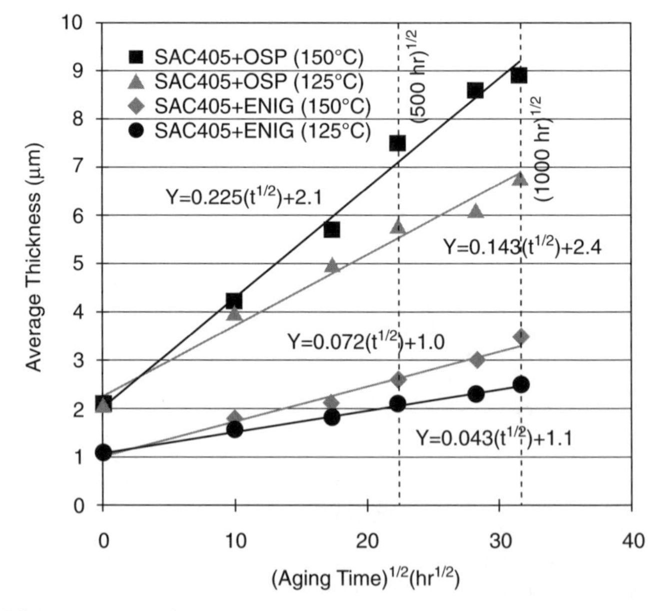

Figure 8.34 Example of IMC growth versus time/temperature for Sn4Ag0.5Cu [11]

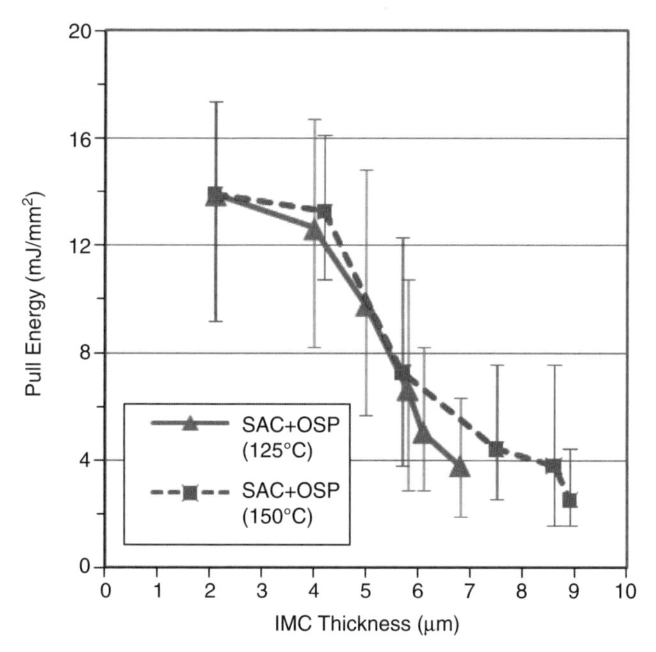

Figure 8.35 Example of Sn4Ag0.5Cu brittle fracture strength (high-speed solder ball pull) versus IMC thickness for OSP surface finish [11]

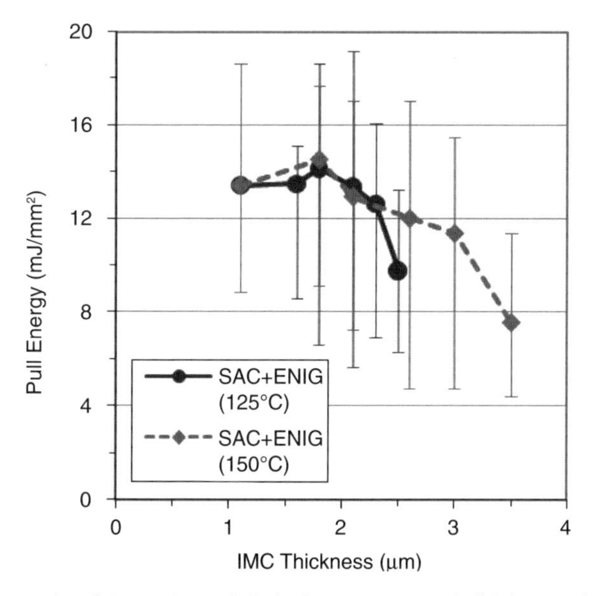

Figure 8.36 Example of Sn4Ag0.5Cu brittle fracture strength (high-speed solder ball pull) versus IMC thickness for ENIG surface finish [11]

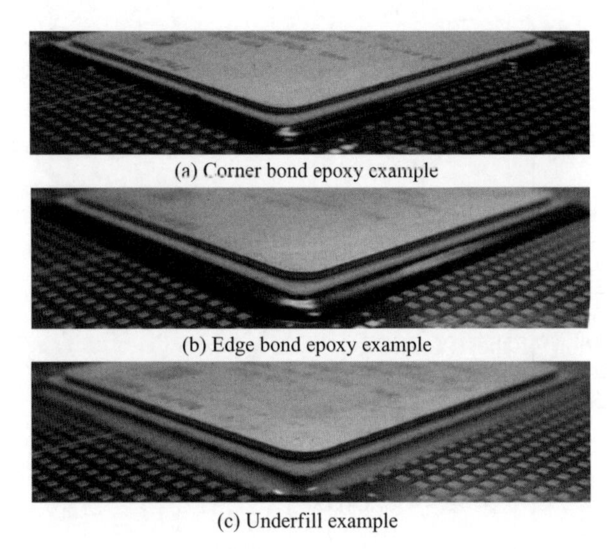

(a) Corner bond epoxy example

(b) Edge bond epoxy example

(c) Underfill example

Figure 8.37 Various examples of epoxy reinforcement to enhance solder joint fracture strength

8.7.5 Epoxy Reinforcement

In addition to modification of surface finish, solder alloy, PWB laminate, and reflow profile as discussed above, other material and design mitigation practices might be implemented. One example would be the use of epoxy to reinforce a component/PWB assembly to improve solder joint brittle fracture strength. The epoxy would normally be applied in one of the three general configurations as described in Figure 8.37: corner bond, edge bond, and underfill. Each of these configurations have different trade-offs relative to strength enhancement, cost, reworkability, productivity, and process compatibility, for example. The drops-to-failure data provided in Figure 8.8 clearly show significant improvements in mechanical shock resistance that epoxy reinforcement can provide. However, care must be taken when making material choices because improper epoxy reinforcement can degrade thermal fatigue reliability.

8.8 CONCLUSIONS

The chapter reviewed the establishment of PWB strain limits for manufacturing, component fracture strength characteristics related to bend, drop, high-speed solder ball shear/pull testing, and other test methods including spherical and cyclic bend testing. It also looked at PWB fracture strength characteristics and strain characterization during board assembly and test, examined optical methods for measuring PWB strain, and discussed strain characterization during unpackaged and packaged drop testing.

Solder joint fracture prediction modeling and some of the methods to improve fracture strength were also reviewed, including surface finish, solder alloy, PWB laminates, thermal exposure, and epoxy reinforcement. Some of the main conclusions from the chapter are listed below:

1. Although lacking a fundamental, analytical basis, the maximum allowable PWB strain guidance presented in this chapter provides a quantitative metric to aid identification of whether observed solder joint brittle fractures are attributable to component/PWB defects or excessive mechanical loading.
2. Publication of the IPC/JEDEC-9702 standard [1] established a board-level monotonic bend test procedure to characterize SMT component brittle fracture strength under conditions similar to expected board assembly, test, and handling operations.
3. Company-specific, in-house drop test methods have provided solder joint fracture strength characterization of components at strain rates of approximately 50,000 to 1,000,000 μStrain/s, consistent with measured values during unpackaged subassembly drop, and packaged assembly shipment conditions.
4. Published data indicate that high-speed solder ball shear/pull testing, as described in the JEDEC JESD22-B117A standard [13] and JEDEC JESD22-B115 standard [14], provide a useful predictor to component manufacturers of board-level solder joint brittle fracture resistance.
5. The IPC/JEDEC-9704 standard [2] has incorporated best practices from a broad selection of electronics manufacturers, and has detailed strain gage test procedures to characterize PWB strain during board assembly, test, and handling operations.
6. For unpackaged/packaged drop testing of actual system assemblies, PWB strain measurements follow the IPC/JEDEC-9704 standard [2], but the drop/impact stress conditions are typically defined by the Telecordia NEBS GR-63-CORE 5.3.2 [23] and IEC 60068-2-31 standards [24,25].
7. Prior experience with Sn37Pb solder attachment to ENIG-plated components and PWBs highlight the potential importance of surface finish with regard to optimization of solder joint brittle fracture strength.
8. The high degree of component mix on large system boards, coupled with proliferation of lead-free solder alloys, creates significant potential for improperly formed solder joints across typical lead-free SMT processing windows.
9. Multiple solder reflow passes or extended high-temperature exposure could result in a significant reduction in brittle fracture strength, depending on specific surface finish and solder alloy.
10. The use of epoxy to reinforce a component/PWB assembly (corner bond, edge bond, or underfill) provides an option to improve solder joint brittle fracture strength.

ACKNOWLEDGMENTS

This chapter has touched on a number of research areas, and has represented the contributions of many individuals and companies. Although it is an incomplete list, I would like to note the following key individuals/mentors and their respective organizations: Mudasir Ahmad (Cisco), Stephen Clark (Dage), Luke Garner (Intel), Dennis Krizman (Celestica), Ricky Lee (HKUST), Frank Liang (Intel), Jorge Martinez-Vargas (Sun), Michael Mello (CalTech), Lei Mercado (Medtronics), Brett Ong (Sun), George Raiser (Medtronics), Sundar Sethuraman (Flextronics), Ken Shaul (Sun), Fubin Song (HKUST), Ahmer Syed (Amkor), Bob Sykes (Dage), and Ee Hua Wong (SIMTech).

Among these listed, a special thanks is expressed to Dr. Fubin Song for his tireless efforts in performing the extensive testing, analysis, and documentation related to much of the shear, pull, and drop testing reported in this chapter.

Final thanks are due to the following organizations that graciously provided permission to publish selected photos and figures in this chapter, reflecting the collaborative efforts associated with much of solder joint reliability related activities: Amkor Technology, Cisco Systems, Dassault Systèmes, Fraunhofer IZM, Hewlett-Packard, Institute of Microelectronics—Singapore, Intel Corp., LSI Corp., and Unovis Solutions.

NOTE

The contents of this chapter (with minor modifications) first appeared as a keynote paper in EuroSimE 2008: "Board-Level Solder Joint Reliability of High Performance Computers under Mechanical Loading," *Proc. 9th International Conference on Thermal, Mechanical and Multi-Physics Simulation and Experiments in Micro-Electronics and Micro-Systems*, Freiburg-im-Breisgau, Germany, April 2008, pp. 672–686.

REFERENCES

1. IPC/JEDEC IPC/JEDEC-9702 Standard, *Monotonic Bend Characterization of Board-Level Interconnects*, June 2004.
2. IPC/JEDEC IPC/JEDEC-9704 Standard, *Printed Wiring Board Strain Gage Test Guideline*, June 2005.
3. C. F. Coombs (ed), *Printed Circuits Handbook*, 6th ed., McGraw-Hill, New York, 2007, pp. 59.1–59.35.
4. JEDEC JESD22-B111 Standard, *Board Level Drop Test Method of Components for Handheld Electronic Products*, July 2003.
5. IPC/JEDEC IPC/JEDEC-9703 Standard, *Mechanical Shock Test Guidelines for Solder Joint Reliability*, Mar. 2009.
6. JEDEC JESD22-B110A Standard, *Subassembly Mechanical Shock*, Nov. 2004.
7. K. Newman, "BGA Brittle Fracture—Alternative Solder Joint Integrity Test Methods," *Proc. 55th Electronic Components and Technology Conf.*, Orlando, FL, June 2005, pp. 1194–1200.

8. F. B. Song, S. W. R. Lee, K. Newman, B. Sykes, and S. Clark, "High Speed Solder Ball Shear and Pull Tests vs. Board Level Mechanical Drop Tests: Correlation of Failure Mode and Loading Speed," *Proc. 57th Electronic Components and Technology Conference*, Reno, NV, June 2007, pp. 1504–1513.
9. F. B. Song, S. W. R. Lee, K. Newman, B. Sykes, and S. Clark, "Brittle Failure Mechanism of SnAgCu and SnPb Solder Balls during High Speed Ball Shear and Cold Ball Pull Test," *Proc. 57th Electronic Components and Technology Conf.*, Reno, NV, June 2007, pp. 364–372.
10. F. B. Song, S. W. R. Lee, S. Clark, B. Sykes, and K. Newman, "Characterization of Failure Modes and Analysis of Joint Strength Using Various Conditions for High Speed Solder Ball Shear and Cold Ball Pull Tests," *Proc. 16th EMPC*, Oulu, Finland, June 2007, pp. 413–418.
11. F. B. Song, S. W. R. Lee, K. Newman, H. Reynolds, B. Sykes, and S. Clark, "Effect of Thermal Aging on High Speed Ball Shear and Pull Tests of SnAgCu Lead-free Solder Balls," *Proc. 9th Electronic Packaging Technology Conf.*, Singapore, Dec. 2007, pp. 463–470.
12. F. B. Song, S. W. R. Lee, K. Newman, S. Clark, and B. Sykes, "Comparison of Joint Strength and Fracture Energy of Lead-free Solder Balls in High Speed Ball Shear/Pull Tests and their Correlation with Board Level Drop Test," *Proc. 9th Electronic Packaging Technology Conf.*, Singapore, Dec. 2007, pp. 450–458.
13. JEDEC JESD22-B117A Standard, *Solder Ball Shear*, Oct. 2006.
14. JEDEC JESD22-B115 Standard, *Solder Ball Pull*, May 2007.
15. F. Liang, R. Williams, and G. Hsieh, "Board Strain States Method and FCBGA Mechanical Shock Analysis," *Proc. 1st Int. Conf. Exhibition on Device Packaging*, Scottsdale, AZ, Mar. 2005.
16. S. K. W. Seah, E. H. Wong, Y. W. Mai, R. Rajoo, and C. T. Lim, "High-Speed Bend Test Method and Failure Prediction for Drop Impact Reliability," *Proc. 56th Electronic Components and Technology Conf.*, San Diego, CA, June 2006, pp. 1003–1008.
17. G. Long, T. Embree, M. L. Mukadam, S. Parupalli, and V. Vasudevan, "Lead Free Assembly Impacts on Laminate Material Properties and Pad Crater Failures," *Proc. of APEX 2007*, Los Angeles, Feb. 2007.
18. M. Ahmad, D. Sink, and J. Burlingame, "Methodology to Characterize Pad Cratering under BGA Pads in Printed Circuit Boards," *Proc. 13th Pan Pacific Microelectronics Symp.*, Kauai, HI, Jan. 2008, pp. 182–188.
19. JEITA EIAJ ET-7407 Standard, Annex 11 and 12, *Peel Test Method for Test Board Land*, Dec. 1999.
20. D. Xie, J. Wang, H. Yu, D. Lau, and D. Shangguan, "Impact Performance of Microvia and Buildup Layer Materials and its Contribution to Drop Test Failures," *Proc. 57th Electronic Components and Technology Conf.*, Reno, NV, June 2007, pp. 391–399.
21. S. Park, C. Shah, J. Kwak, C. Jang, and J. Pitarresi, "Transient Dynamic Simulation and Full-field Test Validation for a Slim-CB of Mobile Phone under Drop/Impact," *Proc. 57th Electronic Components and Technology Conf.*, Reno, NV, June 2007. pp. 914–923.
22. P. Lall, D. Panchagade, D. Iyengar, S. Shantaram, J. Suhling, and H. Schrier, "High Speed Digital Image Correlation for Transient-Shock Reliability of Electronics," *Proc. 57th Electronic Components and Technology Conf.*, Reno, NV, June 2007. pp. 924–939.
23. Telcordia GR-63-CORE Standard, *NEBS Requirements: Physical Protection Environmental Test Methods, Handling Test Methods*, Mar. 2006.

24. IEC 68-2-31 Standard, *Basic Environmental Testing Procedures*, 1969.
25. IEC 68-2-31 Standard Amendment 1, *Basic Environmental Testing Procedures*, 1982.
26. R. Dudek, W. Faust, S. Wiese, M. Röllig, and B. Michel, "Low-Cycle Fatigue of Ag-Based Solders Dependent on Alloying Composition and Thermal Cycle Conditions," *Proc. 9th Electronic Packaging Technology Conf.*, Singapore, Dec. 2007.
27. E. Kaulfersch, S. Rzepka, V. Ganeshan, A. Mueller, and B. Michel, "Dynamic Mechanical Behavior of SnAgCu BGA Solder Joints Determined by Fast Shear Tests and FEM Simulations," *Proc. EuroSIME 2007*, London, Apr. 2007, pp. 172–176.
28. D. Carroll, C. Bates, M. Zampino, and K. Jones, "A Novel Technique for Modeling Solder Joint Failure during System Level Drop Simulations," *Proc. 10th Intersociety Conference on Thermal and Thermomechanical Phenomena in Electronic Systems*, June 2006, pp. 861–868.
29. E. H. Wong, "Dynamics of Board-Level Drop Impact," *ASME J. Electr. Pack.*, vol. 127, Sept. 2005, pp. 200–207.
30. A. Syed, T. S. Kim, Y. M. Cho, C. W. Kim, and M. Y. Yoo, "Alloying Effect of Ni, Co, and Sb in SAC Solder for Improved Drop Performance of Chip Scale Packages with Cu OSP Pad Finish," *Proc. 56th Electronic Components and Technology Conf.*, San Diego, CA, June 2006, pp. 404–411.
31. G. Henshall, M. Roesch, K. Troxel, H. Holder, J. Miremadi, *Manufacturability and Reliability Impacts of Alternate Pb-Free BGA Ball Alloys*, Hewlett-Packard, June 2007, pp. 14–20.

9

LEAD-FREE RELIABILITY IN AEROSPACE/MILITARY ENVIRONMENTS

Thomas A. Woodrow (Boeing) and
Jasbir Bath (Bath Technical Consultancy LLC)

9.1 INTRODUCTION

Most aerospace and military electronics are currently exempt from the European RoHS legislation. However, aerospace and military electronics manufacturers draw on the same supply chain for electronic components and materials as does the broader population of commercial electronics manufacturers. Aerospace and military manufacturers have little leverage within the supply chain since the volume of materials and components purchased by commercial manufacturers greatly exceeds that purchased by aerospace manufacturers. Not surprisingly, suppliers are focusing their attention on the needs of their largest customers, most of whom are demanding RoHS compliant components and materials. Legacy products are rapidly being transitioned to RoHS-compliant versions and new products are being introduced exclusively in RoHS-compliant (lead-free) forms.

Avionics and military electronics are often required to be highly reliable with product lifetimes measured in decades, rather than in years or months. Failure of the electronics to perform can result in mission failure or loss of life. Military and aerospace field environments often include extreme temperatures, pressures, and moisture levels, and high levels of shock and vibration [1,2]. The reliability of lead-free solders and RoHS-compliant materials are still being quantified and aerospace and military electronics manufacturers are reluctant to adopt these materials until their reliability is proven.

Lead-Free Solder Process Development, Edited by Gregory Henshall, Jasbir Bath, and Carol A. Handwerker

Additionally, avionics and military electronics are often repaired when they fail rather than simply replaced. The effect of rework operations on the reliability of lead-free electronics has not been adequately determined. These repair activities often occur many years after initial manufacture, which increases the likelihood of intentionally or accidentally mixing tin-lead and lead-free solders with unknown effects on reliability.

The aerospace and military communities are working together to address these challenges and to produce common and cost-effective solutions. Numerous consortia have been formed to produce the required technical data and to write specifications and guidelines to assist with the transition to lead-free electronics.

9.2 AEROSPACE/MILITARY CONSORTIA

9.2.1 Executive Lead-Free Integrated Process Team (ELF IPT) [3]

The ELF IPT was formed in the fourth quarter of 2005 to address concerns within the US government and the electronics industry regarding the worldwide transition to lead-free electronics. The mission of the ELF IPT was to create a joint government and defense industry position on the implementation of lead-free electronics that minimizes any reliability impact to US Department of Defense (DoD) products.

The objectives of the ELF IPT were as follows:

1. To develop, brief, and maintain government policies, guidelines and standards for the implementation of lead-free materials and processes in the manufacturing of defense electronics.
2. To develop a roadmap for lead-free implementation activities, and to coordinate investment strategies for producing reliability and performance data for lead-free processes and materials that would be used in defense electronics.
3. To draft and recommend implementation of a common Statement of Work (SOW) for use in US government contracts.

The ELF IPT was made up of representatives from the US government and US defense industry (Table 9.1). The role of the government representatives was to provide insight into the defense electronic product acquisition and supportability requirements that may be at risk due to the introduction of lead-free materials. The role of the defense industry representatives was to provide insight into the impact of lead-free materials and processes on defense electronic product reliability, manufacturability and repair/supportability. The ELF IPT was incorporated into the PERM (Pb-Free Electronics Risk Management) Council in 2009.

9.2.2 Pb-Free Electronics Risk Management (PERM) Council [4]

In 2004, the Lead-Free in Aerospace Project (LEAP) was formed to develop a set of guideline documents to assist the aerospace industry in meeting the challenges of

TABLE 9.1 ELF IPT representatives

Defense Industry	Government
AIA	American Competitiveness Institute
BAE Systems	Defense Logistics Agency
The Boeing Company	DoD DMEA
Honeywell	Missile Defense Agency
ITT Industries	NASA/GFSC/KSC
Lockheed Martin	Office of the Secretary of Defense
Northrop Grumman	US Air Force
Raytheon	US Army
Rockwell Collins	US Navy NAVAIR
	US Navy NAVSEA
	US Navy NSWC

lead-free electronics. In 2009, LEAP was restructured to form the PERM Council. The documents written by the PERM Council members were issued initially in the United States by the Government Electronics and Information Technology Association (GEIA, now part of TechAmerica). These documents will be submitted to the International Electrotechnical Commission (IEC) for adoption globally. PERM is jointly sponsored by the Aerospace Industries Association (AIA), Avionics Maintenance Conference (AMC), and the Government Electronics and Information Technology Association (GEIA). PERM membership includes most of the world's major aircraft manufacturers and defense contractors, many mid-tier suppliers, and relevant governmental/customer organizations.

As of the writing of the book chapter, six documents have been completed and issued while one additional document is in the process of being written. The documents are described below:

ANSI-GEIA-STD-0005-1 Performance Standard for Aerospace and High Performance Electronic Systems Containing Lead-Free Solder. This document covers the development and implementation of lead-free control plans (LFCP) to ensure that the plan owners and their customers produce electronic systems that are reliable, affordable, and supportable.

ANSI-GEIA-STD-0005-2 Standard for Mitigating the Effects of Tin Whiskers in Aerospace and High Performance Electronic Systems. This document specifies that users develop and implement tin whisker risk mitigation plans. Aerospace electronics manufacturers and users will be required to specify how they intend to mitigate tin whisker risks based on the level of control required for the given application. Appendixes in the standard provide guidance on addressing risks associated with tin whiskers.

GEIA-HB-0005-1 Program Management/Systems Engineering Guidelines for Managing the Transition to Lead-Free Electronics. This handbook provides assistance to program managers and lead systems engineers for assuring the performance, reliability, airworthiness, safety, and certifiability of product(s), in accordance with ANSI-GEIA-STD-0005-1.

GEIA-HB-0005-2 Technical Guidelines for Aerospace and High Performance Electronic Systems Containing Lead-Free Solder and Finishes. This handbook provides technical guidance for the use of lead-free solder and mixed tin-lead/lead-free alloy systems in high reliability aerospace electronic systems.

ANSI-GEIA-STD-0005-3 Reliability Testing for Aerospace and High Performance Electronics Containing Lead-Free Solder. The purpose of this document is to provide guidance for conducting reliability testing of lead-free aerospace electronics and for interpreting the results from reliability tests.

GEIA-HB-0005-3 Guidelines for Repair and Rework of Lead-Free Assemblies Used in Aerospace and High-Performance Electronic Applications. This handbook provides guidelines for repair and maintenance of lead-free electronics.

GEIA-HB-0005-4 (Document under development) Guidelines for Performing Reliability Predictions for Lead-Free Assemblies used in Aerospace and High-Performance Electronic Applications. This handbook will describe methods for predicting the reliability of lead-free electronics for product certification.

9.2.3 Tin Whisker Alert Group [5]

The Tin Whisker Alert Group was formed in 2002 to collect and exchange information on tin whisker related issues and to explore strategies for mitigating the risks posed by tin whiskers to high-reliability electronics. Members of this group include all major defense contractors, all branches of the Armed Services, NASA, and CALCE (University of Maryland), among others. The group members have conducted studies on the use of conformal coatings to mitigate whisker risks and on the robotic dipping of tin-plated components into molten SnPb solder to remove the tin, and they have developed algorithms for quantifying the risks posed by whiskers to electronics. One of the members (NASA Goddard) hosts a website that contains information on documented failures caused by whiskers, whisker inspection methods, and whisker mitigation strategies (http://nepp.nasa.gov/WHISKER/).

9.2.4 Center for Advanced Life-Cycle Engineering (CALCE) [6]

CALCE at the University of Maryland conducts development and implementation of physics-of-failure (PoF) approaches to reliability, accelerated testing, electronic parts selection and management, and supply-chain management. The CALCE consortium members include all of the major defense contractors. Recently CALCE has conducted numerous thermal cycle studies on lead-free solders. Data from these studies was used to create and validate models that could be used by members to predict the reliability of lead-free solder joints under thermal cycling conditions. In addition CALCE developed models for predicting the performance of lead-free solders under vibration and mechanical shock. CALCE has also been involved in research on tin whisker growth and whisker mitigation and has developed a tool for quantifying the risks due to tin whiskers.

9.2.5 Joint Council on Aging Aircraft/Joint Group on Pollution Prevention (JCAA/JG-PP) Lead-Free Solder Project [7]

This DoD sponsored consortium was founded in 2001 to evaluate lead-free solders and finishes and to determine whether they were suitable for use in high reliability electronics. This consortium was jointly managed by the Joint Council on Aging Aircraft (JCAA) and the Joint Group on Pollution Prevention (JG-PP) and included members from commercial and defense aerospace contractors (BAE Systems, Boeing, Lockheed Martin, Raytheon, Rockwell Collins), NASA Centers (Kennedy Space Center, Jet Propulsion Laboratory, Marshall Space Flight Center, Johnson Space Center, Goddard Space Flight Center, Ames Research Laboratory), NASA contractors (United Space Alliance-Solid Rocket Booster, Boeing-Orbiter), US Air Force, US Army, US Navy, US Marines, US Department of Energy, and more than 20 other companies and universities.

The consortium wrote a test plan called the Joint Test Protocol (JTP) [8] to describe the testing to be done. The testing included thermal cycling, thermal shock, vibration, mechanical shock, combined vibration/thermal cycling, electromigration, SIR (Surface Insulation Resistance), salt fog, and humidity.

Test vehicles were designed and the lead-free solders to be tested were chosen. The solder selection process was documented in the Potential Alternatives Report (PAR) [9].

The main test vehicle was a six-layer circuit board 12.75 inches (324 mm) wide by 9 inches (229 mm) high by 0.090 inches (2.29 mm) thick (Figure 9.1) populated with

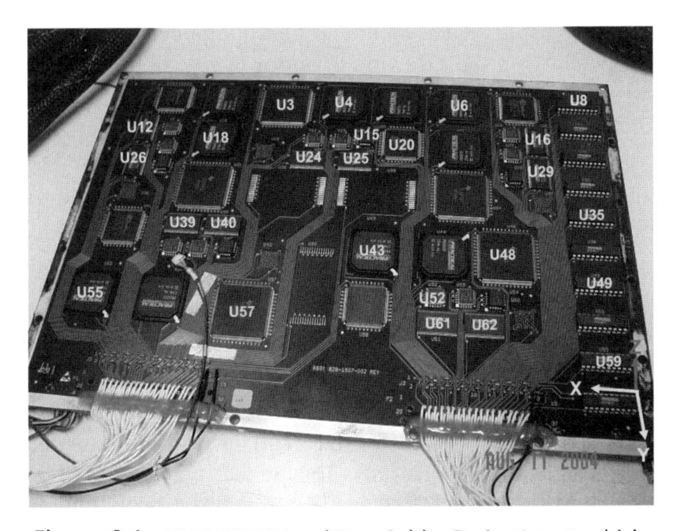

Figure 9.1 JCAA/JG-PP Lead-Free Solder Project's test vehicle

55 components consisting of ceramic leadless chip carriers (CLCCs), plastic leaded chip carriers (PLCCs), Alloy 42 TSOPs, TQFPs, BGAs, and PDIPs.

Four solder alloys were selected for the test:

Sn3.9Ag0.6Cu for reflow and wave soldering (abbreviated as SAC396).

Sn3.4Ag1Cu3.3Bi for reflow soldering (abbreviated as SACB).

Sn0.7Cu0.05Ni for wave soldering (abbreviated as SN100C).

Sn37Pb for reflow and wave soldering (abbreviated as SnPb).

The SAC396 alloy was chosen because extensive testing by iNEMI suggested it was a viable candidate for use in lead-free commercial electronics. The SACB alloy was chosen because it was the best performer in the NCMS study [10]. The SN100C alloy was chosen because it had been used worldwide. Finally, eutectic SnPb was included as the control alloy.

The test vehicles were divided into two types. The first type (named "Manufactured" test vehicles) were made using a laminate with a high glass transition temperature (T_g of 170°C) and an immersion silver board finish. The lead-free "Manufactured" test vehicles were meant to be representative of a printed wiring assembly (PWA) designed for manufacture with lead-free solders and lead-free reflow and wave soldering profiles.

The second type (named "Rework" test vehicles) were made using a laminate with a low glass transition temperature (T_g of 140°C) and a tin-lead HASL board finish. The "Rework" test vehicles were meant to be representative of a typical tin-lead legacy PWA that would have to be reworked using lead-free solders in the future. The "Rework" test vehicles were initially built using tin-lead solder and a tin-lead board finish and using typical tin-lead reflow and wave soldering profiles. Selected components on the "Rework" test vehicles were then removed; residual tin-lead solder was cleaned from the pads using solder wick; and new components attached using a lead-free solder. Components on the "Rework" control test vehicles were reworked with tin-lead solder rather than a lead-free solder.

In general, the lead-free solder test vehicles performed better than the tin-lead controls in thermal cycling but the performance in vibration was mixed (Section 9.4.2). All test results for these evaluations can be found at http://acqp2.nasa.gov/LeadFreeSolderTestingForHighReliability_Proj1.html.

9.2.6 NASA/DoD Lead-Free Electronics Project [7]

This joint project of DoD, NASA, and defense and space contractor representatives has been built on the results from the prior JCAA/JG-PP Lead-Free Solder Project. This new effort has focused on the reliability of reworked tin-lead and lead-free solder alloys and of mixed tin-lead/lead-free materials. The project was launched in 2006 and is scheduled to end in 2011.

The NASA/DoD test vehicle has a very similar design to that used for the JCAA/JG-PP Lead-Free Solder Project. The main test vehicle is a six-layer circuit board 12.75 inches (324 mm) wide by 9 inches (229 mm) high by 0.090 inches (2.29 mm) thick populated with 64 components consisting of ceramic leadless chip carriers (CLCCs), TSOPs, TQFPs, BGAs, QFNs, CSPs, and PDIPs. The board finish is immersion silver. Sn3Ag0.5Cu will be used as the solder paste alloy for attaching the surface mount components and Sn0.7Cu0.05Ni will be used for wave soldering operations. Eutectic SnPb solder will be used on the control test vehicles. Component finishes will include SnPb, Sn, SnBi, and NiPdAu. The BGA balls will be either SnPb or Sn4Ag0.5Cu, and the CSP balls will be either SnPb or Sn1Ag0.5Cu.

Significantly more rework will be done in this study than was done in the JCAA/JG-PP Lead-Free Solder Project. SnPb and lead-free solder joints will be reworked to yield pure SnPb joints, pure lead-free joints, and mixed technology joints.

The testing will include thermal cycling, thermal shock, vibration, mechanical shock, and combined vibration/thermal cycling.

9.2.7 Crane/SAIC Repair Project [11]

The goal of this project will be to evaluate existing data and recommend strategy for the US DoD (Department of Defense) regarding lead-free electronics with emphasis on rework issues within US Navy programs. Issues of concern include solder joint reliability, copper dissolution, mixing of alloys, process temperature effects, alloy identification, tin whiskers, and supply-chain problems. The project team members include NAVSEA Crane, SAIC, Purdue (West Lafayette), Purdue (Calumet), and Raytheon.

9.3 LEAD-FREE CONTROL PLANS FOR AEROSPACE/MILITARY ELECTRONICS

The ANSI-GEIA-STD-0005-1 and ANSI-GEIA-STD-0005-2 documents require each supplier of electronic products and systems to the military, aerospace, and other high-performance industries to develop a lead-free control plan (LFCP). Each LFCP describes the policies, processes, and documents that the supplier will use to ensure that its products will meet all performance, reliability, airworthiness, safety, and certifiability requirements throughout the product's specified performance life.

Creation of a LFCP would not necessarily be a plan to make the transition to lead-free solder, although this would be considered an option. The LFCP would be viewed as the supplier's response to the transition that has already been taking place and that is impacting the supplier's products. Those suppliers who plan to prohibit the use of lead-free materials in their products should have in place a LFCP that references processes for monitoring incoming materials and for removing or replacing lead-free materials.

9.4 AEROSPACE/MILITARY LEAD-FREE RELIABILITY CONCERNS

9.4.1 Thermal Fatigue Resistance

Numerous studies have been conducted by the commercial electronics industry to determine the reliability of lead-free solder joints during thermal cycling, but very few studies have been conducted by the manufacturers of IPC Class 3 (high-performance) aerospace/military electronics. The most notable exception is the JCAA/JG-PP Lead-Free Solder Project (Section 9.2.5). Two thermal cycles were conducted in this study: –20°C to +80°C and –55°C to +125°C. The data are intended to be used by the modeling community to determine acceleration factors and to validate their thermal cycling models [12,13].

The thermal cycling chamber that was used for the –20°C to +80°C test is shown in Figure 9.2. The test vehicles were held vertically in racks that allow airflow between the vehicles. Figure 9.3 shows actual air and test vehicle temperatures recorded during the –20°C to +80°C test. Dwell times of 30 minutes (hot dwell) and 10 minutes (cold dwell) were used for both the –20°C to +80°C and the –55°C to +125°C thermal cycle tests and the ramp rates were approximately 10°C/min.

During the JCAA/JG-PP thermal cycle tests, the 55 components on each test vehicle and the 1206 chip resistors on break-off coupons were individually monitored using event detectors capable of detecting a two-hundred nanosecond event with a resistivity of 300 ohms or greater (Figure 9.4). Data collection software was used to record the cycle when each event (solder joint failure) occurred.

For those component types that had a significant number of failures, Weibull plots of the failure data were created to determine the beta (slope) and the characteristic lifetime (time to fail 63.2% of the population, also called alpha or eta) for each

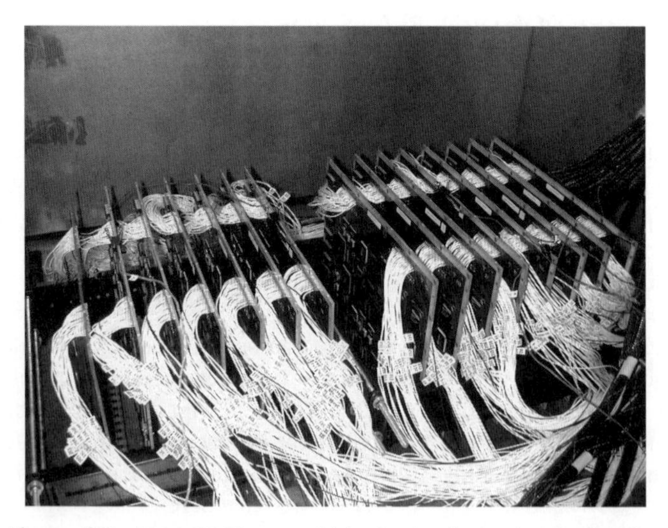

Figure 9.2 JCAA/JG-PP test vehicles in the thermal cycle chamber

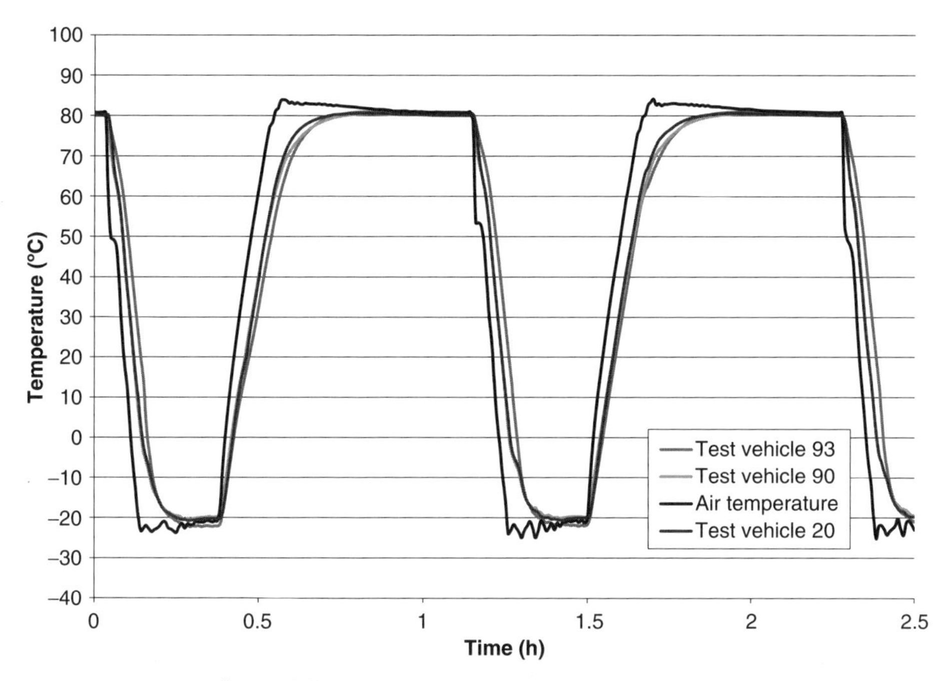

Figure 9.3 JCAA/JG-PP thermal cycle, −20°C to +80°C

Figure 9.4 JCAA/JG-PP event detectors and data collection system

component type. The number of cycles required to fail 50% of each component population attached with eutectic SnPb or with SAC396 solder are shown in Figures 9.5 and 9.6. The −20°C to +80°C test data clearly showed that SAC solder outperformed SnPb solder for every component type tested.

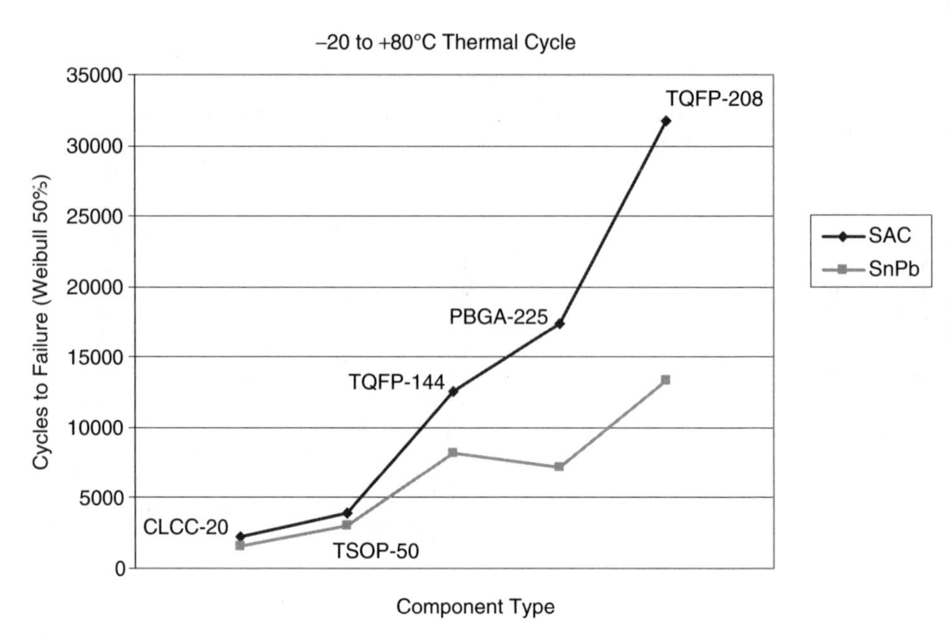

Figure 9.5 Comparison of eutectic SnPb and SAC396 solders on various components during the JCAA/JG-PP –20°C to +80°C thermal cycle test [12]

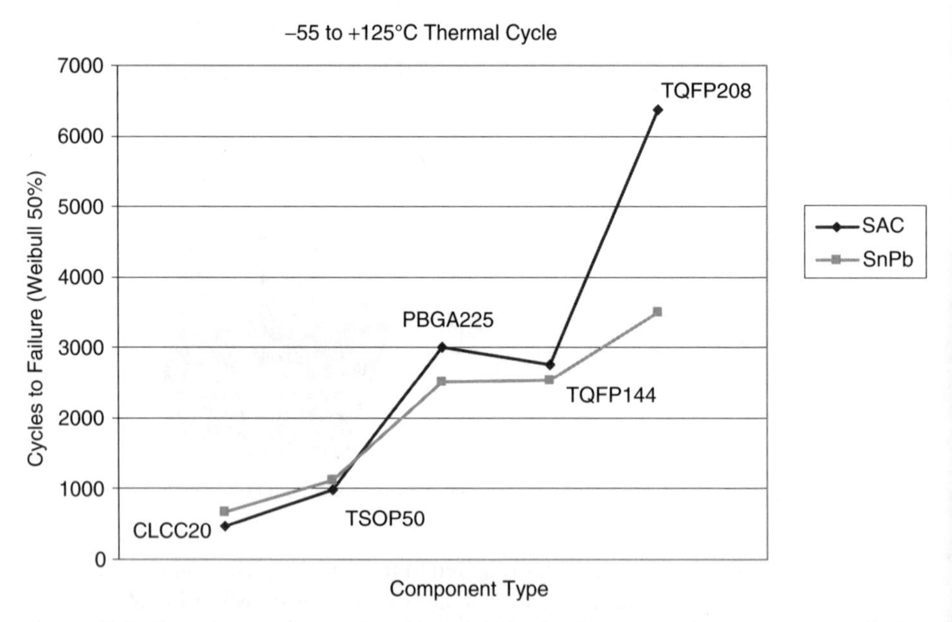

Figure 9.6 Comparison of eutectic SnPb and SAC396 solders on various components during the JCAA/JG-PP –55°C to +125°C thermal cycle test [13]

The −55°C to +125°C data showed the same general trend where SAC396 outperformed SnPb. However, the trend was reversed for the highly noncompliant ceramic CLCCs and alloy 42 lead-frame TSOPs. The combination of noncompliancy and highly accelerated thermal cycle conditions resulted in large solder joint strains for these component types. Clech [14] showed that if the solder strain exceeded 6.2%, SnPb would outperform SnAgCu solder while the opposite was true at lower solder strains.

Based on these test results, we could say with some confidence that the thermal fatigue resistance of the SnAgCu family of solder alloys (with 3–4 wt% Ag) are as good or better than that of eutectic SnPb solder for most aerospace/military thermal use conditions.

9.4.2 Vibration Fatigue Resistance

Very few studies have been conducted by the manufacturers of IPC Class 3 (high-performance) aerospace/military electronics to determine the reliability of lead-free solders in vibration environments. One of the exceptions is the JCAA/JG-PP Lead-Free Solder Project (Section 9.2.5) for which a large-scale vibration study was conducted [15].

An aluminum fixture was built to hold up to 15 test vehicles at one time. Slots were cut into the fixture to accept wedge locks that were mounted on both ends of the test vehicles. The fixture was mounted on an electrodynamic shaker for testing in the *y*-axis. For testing in the *x*- and *z*-axes, the fixture was mounted on a slip table that in turn was connected to the electrodynamic shaker.

One set of JCAA/JG-PP test vehicles mounted in the test fixture are shown in Figure 9.7. Event detectors were used for detecting solder joint failures during the test. A calibrated accelerometer was mounted on each test vehicle to record the resonance frequencies and the response of each test vehicle.

The shaker input into the test vehicles is shown in Figure 9.8 with the typical response of a test vehicle shown in Figure 9.9 (both during a 14.0 G_{rms} run). Note that

Figure 9.7 JCAA/JG-PP test vehicles in the vibration fixture

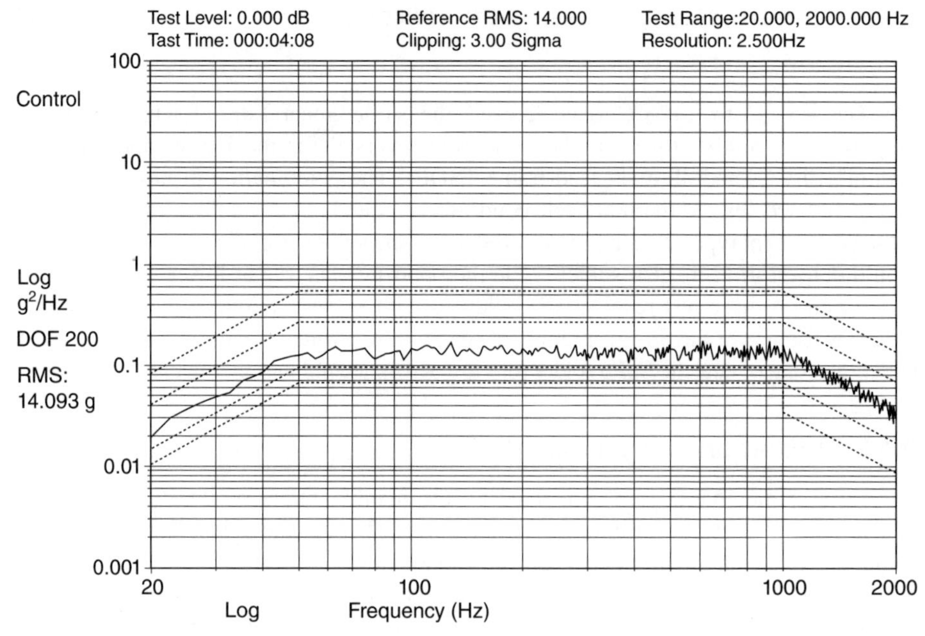

Figure 9.8 JCAA/JG-PP vibration input (14.0 G_{rms}, z-axis)

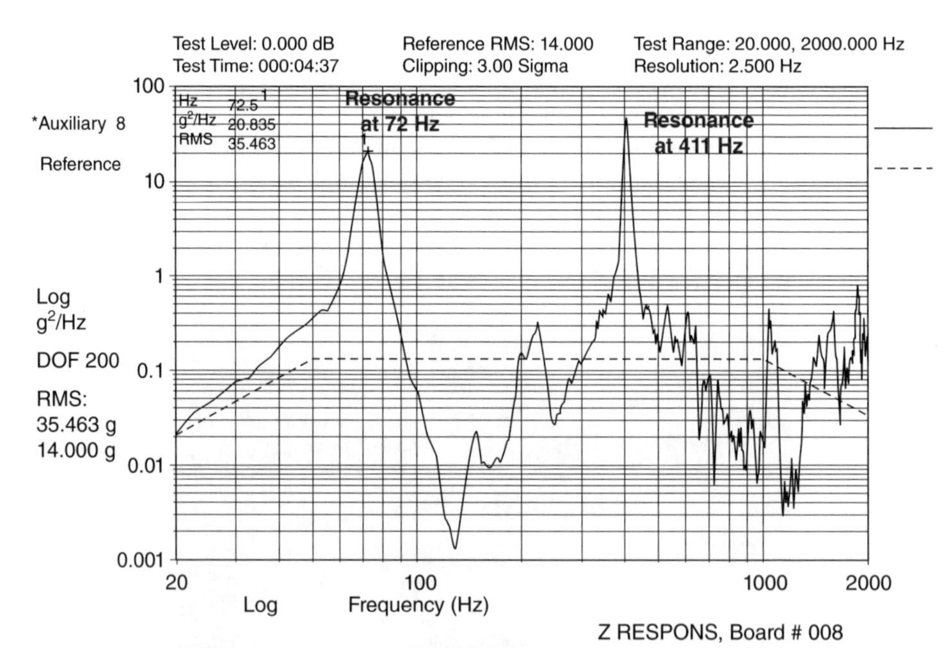

Figure 9.9 JCAA/JG-PP test vehicle response (14.0 G_{rms}, z-axis)

the response of the test vehicle differed greatly from the input power spectral density (PSD) spectrum, with the major test vehicle resonances occurring at 72 and 411 Hz (Figures 9.10 and 9.11). The first bending mode at 72 Hz caused the most board flexure that in turn was responsible for the majority of solder joint failures. Components located down the center line of the circuit board tended to fail first because that is where the largest radius of circuit board curvature was produced.

The bending modes induced in the test vehicle during the testing resulted in regions of high stress and regions of low stress across the surface of the circuit board. This meant that only identical components in identical locations on identical test vehicles could be directly compared. For the comparisons to be made, the test solder had to be used on one set of test vehicles and the control solder on a second set of test vehicles.

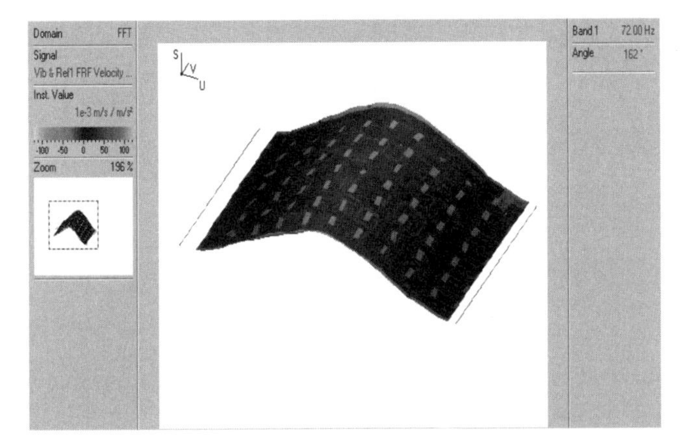

Figure 9.10 JCAA/JG-PP test vehicle bending mode at 72 Hz

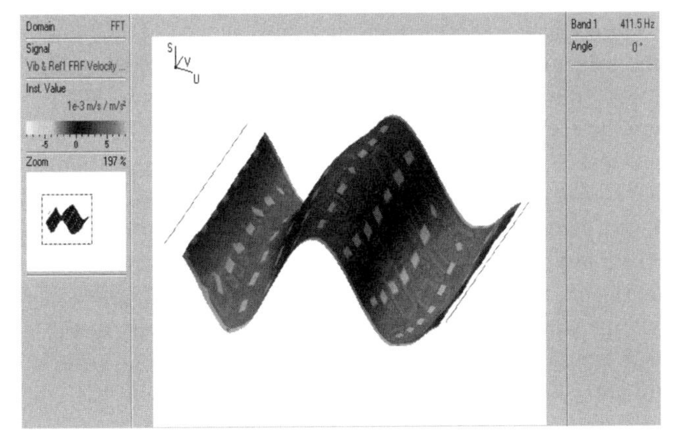

Figure 9.11 JCAA/JG-PP test vehicle bending mode at 411 Hz

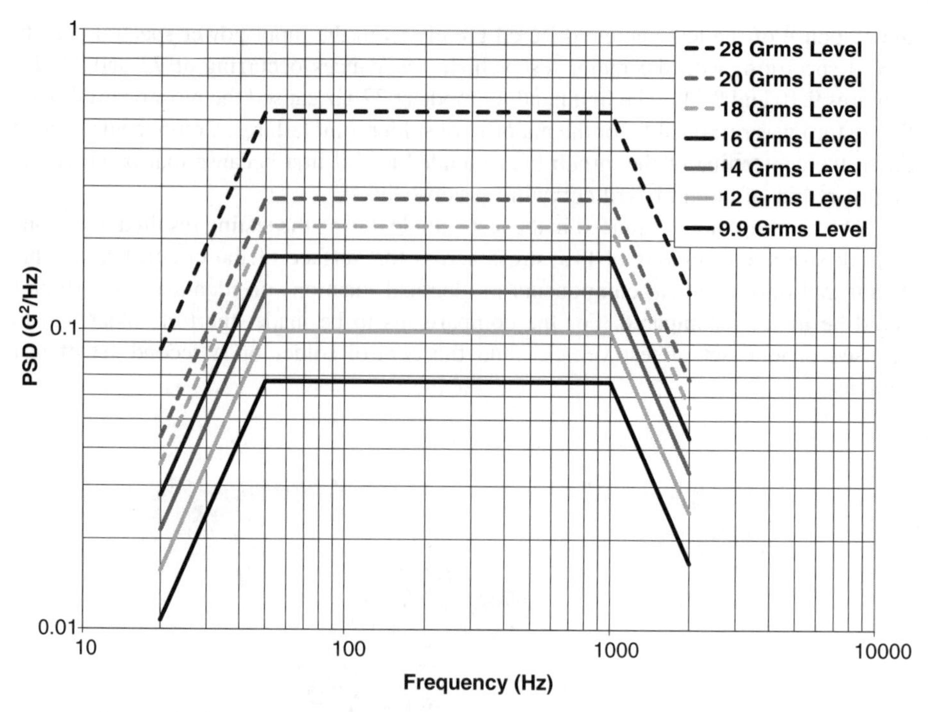

Figure 9.12 Vibration test levels used for the JCAA/JG-PP step stress test

Each group of JCAA/JG-PP test vehicles was subjected sequentially to the seven random vibration levels shown in Figure 9.12 and Table 9.2. The test consisted of one hour of vibration at 9.9 G_{rms} in the *y*-axis, followed by one hour of vibration at 9.9 G_{rms} in the *x*-axis, followed by one hour of vibration at 9.9 G_{rms} in the *z*-axis (perpendicular to the plane of the test vehicle). Then the test vehicles were subjected to an additional vibration step stress testing in the *z*-axis only, starting with one hour of vibration at 12.0 G_{rms}. The vibration levels were then increased in 2.0 G_{rms} increments, shaking at each level in the *z*-axis for one hour until 20.0 G_{rms} were reached. A final 28.0 G_{rms} run completed the test. This meant that the stress experienced by the solder joints increased substantially with every 60 minute increment of test time.

This type of test is called a step stress test because the severity of the vibration is increased in steps. With a step stress test the time required to fail a majority of components is reduced, but differences in the solders being tested can still be determined. If the JCAA/JG-PP test had been run at a constant low level of vibration (e.g., 9.9 G_{rms}), only a few components would have failed in the first eight hours of testing and hundreds of hours of test time would have been required to fail a majority of the components on the JCAA/JG-PP test vehicle. In contrast, if the test had been run at a constant high level of vibration (e.g., 28 G_{rms}), some of the components would have failed in the first few seconds of testing, and any discrimination between solder types would have been lost for those components.

TABLE 9.2 JCAA/JG-PP vibration test levels

Level 1	*Level 2*	*Level 3*
20 Hz @ 0.0107 G^2/Hz	20 Hz @ 0.0157 G^2/Hz	20 Hz @ 0.0214 G^2/Hz
20–50 Hz @ +6.0 dB/octave	20–50 Hz @ +6.0 dB/octave	20–50 Hz @ +6.0 dB/octave
50–1000 Hz @ 0.067 G^2/Hz	50–1000 Hz @ 0.0984 G^2/Hz	50–1000 Hz @ 0.134 G^2/Hz
1000–2000 Hz @ –6.0 dB/octave	1000–2000 Hz @ –6.0 dB/octave	1000–2000 Hz @ –6.0 dB/octave
2000 Hz @ 0.0167 G^2/Hz	2000 Hz @ 0.0245 G^2/Hz	2000 Hz @ 0.0334 G^2/Hz
Composite = 9.9 G_{rms}	**Composite = 12.0 G_{rms}**	**Composite = 14.0 G_{rms}**
Level 4	*Level 5*	*Level 6*
20 Hz @ 0.0279 G^2/Hz	20 Hz @ 0.0354 G^2/Hz	20 Hz @ 0.0437 G^2/Hz
20–50 Hz @ +6.0 dB/octave	20–50 Hz @ +6.0 dB/octave	20–50 Hz @ +6.0 dB/octave
50–1000 Hz @ 0.175 G^2/Hz	50–1000 Hz @ 0.2215 G^2/Hz	50–1000 Hz @ 0.2734 G^2/Hz
1000–2000 Hz @ –6.0 dB/octave	1000–2000 Hz @ –6.0 dB/octave	1000–2000 Hz @ –6.0 dB/octave
2000 Hz @ 0.0436 G^2/Hz	2000 Hz @ 0.0552 G^2/Hz	2000 Hz @ 0.0682 G^2/Hz
Composite = 16.0 G_{rms}	**Composite = 18.0 G_{rms}**	**Composite = 20.0 G_{rms}**
Level 7		
20 Hz @ 0.0855 G^2/Hz		
20–50 Hz @ +6.0 dB/octave		
50–1000 Hz @ 0.5360 G^2/Hz		
1000–2000 Hz @ –6.0 dB/octave		
2000 Hz @ 0.1330 G^2/Hz		
Composite = 28.0 G_{rms}		

Because of the large amount of data collected, not all of the data will be presented here. Examples of the test results for selected components are given in Tables 9.3 through 9.11. Since only identical components in identical locations on identical test vehicles could be directly compared, each table shows only the test data for components with the same reference designation (i.e., components occupying the same location on each of the test vehicles). Since some of the "Manufactured" test vehicles were assembled with lead-free solders and some with tin-lead solder, a direct comparison could be done to determine which solder type performed better under vibration. The relevant tables show the solder/component finish combinations that were tested and the number of test levels that each survived. A "60" number means that that particular component survived the entire 60-minute duration of the test at that test level.

A brief summary of the SnPb and SnAgCu data for each component type is given in the following sections. In general, the eutectic SnPb solder joints outperformed the SnAgCu solder joints for most of the component types tested. However, there was one case where the lead-free solders outperformed SnPb solder, namely Sn0.7Cu0.05Ni and Sn3Ag0.5Cu alloys used to wave solder the PDIP through-hole components at location U8 (Table 9.6).

TABLE 9.3 JCAA/JG-PP vibration data for the BGA component at U43

Test Vehicle ID	Solder Paste/ Solder Balls	Time at Each Level (mins)								
		y-axis	*x*-axis	*z*-axis	*z*-axis	*z*-axis	*z*-axis	*z*-axis	*z*-axis	*z*-axis
		9.9 G_{rms}	9.9 G_{rms}	9.9 G_{rms}	12.0 G_{rms}	14.0 G_{rms}	16.0 G_{rms}	18.0 G_{rms}	20.0 G_{rms}	28.0 G_{rms}
79	SAC396/SAC405	60	60	5						
77	SAC396/SAC405	60	60	6						
75	SAC396/SAC405	60	60	10						
76	SAC396/SAC405	60	60	23						
78	SAC396/SAC405	60	60	60	10					
6	Sn37Pb/Sn37Pb	60	60	54						
5	Sn37Pb/Sn37Pb	60	60	60	3					
8	Sn37Pb/Sn37Pb	60	60	60	16					
9	Sn37Pb/Sn37Pb	60	60	60	16					
7	Sn37Pb/Sn37Pb	60	60	60	20					

Note: Stress on the solder joints was increased substantially every 60 minutes by stepping up the power spectral density (PSD) input. Therefore the first 60 test minutes were less damaging than the second 60 test minutes, which were less damaging than the third 60 test minutes, and so on.

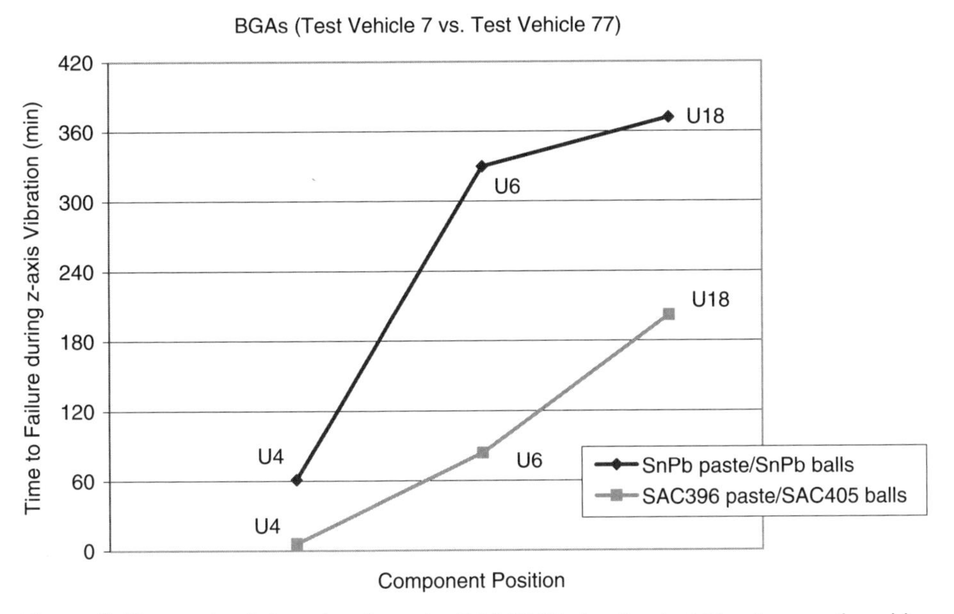

Figure 9.13 BGA225 failure data from the JCAA/JG-PP vibration test. The stress on the solder joints was increased substantially every 60 minutes by stepping up the power spectral density (PSD) input. Therefore the first 60 test minutes were less damaging than the second 60 test minutes, which were less damaging than the third 60 test minutes, and so on

BGA-225 ("Manufactured" Test Vehicles) SnPb BGA balls assembled with SnPb solder paste always outperformed SAC405 BGA balls assembled with SAC396 solder paste regardless of location on the test vehicle (Tables 9.3 and 9.4 and Figure 9.13). BGAs are large stiff components that make the solder joints sensitive to the changes in the curvature of the test vehicle during vibration and accentuates differences in the material properties of the solders. Figure 9.13 compares the BGAs locations at U4, U6, and U18 on Test Vehicle 7 (assembled with Sn37Pb solder paste) and Test Vehicle 77 (assembled with SAC396 solder paste).

Note that the BGA at U4 was in a higher strain environment that caused those BGAs to fail earlier in the test. The BGAs at U18 were in a lower strain environment, and those BGAs failed later in the test. Similarly the study found that SnPb BGA balls assembled with SnPb solder paste generally outperformed SnPb balls assembled with SAC396 solder paste.

CLCC-20s ("Manufactured" Test Vehicles) Overall, a large percentage of the CLCC components did not fail during the test. In those cases where there were enough failures to rank the solders, the combination of SnPb paste/SnPb component part finish outperformed SnAgCu paste/SnAgCu component part finish (Table 9.5).

PDIP-20s ("Manufactured" Test Vehicles) When combined with a NiPdAu component part finish, the Sn0.7Cu0.05Ni wave solder alloy (SN100C) was the best performer, followed by the SAC396 alloy and then SnPb (Table 9.6).

TABLE 9.4 JCAA/JG-PP vibration data for the BGA component at U55

Test Vehicle ID	Solder Paste/ Solder Balls	Time at Each Level (mins)								
		y-axis	x-axis	z-axis	z-axis	z-axis	z-axis	z-axis	z-axis	z-axis
		$9.9 G_{rms}$	$9.9 G_{rms}$	$9.9 G_{rms}$	$12.0 G_{rms}$	$14.0 G_{rms}$	$16.0 G_{rms}$	$18.0 G_{rms}$	$20.0 G_{rms}$	$28.0 G_{rms}$
75	SAC396/SAC405	60	60	8						
77	SAC396/SAC405	60	60	11						
79	SAC396/SAC405	60	60	13						
76	SAC396/SAC405	60	60	37						
78	SAC396/SAC405	60	60	60	13					
8	Sn37Pb/Sn37Pb	60	60	60	18					
9	Sn37Pb/Sn37Pb	60	60	60	20					
5	Sn37Pb/Sn37Pb	60	60	60	60	33				
6	Sn37Pb/Sn37Pb	60	60	60	60	60	5			
7	Sn37Pb/Sn37Pb	60	60	60	60	60	20			

Note: Stress on the solder joints was increased substantially every 60 minutes by stepping up the power spectral density (PSD) input. Therefore the first 60 test minutes were less damaging than the second 60 test minutes, which were less damaging than the third 60 test minutes, and so on.

TABLE 9.5 JCAA/JG-PP vibration data for the CLCC component at U52

Test Vehicle ID	Solder Paste/ Finish	Time at Each Level (mins)								
		y-axis	*x*-axis	*z*-axis	*z*-axis	*z*-axis	*z*-axis	*z*-axis	*z*-axis	*z*-axis
		$9.9\,G_{rms}$	$9.9\,G_{rms}$	$9.9\,G_{rms}$	$12.0\,G_{rms}$	$14.0\,G_{rms}$	$16.0\,G_{rms}$	$18.0\,G_{rms}$	$20.0\,G_{rms}$	$28.0\,G_{rms}$
76	SAC396/SAC396	60	60	60	60	44				
79	SAC396/SAC396	60	60	60	60	60	17			
77	SAC396/SAC396	60	60	60	60	60	60	4		
78	SAC396/SAC396	60	60	60	60	60	60	32		
75	SAC396/SAC396	60	60	60	60	60	60	37		
7	Sn37Pb/Sn37Pb	60	60	60	60	60	60	37		
9	Sn37Pb/Sn37Pb	60	60	60	60	60	60	60	31	
6	Sn37Pb/Sn37Pb	60	60	60	60	60	60	60	60	5
8	Sn37Pb/Sn37Pb	60	60	60	60	60	60	60	60	21
5	Sn37Pb/Sn37Pb	60	60	60	60	60	60	60	60	24

Note: Stress on the solder joints was increased substantially every 60 minutes by stepping up the power spectral density (PSD) input. Therefore the first 60 test minutes were less damaging than the second 60 test minutes, which were less damaging than the third 60 test minutes, and so on.

TABLE 9.6 JCAA/JG-PP vibration data for the PDIP component at U8

Test Vehicle ID	Solder/Finish	Time at Each Level (mins)								
		y-axis	*x*-axis	*z*-axis	*z*-axis	*z*-axis	*z*-axis	*z*-axis	*z*-axis	*z*-axis
		9.9 G_{rms}	9.9 G_{rms}	9.9 G_{rms}	12.0 G_{rms}	14.0 G_{rms}	16.0 G_{rms}	18.0 G_{rms}	20.0 G_{rms}	28.0 G_{rms}
77	SAC396/NiPdAu	60	60	5						
76	SAC396/NiPdAu	60	60	60	34					
78	SAC396/NiPdAu	60	60	60	60	1				
79	SAC396/NiPdAu	60	60	60	60	60	3			
75	SAC396/NiPdAu	60	60	60	60	60	60	18		
115	Sn0.7Cu0.05Ni/NiPdAu	60	60	60	60	60	27			
114	Sn0.7Cu0.05Ni/NiPdAu	60	60	60	60	60	60	26		
116	Sn0.7Cu0.05Ni/NiPdAu	60	60	60	60	60	60	28		
117	Sn0.7Cu0.05Ni/NiPdAu	60	60	60	60	60	60	60	1	
118	Sn0.7Cu0.05Ni/NiPdAu	60	60	60	60	60	60	60	17	
8	Sn37Pb/NiPdAu	60	60	14						
5	Sn37Pb/NiPdAu	60	60	20						
6	Sn37Pb/NiPdAu	60	60	20						
7	Sn37Pb/NiPdAu	60	60	22						
9	Sn37Pb/NiPdAu	60	60	60	8					

Note: Stress on the solder joints was increased substantially every 60 minutes by stepping up the power spectral density (PSD) input. Therefore the first 60 test minutes were less damaging than the second 60 test minutes, which were less damaging than the third 60 test minutes, and so on.

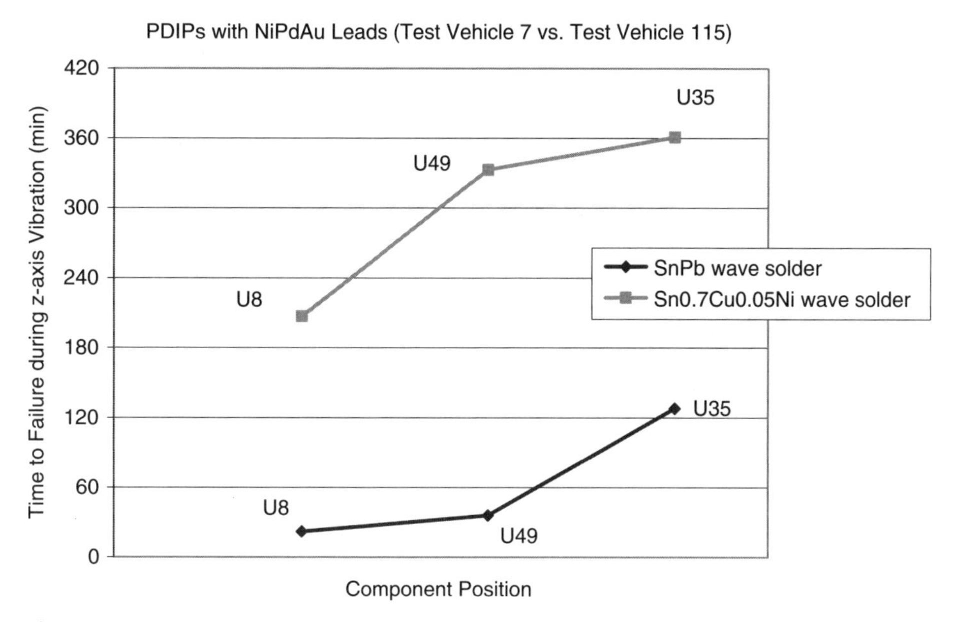

Figure 9.14 PDIP component failure data from the JCAA/JG-PP vibration test. The stress on the solder joints was increased substantially every 60 minutes by stepping up the power spectral density (PSD) input. Therefore the first 60 test minutes were less damaging than the second 60 test minutes, which were less damaging than the third 60 test minutes, and so on

Figure 9.14 compares PDIP components at locations U8, U35, and U49 on Test Vehicle 7 (wave soldered with Sn37Pb) and Test Vehicle 115 (wave soldered with Sn0.7Cu0.05Ni).

When combined with a tin component finish, the Sn0.7Cu0.05Ni wave solder alloy was still the best performer. SnPb appeared to perform better (relative to the SN100C alloy) when combined with a tin component part finish rather than a NiPdAu component finish.

The PDIP component solder joints tended to form circumferential cracks around the corner leads closest to the edge of the test vehicle. Figure 9.15 shows a photograph of a cracked PDIP corner solder joint (SnPb solder with NiPdAu component finish).

TSOP-50s ("Manufactured" Test Vehicles) The TSOPs oriented parallel to the shorter side of the test vehicle at locations U12, U16, U26, and U29 tended to fall off during the testing. The TSOPs oriented perpendicular to the shorter side of the test vehicle located at U24, U25, U39, U40, U61, and U62 did not fall off during the testing. The general failure mechanism for those TSOPs that fell off was for all of the leads on one side to come loose from the solder (Figure 9.16). The TSOP would now be free to rotate, which caused the leads on the opposite side of the TSOP to break off. The effect of orientation in determining which TSOPs fell off could have been due to the greater radius of curvature change experienced by the foot of a TSOP lead when oriented perpendicular to the shorter edge of the vehicle, as opposed to the smaller radius of

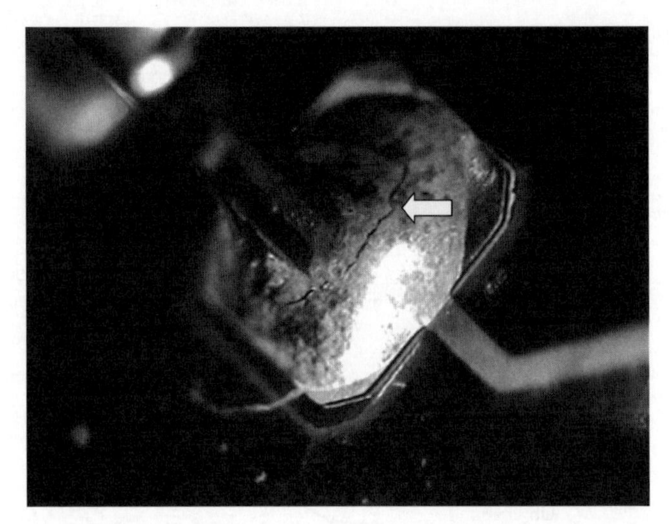

Figure 9.15 PDIP component with cracked solder joint. SnPb solder with NiPdAu component finish (100×)

Figure 9.16 TSOP component with leads lifted from the solder during testing (100×)

curvature change experienced when oriented parallel to the vehicle edge. The greater radius of curvature change would result in more damage to the solder joints.

Interpretation of the TSOP data was not straightforward. The orientation of the TSOPs could have played a role in how well the solders performed and in their relative ranking. For example, with the TSOPs at location U12 (orientated parallel to the short vehicle edge) the SnPb outperformed the SnAgCu soldered joints (Table 9.7). However, for the TSOPs at U25 (orientated perpendicular to the short vehicle edge), SnAgCu

TABLE 9.7 JCAA/JG-PP vibration data for the TSOP at U12

Test Vehicle ID	Solder Paste/ Finish	Time at Each Level (mins)								
		y-axis	*x*-axis	*z*-axis	*z*-axis	*z*-axis	*z*-axis	*z*-axis	*z*-axis	*z*-axis
		$9.9\,G_{rms}$	$9.9\,G_{rms}$	$9.9\,G_{rms}$	$12.0\,G_{rms}$	$14.0\,G_{rms}$	$16.0\,G_{rms}$	$18.0\,G_{rms}$	$20.0\,G_{rms}$	$28.0\,G_{rms}$
79	SAC396/SnCu	60	60	60	49					
77	SAC396/SnCu	60	60	60	60	60	4			
78	SAC396/SnCu	60	60	60	60	60	38			
75	SAC396/SnCu	60	60	60	60	60	39			
76	SAC396/SnCu	60	60	60	60	60	60	39		
8	Sn37Pb/Sn37Pb	60	60	60	60	22				
7	Sn37Pb/Sn37Pb	60	60	60	60	58				
9	Sn37Pb/Sn37Pb	60	60	60	60	60	60	60	60	6
5	Sn37Pb/Sn37Pb	60	60	60	60	60	60	60	60	13
6	Sn37Pb/Sn37Pb	60	60	60	60	60	60	60	60	33

Note: Stress on the solder joints was increased substantially every 60 minutes by stepping up the power spectral density (PSD) input. Therefore the first 60 test minutes were less damaging than the second 60 test minutes, which were less damaging than the third 60 test minutes, and so on.

outperformed the SnPb-soldered joints. These patterns were repeated for the same components on the "Rework" test vehicles, which supported the hypothesis that there could be an orientation effect.

PLCC-20s ("Manufactured" Test Vehicles) Most of the PLCCs components did not fail during the vibration testing. Only the PLCC at location U15 had enough failures to allow ranking of the solders, which suggests that the PLCCs are relatively resistant to solder joint damage in high-vibration environments. At location U15, the SnPb outperformed the SnAgCu soldered joints. No cracked or missing leads were noted.

TQFPs ("Manufactured" Test Vehicles) Most of the TQFP-144s and the TQFP-208s had broken and/or missing leads at the end of the test (Figure 9.17 and Tables 9.8 and 9.9). Since it appeared that most of the recorded failures were due to lead failure rather than solder joint failure, no comparisons of solder performance could be made.

BGA-225s ("Rework" Test Vehicles) Only eight components were reworked on each of the "Rework" test vehicles (two BGAs, two TSOPs, two PDIPs, and two TQFP-208s). The majority of the components on the "Rework" test vehicles were not reworked, but they did provide useful information on the use of SnPb solder with lead-free component finishes.

The "Rework" test vehicles were initially assembled with SnPb solder. During Rework with the lead-free solders, the old component was removed; the pads were wicked clean of most but not all of the SnPb solder; and a new component was attached

Figure 9.17 TQFP208 component with broken and missing leads after vibration testing (100×)

TABLE 9.8 JCAA/JG-PP Vibration Data for the TQFP-144 component at U20

Test Vehicle ID	Solder Paste/Finish	Time at Each Level (mins)								
		y-axis	*x*-axis	*z*-axis	*z*-axis	*z*-axis	*z*-axis	*z*-axis	*z*-axis	*z*-axis
		$9.9 G_{rms}$	$9.9 G_{rms}$	$9.9 G_{rms}$	$12.0 G_{rms}$	$14.0 G_{rms}$	$16.0 G_{rms}$	$18.0 G_{rms}$	$20.0 G_{rms}$	$28.0 G_{rms}$
79	SAC396/Sn	60	60	60	27*					
77	SAC396/Sn	60	60	60	60	14*				
75	SAC396/Sn	60	60	60	60	48*				
76	SAC396/Sn	60	60	60	60	57*				
78	SAC396/Sn	60	60	60	60	60	15*			
5	Sn37Pb/Sn	60	60	60	60	48*				
6	Sn37Pb/Sn	60	60	60	60	60	16*			
7	Sn37Pb/Sn	60	60	60	60	60	34*			
8	Sn37Pb/Sn	60	60	60	60	60	36*			
9	Sn37Pb/Sn	60	60	60	60	60	43*			

Note: An asterisk denotes broken or cracked leads. Stress on the solder joints was increased substantially every 60 minutes by stepping up the power spectral density (PSD) input. Therefore the first 60 test minutes were less damaging than the second 60 test minutes, which were less damaging than the third 60 test minutes, and so on.

TABLE 9.9 JCAA/JG-PP vibration data for the TQFP-208 component at U48

Test Vehicle ID	Solder Paste/ Finish	Time at Each Level (mins)								
		y-axis	*x*-axis	*z*-axis	*z*-axis	*z*-axis	*z*-axis	*z*-axis	*z*-axis	*z*-axis
		$9.9\,G_{rms}$	$9.9\,G_{rms}$	$9.9\,G_{rms}$	$12.0\,G_{rms}$	$14.0\,G_{rms}$	$16.0\,G_{rms}$	$18.0\,G_{rms}$	$20.0\,G_{rms}$	$28.0\,G_{rms}$
76	SAC398/NiPdAu	60	60	60	60	60	60	60	4*	
77	SAC398/NiPdAu	60	60	60	60	60	60	60	4*	
79	SAC398/NiPdAu	60	60	60	60	60	60	60	14*	
78	SAC398/NiPdAu	60	60	60	60	60	60	60	32*	
75	SAC398/NiPdAu	60	60	60	60	60	60	60	50*	
7	Sn37Pb/NiPdAu	60	60	60	60	60	60	60	60	19
5	Sn37Pb/NiPdAu	60	60	60	60	60	60	60	60	25
8	Sn37Pb/NiPdAu	60	60	60	60	60	60	60	60	29
9	Sn37Pb/NiPdAu	60	60	60	60	60	60	60	60	31*
6	Sn37Pb/NiPdAu	60	60	60	60	60	60	60	60	38*

Note: An asterisk denotes broken or cracked leads. Stress on the solder joints was increased substantially every 60 minutes by stepping up the power spectral density (PSD) input. Therefore the first 60 test minutes were less damaging than the second 60 test minutes, which were less damaging than the third 60 test minutes, and so on.

using a lead-free solder. Therefore all solder joints on the "Rework" vehicles contained some lead, even the components that were reworked. In addition the effect of lead contamination and the effect (if any) of the heat of the rework operation on the reliability of the solder joints in this test were not separable.

For BGAs that were not reworked, SnPb BGA balls assembled with SnPb solder paste always outperformed SAC405 BGA balls assembled with SnPb solder paste. For BGAs that were reworked, Sn37Pb BGA balls assembled using flux only outperformed Sn4Ag0.5Cu balls assembled using flux only (Table 9.10). In the latter case the final SnAgCu BGA solder joints contained approximately 0.3% Pb contamination as determined by inductively coupled plasma (ICP) spectroscopy.

PDIP-20s ("Rework" Test Vehicles) For the PDIP components with NiPdAu finish that were reworked at location U59, SnPb solder was the best performer (Table 9.11). This is in sharp contrast to the results from the "Manufactured" vehicles where the unreworked Sn0.7Cu0.05Ni wave solder alloy (combined with a NiPdAu component finish) was the best performer (Table 9.6). These results could be partly due to the negative effect that small amounts of Pb have on the reliability of Sn0.7Cu [16]. Chemical analysis of two reworked PDIP solder joints showed that the residual lead contamination could vary from 0.38% to 2.98% Pb. The amount of residual lead in the PDIP joints was probably dependent upon how well the operator was able to remove residual SnPb solder from the plated through-hole during rework.

TSOP-50s ("Rework" Test Vehicles) As with the "Manufactured" vehicles, the TSOPs oriented parallel to the shorter side of the "Rework" test vehicle (at locations U12, U16, U26, and U29) tended to fall off during the testing. The TSOPs oriented perpendicular to the shorter side of the test vehicle (at locations U24, U25, U39, U40, U61, and U62) did not fall off during the testing. The general failure mechanism for the TSOPs was the same as already mentioned for the "Manufactured" test vehicles.

The performance ranking of solders for the TSOP components that were reworked (at locations U12 and U25) mirrored that observed for the corresponding TSOPs on the "Manufactured" test vehicles. For example, with the U12 TSOP50 (oriented parallel to the short test vehicle edge), Sn37Pb outperformed SnAgCu reworked joints. However, for the TSOP50 at U25 (oriented perpendicular to the short vehicle edge), SnAgCu outperformed SnPb reworked joints.

Most of the TSOPs on the "Rework" test vehicles were assembled using only SnPb solder. However, TSOPs with a SnPb component finish were used on the "control" SnPb solder test vehicles while a SnCu TSOP component finish was used on the balance of the test vehicles. The test data comparing the two component finishes can be found in the Joint Test Report. In general, there was no distinct difference in the survival times of the SnPb and SnCu TSOP component finishes when combined with SnPb solder.

TQFPs ("Rework" Test Vehicles) The TQFP-208s that were reworked were at locations U3 and U57. For the TQFP components reworked at U57, it appeared that most of the recorded failures were due to lead failures rather than solder joint failures meaning that no comparisons of solder performance could be made.

TABLE 9.10 JCAA/JG-PP vibration data for the Reworked BGA at U4

Test Vehicle ID	Solder Paste/ BGA Balls	Time at Each Level (mins)								
		y-axis	*x*-axis	*z*-axis	*z*-axis	*z*-axis	*z*-axis	*z*-axis	*z*-axis	*z*-axis
		$9.9 G_{rms}$	$9.9 G_{rms}$	$9.9 G_{rms}$	$12.0 G_{rms}$	$14.0 G_{rms}$	$16.0 G_{rms}$	$18.0 G_{rms}$	$20.0 G_{rms}$	$28.0 G_{rms}$
180	flux only/SAC405	60	60	3						
153	flux only/SAC405	60	60	4						
185	flux only/SAC405	60	60	4						
183	flux only/SAC405	60	60	5						
154	flux only/SAC405	60	60	6						
155	flux only/SAC405	60	60	6						
156	flux only/SAC405	60	60	6						
157	flux only/SAC405	60	60	7						
182	flux only/SAC405	60	60	8						
184	flux only/SAC405	60	60	10						
50	flux only/Sn37Pb	60	60	6						
49	flux only/Sn37Pb	60	60	8						
43	flux only/Sn37Pb	60	60	17						
46	flux only/Sn37Pb	60	60	19						
47	flux only/Sn37Pb	60	60	22						

Note: These components were reworked. Stress on the solder joints was increased substantially every 60 minutes by stepping up the power spectral density (PSD) input. Therefore the first 60 test minutes were less damaging than the second 60 test minutes, which were less damaging than the third 60 test minutes, and so on.

TABLE 9.11 JCAA/JG-PP vibration data for the Reworked PDIP component at U59

Test Vehicle ID	Solder/Finish	Time at Each Level (mins)								
		y-axis	x-axis	z-axis	z-axis	z-axis	z-axis	z-axis	z-axis	z-axis
		$9.9\,G_{rms}$	$9.9\,G_{rms}$	$9.9\,G_{rms}$	$12.0\,G_{rms}$	$14.0\,G_{rms}$	$16.0\,G_{rms}$	$18.0\,G_{rms}$	$20.0\,G_{rms}$	$28.0\,G_{rms}$
153	SAC396/NiPdAu	60	60	14						
155	SAC396/NiPdAu	60	60	25						
157	SAC396/NiPdAu	60	60	28						
156	SAC396/NiPdAu	60	60	60	60	22				
154	SAC396/NiPdAu	60	60	60	60	49				
182	Sn0.7Cu0.05Ni/NiPdAu	60	60	45						
185	Sn0.7Cu0.05Ni/NiPdAu	60	60	58						
183	Sn0.7Cu0.05Ni/NiPdAu	60	60	60	16					
180	Sn0.7Cu0.05Ni/NiPdAu	60	60	60	60	36				
184	Sn0.7Cu0.05Ni/NiPdAu	60	60	60	60	60	60	60	60	60
49	Sn37Pb/NiPdAu	60	60	60	60	60	39			
46	Sn37Pb/NiPdAu	60	60	60	60	60	60	60	60	1
50	Sn37Pb/NiPdAu	60	60	60	60	60	60	60	60	30
47	Sn37Pb/NiPdAu	60	60	60	60	60	60	60	60	60
43	Sn37Pb/NiPdAu	60	60	60	60	60	60	60	60	60

Note: These components were reworked. Stress on the solder joints was increased substantially every 60 minutes by stepping up the power spectral density (PSD) input. Therefore the first 60 test minutes were less damaging than the second 60 test minutes, which were less damaging than the third 60 test minutes, and so on.

The TQFP component at U3 was unusual in that 7 out of 15 components had electrical opens before the test began. At least two of the seven bad components failed during normal handling between the time the test vehicles were received and when the vibration test was started. In addition 11 out of 15 U3 TSOP components fell off from the "Rework" vehicles during test (compared to 0 out of 15 during testing of the "Manufactured" vehicles). Visual examination of the board pads that had lost a component revealed that many of the board pads no longer had any significant solder on them (Figure 9.18). The TQFP at location U3 and the adjacent BGAs (U4 and U18) were removed at the same time during rework. It was believed that replacement of the BGAs prior to replacement of the TQFP component at U3 affected the U3 board pads, resulting in a weak pad/solder interface. In contrast, the other TQFP that was reworked (at U57) did not exhibit premature electrical failure during normal handling and did not come off of the vehicle during testing.

Little data are available in the open literature on the relative reliability of SnPb and lead-free solders in vibration. For the data that do exist, SnPb solder generally outperforms SnAgCu solder [17,18,19].

Arnold [19] conducted vibration testing on 2512 chip resistors, Alloy 42 TSOPs, and CSP soldered components. The components were attached to the test vehicles using Sn37Pb, Sn3Ag0.5Cu, or Sn0.7Cu0.05Ni solder. The test vehicles were then tested to failure on an electrodynamic shaker using 30 G sinusoidal dwell vibration at the first resonance frequency of the vehicles. For the chip resistors and the CSPs, the relative ranking of the solders was Sn37Pb was better than Sn0.7Cu0.05Ni which was better than Sn3Ag0.5Cu. With the Alloy 42 TSOPs, the ranking was Sn37Pb > Sn3Ag0.5Cu > Sn0.7Cu0.05Ni.

Figure 9.18 TQFP208 component at location U3 showing failures at the reworked board pad interface (75×)

The existing test data would imply that the use of the SnAgCu family of solder alloys (3 to 4 wt% Ag) in aerospace and military electronics could result in an increase in vibration failures unless measures are taken to reduce printed wiring assembly (PWA) flexure. These measures would include the use of stiffeners on the circuit boards or the use of vibration isolators on the boxes holding the circuit assemblies.

9.4.3 Mechanical Shock Resistance

Very few published studies have been conducted on the reliability of lead-free solders under mechanical shock conditions by the manufacturers of IPC Class 3 (high-performance) aerospace/military electronics. One exception is the JCAA/JG-PP Lead-Free Solder Project (Section 9.2.5).

A mechanical shock test was performed in accordance with MIL-STD-810F, Method 516.5 [20]. The test vehicles were mounted on an electrodynamic shaker, and three shocks were applied in each direction along each of the three orthogonal axes for each test level. The three test levels used were the Functional Test for Flight Equipment, the Functional Test for Ground Equipment, and the Crash Hazard Test for Ground Equipment (Figure 9.19). All components soldered with eutectic SnPb and SAC396 passed the test.

Numerous studies have been conducted by the commercial electronics industry to determine the reliability of lead-free solder joints exposed to mechanical shock. In general, these studies indicate that the common lead-free solders (e.g., SAC387 and SAC305) are less reliable than eutectic SnPb solder in drop shock.

Liu et al. conducted extensive drop testing of BGAs soldered to test vehicles with SnPb and lead-free alloys [21]. The BGAs were tested using a standard drop test

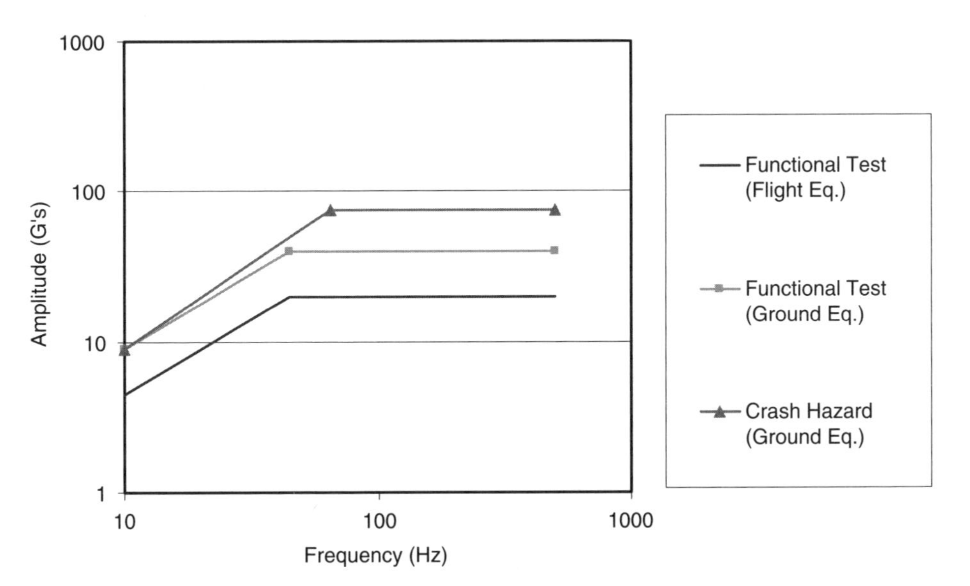

Figure 9.19 SRS levels for the JCAA/JG-PP mechanical shock test

fixture and a drop height of 0.5 meters. The prototype BGAs had a 3 by 3 array of solder balls and a single BGA was soldered to the center of an 80-mm by 80-mm circuit board. Under these test conditions-SAC387 and SAC305 solders survived an average of 1.1 and 1.2 drops respectively compared to 23.8 drops for eutectic SnPb. Reducing the percentage of silver in the SnAgCu alloy improved the drop performance but not to the levels seen with SnPb. For example, SAC105 failed at a mean number of drops of 5.1 in the tests, which was still inferior to the performance of SnPb. The addition of dopants such as manganese (Mn) or cerium (Ce) greatly improved the drop shock performance of SnAgCu alloys by improving the adhesion between the solder and the BGA pad.

The drop shock failures in SnAgCu BGAs tended to be at the BGA/SnAgCu solder interface while the failure mode for SnPb was within the bulk solder or was a mixed failure mode (at the interface and in the bulk solder). Reiff et al. [22] showed that SAC387 deformed much less under mechanical shock than SnPb, resulting in greater stresses at the BGA/SnAgCu solder interface, which could explain the shift in failure mode.

Some failure modes have become more prevalent with the increased use of lead-free solders in electronics, one of these failure modes being "pad cratering." The greater stiffness of lead-free BGA balls combined with the increased brittleness of some lead-free compatible PCB dielectrics would allow the transfer of more strain to the PCB during drop shock. The increased strain would cause a cohesive failure of the printed wiring board (PWB) underneath the BGA corner pads, termed "pad cratering" [23]. The resulting crack can also extend into the copper of the traces and cause complete failure of the trace.

The lower reliability of the common lead-free solders in drop shock conditions and the increase of unusual failure mechanisms, as observed with some lead-free solders exposed to mechanical shock, have been a source of concern for the military and aerospace industries. For those cases where lead-free solders must be used in electronics exposed to high-shock conditions, appropriate mitigation measures should be incorporated into the design (shock isolators, stiffeners, etc.) to ensure that the stresses experienced by the lead-free solder joints are minimized.

9.4.4 Tin Whiskers and Tin Pest

The worldwide transition to lead-free electronics is forcing most major suppliers of components to convert their product lines from tin-lead to lead-free finishes. Their predominant choice for a lead-free component finish appears to be pure tin. The propensity of pure tin plating to form tin whiskers has been known for many years [24,25]. Tin whiskers have been found to form on a wide variety of tin-plated component types under a range of environmental conditions [26]. These whiskers are comprised of nearly pure tin and are therefore electrically conductive and can cause shorting of electronics. The growth of whiskers has caused, and continues to cause, reliability problems for electronic systems that employ components that are plated with tin. Manufacturers of high-reliability systems and government users have not been immune to these difficulties [24,27]. Field failures attributable to tin whiskers have cost individual programs many millions of US dollars and have caused significant customer dissatisfaction. Table 9.12 documents some of the recent failures of aerospace/military electronics caused by metal whiskers.

Annealing of the tin plating or placing a nickel barrier layer under the tin plating have been proposed as mitigation strategies, but neither has been shown to be 100% effective in preventing whisker growth [28,29]. The use of conformal coatings appears attractive as a tin whisker mitigation strategy, since this is a parameter that can be controlled by the OEMs. However, it has been shown that none of the commonly used aerospace conformal coatings can totally prevent penetration by tin whiskers

TABLE 9.12 Documented metal whisker failures in aerospace and military equipment

Year	Application	User	Whiskers On?
1986	F-15 radar	Military	Hybrid package lid
1986	Phoenix Missile	Military	Electronics enclosure
1987	Military/aerospace PWB	Military/aerospace	PWB traces
1988	Missile Program "A"	Military	Relays
1992	Missile Program "C"	Military	Transistor package + standoff
1993	Government electronics	Government systems	Transistor, diode, lug
1996	Military aerospace	Military aerospace	Relays
1998	Aerospace electronics	Space	Hybrid package lid
1998	DBS-1 (Side 1)	Space	Relays
1998	GALAXY IV (Side 2)	Space (complete loss)	Relays
1998	GALAXY VII (Side 1)	Space	Relays
1998	Military/aerospace	Military/aerospace	Plastic film capacitor
1998	PAS-4 (Side 1)	Space	Relays
1999	SOLIDARIDAD I (Side 1)	Space	Relays
2000	GALAXY VII (Side 2)	Space (complete loss)	Relays
2000	Missile Program "D"	Military	Terminals
2000	SOLIDARIDAD I (side 2)	Space (complete loss)	Relays
2001	GALAXY IIIR (side 1)	Space	Relays
2001	Hi-Rel	Hi-Rel	Ceramic chip caps
2001	Space ground test eqpt.	Ground support	Bus rail
2002	DirecTV 3 (side 1)	Space	Relays
2002	GPS receiver	Aeronautical	RF enclosure
2002	Military aerospace	Military aerospace	Mounting hardware (nuts)
2002	Military aircraft	Military	Relays
2003	Missile Program "E"	Military	Connectors
2003	Missile Program "F"	Military	Relays
2004	Military	Military	Waveguide
2005	OPTUS B1	Space	Relays
2006	GALAXY IIIR (side 2)	Space	Relays

Source: NASA.

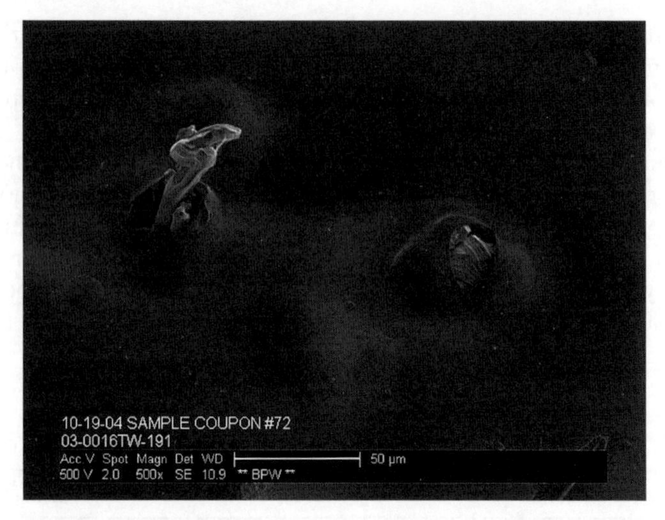

Figure 9.20 Tin whiskers penetrating a silicone conformal coating [30]

growing directly beneath them (Figure 9.20) [30,31]. However, even if a whisker can puncture the coating directly above it, the real value of conformal coatings is to prevent whiskers growing on one component lead from electrically shorting to an adjacent component lead.

In addition to forming tin whiskers, tin can undergo an allotropic transformation from beta-tin (body-centered tetragonal) into alpha-tin (diamond cubic) at temperatures below 13°C [32]. This transformation is called "tin pest." The change is accompanied by an increase in volume of 26%, which results in the disintegration of the tin. The maximum rate of the allotropic transformation appears to occur between –30°C and –35°C [33]. Yoshiharu et al. [34] noted tin pest on Sn0.5Cu specimens with prolonged exposure at a temperature of –18°C. Tin pest was also observed in Sn3.5Ag and Sn9Zn specimens. Hedges [32] suggests that impurities such as lead, bismuth, and antimony can suppress the allotropic transformation.

Sweatman et al. [35] showed that trace impurities in the tin used to make solder pastes can help inhibit the formation of tin pest. Lead, bismuth, silver, zinc, indium, and phosphorus inhibited tin pest formation. Gold, aluminum, copper, germanium, nickel, antimony, gallium, and especially iron tended to promote tin pest formation. The only element found that could totally suppress tin pest formation at the 0.01% addition level was lead. The study suggested that commercial grade tin-rich solders were unlikely to form tin pest because the trace impurities in the solders would inhibit the transformation.

This does not mean to say that tin pest had not been observed in aerospace/military equipment. Recently tin pest was found on a tin-plated copper beryllium grounding strap on the International Space Station (Figures 9.21 and 9.22). The presence of tin pest was confirmed by X-ray crystallographic analysis [36].

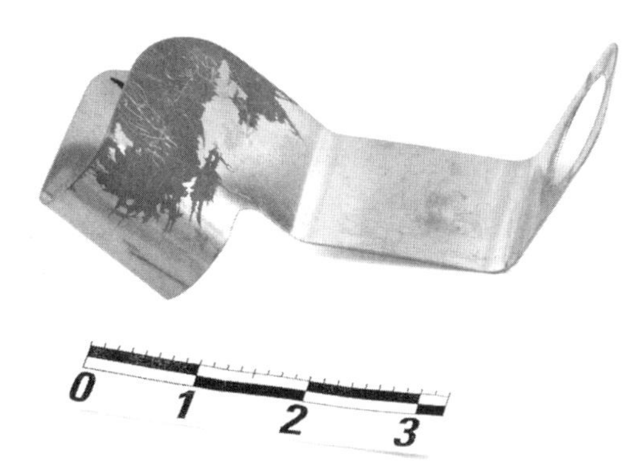

Figure 9.21 Tin pest found on an International Space Station grounding strap. Source: NASA

Figure 9.22 Close-up of tin pest found on an International Space Station grounding strap (10×). Source: NASA

9.5 SUMMARY AND CONCLUSIONS

The data generated to date indicate that the SnAgCu family of alloys (with 3- to 4-wt% Ag) is as reliable as or more reliable than eutectic SnPb solder in most thermal cycling conditions. However, the reliability of SnAgCu alloys in vibration and mechanical shock can be significantly less than that of eutectic SnPb solder, especially when used on large area array packages. New lead-free solder alloys are being developed that have better performance in vibration and mechanical shock, but the thermal cycle properties of these new alloys have not yet been well defined. In addition the increasing use of tin plating on electronic components could lead to an epidemic of tin whisker related failures on electronic assemblies. The uncertainties associated with the use of lead-free materials in high-reliability electronics pose severe challenges for the aerospace and military communities. The best strategy for the short term would be to avoid the use of these materials, if possible. For those cases where lead-free materials must be used, appropriate mitigation measures should be employed to minimize the effects of shock and vibration on solder joints and to control whisker formation and whisker growth on tin-plated components.

REFERENCES

1. IPC-9701A Standard, *Performance Test Methods and Qualification Requirements for Surface Mount Solder Attachments*, IPC, Bannockburn, IL, Feb. 2006.
2. MIL-STD-810F Standard, *Method 514.5, Vibration*, Jan. 2000.
3. For more information on ELF IPT, contact Gerald Aschoff at gerald.r.aschoff@boeing.com.
4. For more information on PERM, contact Lloyd Condra at lloyd.w.condra@boeing.com.
5. For more information on the Tin Whisker Alert Group, contact William Rollins at wprollins@raytheon.com.
6. For more information on CALCE, contact Michael Osterman at osterman@calce.umd.edu.
7. For more information on JCAA/JG-PP and NASA/DoD projects, contact Kurt Kessel at kurt.r.kessel@nasa.gov.
8. J-01-EM-026-P1, *Draft Joint Test Protocol for Validation of Alternatives to Eutectic Tin-Lead Solders Used in Manufacturing and Rework of Printed Wiring Assemblies*, Joint Group on Pollution Prevention (JG-PP), Feb. 14, 2003 (rev. April 2004).
9. "Potential Alternatives Report for Validation of Alternatives to Eutectic Tin-Lead Solders used in Electronics Manufacturing and Repair, Final," Contract No. DAAE30-98-C-1050, Task 272, Concurrent Technologies Corporation, Johnstown, PA, May 27, 2003.
10. NCMS Report 0096RE01, "Lead-Free, High-Temperature Fatigue-Resistant Solder, Final Report," National Center for Manufacturing Sciences, Aug. 2001.
11. For more information on the Crane/SAIC Repair Project, contact Dennis Fritz at DENNIS.D.FRITZ@saic.com.
12. T. A. Woodrow, "JCAA/JG-PP Lead-Free Solder Project: –20 to +80°C Thermal Cycle Test," *SMTA Int. Conf. Proc.*, Rosemont, IL, Sept. 24–28, 2006. http://acqp2.nasa.gov/LeadFreeSolderTestingForHighReliability_Proj1.html.
13. D. Hillman and R. Wilcoxon, "JCAA/JG-PP No-Lead Solder Project: –55 to 125°C Thermal Cycle Testing Final Report, Rev. B," May 28, 2006. http://acqp2.nasa.gov/LeadFreeSolderTestingForHighReliability_Proj1.html.
14. J.-P. Clech, "Lead-Free and Mixed Solder Joint Reliability Trends," IPC, *Proc. IPC Printed Circuits Expo, APEX and Designers Summit 2004*, Feb. 24–26, Anaheim, CA.
15. T. A. Woodrow, "JCAA/JG-PP Lead-Free Solder Project: Vibration Test," Boeing Electronics Materials and Processes Report–582, Rev. A, Jan. 9, 2006. http://acqp2.nasa.gov/LeadFreeSolderTestingForHighReliability_Proj1.html.
16. T. Woodrow, "The Effects of Trace Amounts of Lead on the Reliability of Six Lead-Free Solders," IPC, *Proc. 3rd Int. Conf. Lead-Free Components and Assemblies*, San Jose, CA, April 23–24, 2003.
17. N. Barry, I. P. Jones, T. Hirst, I. M. Fox, and J. Robins, "High-Cycle Fatigue Testing of Pb-Free Solder Joints," *Soldering Surf. Mount Technol.*, vol. 19, No. 2, pp. 29–38, 2007.
18. D. D. Maio and C. Hunt, "High Frequency Vibration Tests of Sn-Pb and Lead-Free Solder Joints," National Physical Laboratory Report MAT 2, Aug. 2007.

19. J. Arnold, "Accelerated Reliability Testing of Ni-Modified SnCu and SAC305," DfR Solutions, Presented at the LEAP Working Group Meeting, June 13, 2008.
20. American Competitiveness Institute, "Mechanical Shock Environmental Testing of Tin-Lead and Lead-Free Circuit Boards, Final Report," JCAA/JG-PP Lead-Free Solder Project, June 7, 2006. http://acqp2.nasa.gov/projects/LeadFreeSolderTestingForHighReliability_Proj1.html.
21. W. Liu and N.-C. Lee, "The Effects of Additives to SnAgCu Alloys on Microstructure and Drop Impact Reliability of Solder Joints," *JOM*, July, pp. 26–31, 2007.
22. D. Reiff and E. Bradley, "A Novel Mechanical Shock Test Method to Evaluate Lead-Free BGA Solder Joint Reliability," ECTC, *Proc. ECTC2005 Conf.*, Lake Buena Vista, FL, May 31–June 3, 2005, pp. 1519–1525.
23. M. Ahmad, D. Senk, and J. Burlingame, "Methodology to Characterize Pad Cratering Under BGA Pads in Printed Circuit Boards," SMTA Pan Pacific Microelectronics Symp., January 22–24 2008.
24. J. A. Brusse, G. J. Ewell, and J. P. Siplon, "Tin Whiskers: Attributes and Mitigation," CARTS 2002, *22nd Capacitor and Resistor Technology Symp. Proc.*, March 25–29, 2002, pp. 67–80.
25. G. T. Galyon, "Annotated Tin Whisker Bibliography," *NEMI*, Feb. 2003, pp. 1–21.
26. B. D. Dunn, "Whisker Formation on Electronic Materials," *Circuit World*, vol. 2, no. 4, pp. 32–40, 1976.
27. M. E. McDowell, "Tin Whiskers: A Case Study," IEEE, *Proc. Aerospace App. Conf.*, 1993, pp. 207–215.
28. M. Osterman, "Assessing the Risk Posed by Tin Whiskers," SMTA Capital Vendor Show (keynote address), Columbia, MD, Sept. 7, 2006.
29. L. Panashchenko, S. Mathew, S. Han, M. Osterman, and M. Pecht, "Tin Whisker Growth Measurements and Observations," 2nd Int. Symp. Tin Whiskers, Tokyo, April 24–25, 2008.
30. T. A. Woodrow and E. A. Ledbury, "Evaluation of Conformal Coatings as a Tin Whisker Mitigation Strategy," IPC/JEDEC, *Proc. IPC/JEDEC 8th Int. Conf. Lead-Free Electronic Components and Assemblies*, San Jose, CA, April 18–20, 2005.
31. T. A. Woodrow and E. A. Ledbury, "Evaluation of Conformal Coatings as a Tin Whisker Mitigation Strategy, Part II," SMTA, *Proc. SMTA Int. Conf.*, Rosemont, IL, Sept. 24–28, 2006.
32. H. E. Hedges, *Tin and Its Alloys*, London: Arnold, 1960, pp. 51–53.
33. J. H. Becker, "On the quality of Gray Tin Crystals and Their Rate of Growth," *J. Appl. Phys.*, vol. 29, no. 7, pp. 1110–1121, 1958.
34. K. Yoshiharu, C. Gagg, and W. J. Plumbridge, "Tin Pest in Lead-free Solders," *Soldering Surf. Mount Technol.*, vol. 13, no. 1, pp. 39–40, 2000.
35. K. Sweatman, S. Suenaga, and T. Nishimura, "Suppression of Tin Pest in Lead-Free Solders," IPC/JEDEC 8th Int. Conf. Lead-Free Electronic Components and Assemblies, San Jose, CA, April 18–20, 2005.
36. Private communication from Christina Reich (Boeing), Clinton Wylie (Boeing), and Victoria Salazar (NASA-KSC).

10

LEAD-FREE RELIABILITY IN AUTOMOTIVE ENVIRONMENTS

Richard D. Parker (Delphi Electronics and Safety)

10.1 INTRODUCTION TO ELECTRONICS IN AUTOMOTIVE ENVIRONMENTS

The automotive market segment has been increasing its utilization of electronics for many decades. The legislative drive to increase safety, reduce emissions, and improve the "green footprint" of the vehicle has resulted in much of the current level of integration of electronics for the vehicle. The desire that consumers have for entertainment and navigation devices in the car are also contributing to the influx of sophisticated electronics. Consumers want to be connected in their vehicles, and this has driven totally new types of electronics into the vehicle.

Most people would be surprised at the number of different electronic modules that are present in any given car, from the lower level of options in an entry level vehicle to complex systems in the high end luxury vehicle. Figure 10.1 shows an example of many of the systems and subsystems that are currently employed or that are upcoming trends (depending on the car make and model). This graphic does not even address the requirements of the hybrid electric vehicles that add significant complexity to a vehicle's electronics.

Present trends also suggest that more electronics will be integrated into the vehicle with increasing complexities as we move forward in time. New requirements

Lead-Free Solder Process Development, Edited by Gregory Henshall, Jasbir Bath, and Carol A. Handwerker

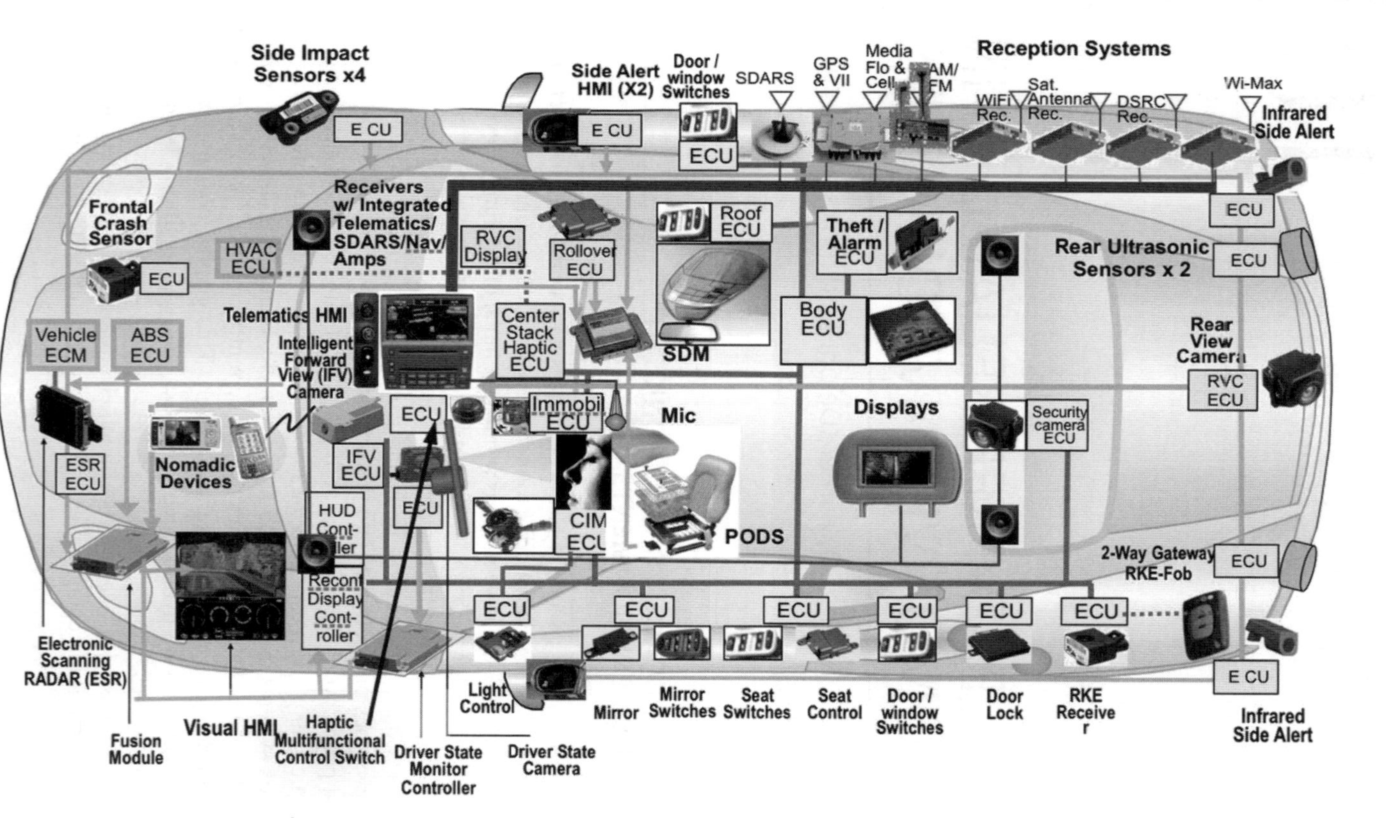

Figure 10.1 Location and types of electronics found in new vehicle models

for safety will continue to be added to the vehicle's systems. The movement toward "greener" transportation will drive more sophisticated engine management systems.

The proliferation of electronics in the automobile has been made possible in part by the success of SnPb soldering technologies. This mature technology has provided electronics with needed reliability, at moderate costs, for the automotive environment. Legislative requirements have now changed and have restricted the use of lead to solder electronics. The use of lead-free soldering technology is too new to have sufficient design and reliability data to draw upon, thus presenting an element of uncertainty and risk for high reliability automotive applications.

10.2 PERFORMANCE RISKS AND ISSUES

Consumers expect products they purchase to work for the entire period they intend to own them. This could not be more significant than in the automotive market. People depend on their vehicles to perform, usually without understanding the complexities of the task. In the context of the electronics in the vehicle, reliability is a prime concern of the manufacturer.

All automotive manufacturers have their own requirements for reliability, thus their own requirements for testing of their electronics. This testing requires many steps and is time-consuming, taking as much as a year to complete. In the context of the lead-free transition, it means that every product that switches to lead-free assemblies must be re-evaluated and certified. This drives cost and risk into the equation, and as a result the manufacturer has a desire to wait until necessary to make this change. The timing of this change will be discussed in the next section.

Reliability risks come into play in three areas. The first and most obvious is the solder joint reliability, discussed in detail in Section 10.5. The issues around this topic have been discussed in previous chapters. Solder joint reliability manifests itself in both final product life cycle and during product assembly. So thermal history and vibration history are key factors. The second biggest risk factor is in the component's thermal stability during the lead-free processing. Last, the long-term field reliability risk is different than for SnPb due to issues such as tin whisker induced short circuits, and joint reliability both with long-term warranty implications. These risks will, by their nature, tend to postpone introduction of lead-free solders to the automotive arena until the last moment possible. That way the experiences and lead-free developments of the consumer electronics industry can be applied to automotive programs, reducing learning curves and reliability risks for the lead-free product transition.

Reliability issues become more complicated by the introduction of more BGA and CSP devices into automotive electronic designs. The lead-free solder balls are available in many SnAgCu (SAC) alloys, but their reliability in harsh environments has not been proved. While the industry is beginning to get a handle on the near eutectic Sn3–4AgCu alloys, which perform similar to SnPb in thermal cycling tests, data are lacking for the new low-silver SnAgCu alloys (e.g., Sn1Ag0.5Cu). There are indications from the consumer electronics industry that the high-silver content alloys (near-eutectic SnAgCu alloys) could have problems surviving high shock/vibration testing (1). Some BGA and CSP packages are now available with a lower silver content (e.g., Sn1Ag0.5Cu) to address this concern but thermal cycle fatigue may be reduced. Currently there are insufficient data available that can be used to evaluate BGA solder joint reliability in harsh automotive environments.

Another concern is managing a lead-free soldering process. The lead-free solder alloys require higher processing temperatures than the incumbent SnPb solders. The process window is also considerably narrower and requires more stringent thermal management and control. The electrical components being assembled are also being thermally stressed closer to their design limits, which can result in more risk of damage during the soldering processes. Component warpage, moisture sensitivity issues, and the volatilization of a component's chemistry (e.g., electrolytic capacitors) have to be managed.

10.3 LEGISLATION DRIVING LEAD-FREE AUTOMOTIVE ELECTRONICS

Since the beginning of the lead-free transition in Europe, the automotive segment was considered a separate concern. The automotive environmental needs fell under legislation entitled, Annex II of Directive 2000/53/EC of the European Parliament and of the Council on End-of-Life Vehicles (ELV), and was enacted in June 2002. This legislation documented the restrictions of lead in the automobile but allowed exemptions for lead in electrical systems and electronics. This was a reasonable action, since at that time the reliability of lead-free electronics was still in question, especially for high-reliability or safety-related applications.

The automotive industry treated this legislation as if it were global, not just for the European market. The rationale was that automotive manufacturers did not want to design vehicles for specific markets. In general, the most restrictive environmental rules would have to be applied, at least at the electronic module level, to all products.

Much of the rest of the world's electronics have fitted into the EU RoHS legislation and timing requirements for restrictions on the use of lead in electronic systems. This document has largely driven the electronics component manufacturing industry. The volume needs of this whole market have dwarfed the automotive electronics industry. In 2008 the automotive electronics market accounted for less than 9.4% of the world's total electronics market revenue (2). Therefore, for most components, the automotive electronics market would use the same components as the consumer electronics manufacturers. Special testing and binning could be used to select certain performance requirements, but in the context of lead-free requirements, the components for automotive products would be made using the same processes and the same material sets as for other segments of the electronics industry.

The overall transition plans for most automotive manufacturers were constructed with the expectation that legislation would not force lead-free transitions and that a slow, reliability data driven transition could occur. Most manufacturers had planned a transition period lasting many years.

In August 2008 the Annex II Revision of the ELV Directive (2000/53/EC) was approved by the European Parliament to change the transition timing. This change would mandate much of the new automotive electronics to be lead-free by January 1, 2011 [3]. This legislation had gone one step beyond the RoHS legislation in that it did not include an exemption for high lead (>85%Pb) solders. These alloys are typically used as silicon die attach in thermally challenged packages or power device applications. This difference would put power electronics at risk because there are no drop-in replacements for high-PbSn solder alloys. Therefore the reliability of the package would be affected. This change in legislation was under review for several months to postpone the effectivity date and to bring exemptions in line with those in RoHS. The EU Commission granted an exemption (8e) on February 23, 2010, to allow high-lead solders (>85%Pb) to be used, bringing the ELV legislation in line with RoHS. In addition the EU Commission pushed out legislation on the use of lead in solders to attach components to electronic circuit boards in automotive electronics to January 1, 2016 (exemption 8a).

10.4 RELIABILITY REQUIREMENTS FOR AUTOMOTIVE ENVIRONMENTS

Outside of the automotive electronics industry, there may be thought that there is only one thermal environment for a vehicle. However, there are many thermal environments in a car; in fact there are at least four temperature cycling regimes that are used to model the thermal environments in the typical automobile. These relate to the physical locations on the vehicle that the electronics would operate in and the thermal exposure that would be present at that location. The environment that a vehicle operates can be harsh for electronics. Furthermore the automotive reliability requirements are very stringent. Each automobile manufacturer has its own specific set of validation requirements, but these temperatures can be generalized in the groupings shown in Table 10.1.

TABLE 10.1 Automotive temperature environments

General Location of Electronics	Temperature Conditions	Typical Cycle Counts[a]
General passenger compartment electronics	–40°C / +85°C	500, 1000, or 3000
Diesel engine controller, High heat dissipation passenger compartment	–40°C / +105°C	1000 or 2000
Gas engine controller, Power train controller. General under hood electronics	–40°C / +115°C	500, 1000, 2500, or 3000
On engine electronics	–40°C / +125°C	1000, 1500, 2000, or 2500

[a]Exact thermal cycle count is dependent on customer requirements.

The environmental specifications for the coldest and hottest temperature conditions are also shown. The temperature ranges reflect the actual temperatures that the electronics can be exposed to in these locations. The actual temperatures in the vehicle are influenced by the outside temperature conditions and the state of operation the vehicle is in. The table includes some thermal cycle testing requirements that are typical for each product location in the automobile.

The different temperature regimes must be taken into account in the design cycle for lead-free electronics. Since the temperature conditions are far more severe than for most electronics, the lead-free design experiences from the consumer electronics industry do not usually apply. The learning curve had to be developed specifically for these types of electronics. The higher temperature environments, in particular, need to be studied with extreme care to properly characterize failure modes with the lead-free solder alloys.

To add to this challenge, the development cycle times for automotive electronics are long, typically 3 to 4 years. Product program life is also long, typically 3 to 5 years or more. Beyond that, the product reliability requirements (warranties) are long, in the range of 3 to 10 years, depending on the customer. During the product's lifetime, single-digit PPM failure rates may be of considerable concern, especially if safety functions are compromised. Together, these constraints put a large burden on automotive electronics manufacturers to design high-reliability assemblies and to be able to predict solder joint life over periods of many years of varying environmental exposures.

The thermal cycle reliability testing for automotive electronics will likely be specific to each manufacturer, but it is common to require testing to beyond 2000 cycles. Referring back to Table 10.1, notice how demanding these types of tests can be. It is also typical to test assemblies in a powered state, which can add up to 20°C to these temperatures due to the product's self-heating.

Vibration requirements are more severe in a vehicle than for most consumer products, and are specific to locations within the structure. Vibration fatigue failures come into play in the reliability of lead-free solder joints, more so than with SnPb joints, because of the higher elastic modulus of the SnAgCu family of solder alloys. Vibration testing requirements and the mounting locations within a vehicle are also unique to each automobile manufacturer.

10.5 FAILURE MODES OF LEAD-FREE JOINTS

Solder joint failures in automotive electronics would not be different than the failures in other types of electronics. However, the stress levels and forces that can break solder joints are higher in automotive electronics because of the larger temperature cycling and vibration exposures that exist during road travel. The change to lead-free processing and components would necessitate reliability testing of the various components used. This testing has revealed that some of the components that were used in previous "leaded" designs cannot be used in lead-free designs at the same reliability levels.

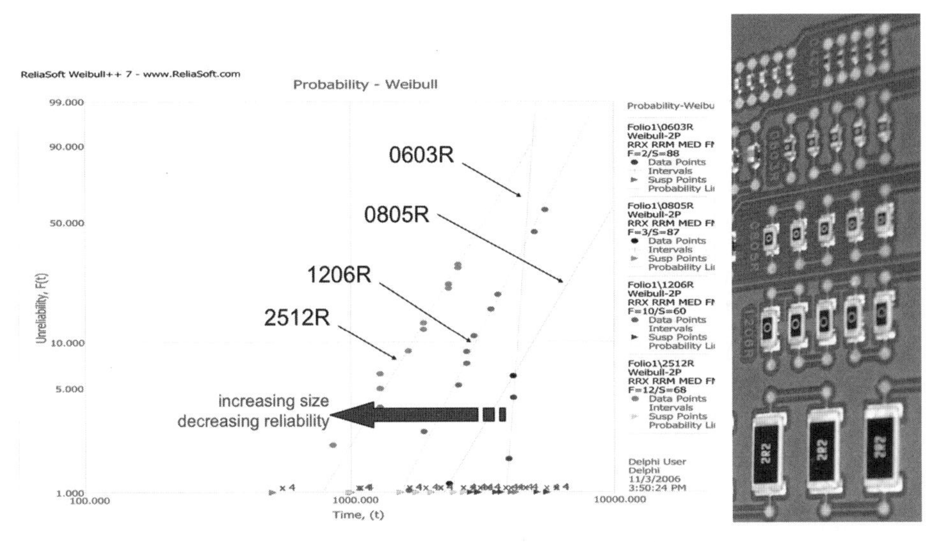

Figure 10.2 SMT reliability as a function of component size for automotive temperature cycles (in hours). Strain per cycle increased as component chip size increased from 0603 (1608 metric) to 0805 (2012 metric) to 1206 (3216 metric) to 2512 (6332 metric)

It is well known that the SnAgCu solder alloys are strain sensitive for thermal cycling reliability. Surface mount chip resistors are a good example of this relationship. Figure 10.2 shows the relative reliability of four common sizes of SMT resistors, ranging from 0603 (1608 metric) up to 2512 (6332 metric). The Weibull plot curves for the –40°C to +125°C thermal cycle clearly show the reduction in reliability as the size (and therefore the strain per cycle) of the SMT component is increased.

Figure 10.3 shows micrographs comparing solder joints for three different sizes of SMT chip resistors thermal cycle tested from –40°C to +125°C. Samples were taken out of the thermal cycling chamber at 1000 cycles and 3000 cycles for each SnPb and lead-free Sn3Ag0.5Cu solder alloy tested and cross-sectioned to look for solder joint cracks. The damage to the Pb-free solder joints is dramatic even for the 0603 (1608 metric) chip size, showing significant joint cracks at 1000 cycles. As the component chip size increased, the level of cracking increased. More important, the level of cracking in the lead-free joints was greater than that for the SnPb joints for all chip sizes tested. Thus product designs that used these sizes of SMT devices in the past may need to be changed to smaller chip resistor sizes for lead-free designs in order to achieve the required reliability.

This reduction in solder joint reliability for lead-free solder joints under high-stress conditions is a function of the mechanical properties of the new SnAgCu solder alloys. It is manifested when there is a large difference in the coefficient of thermal expansion

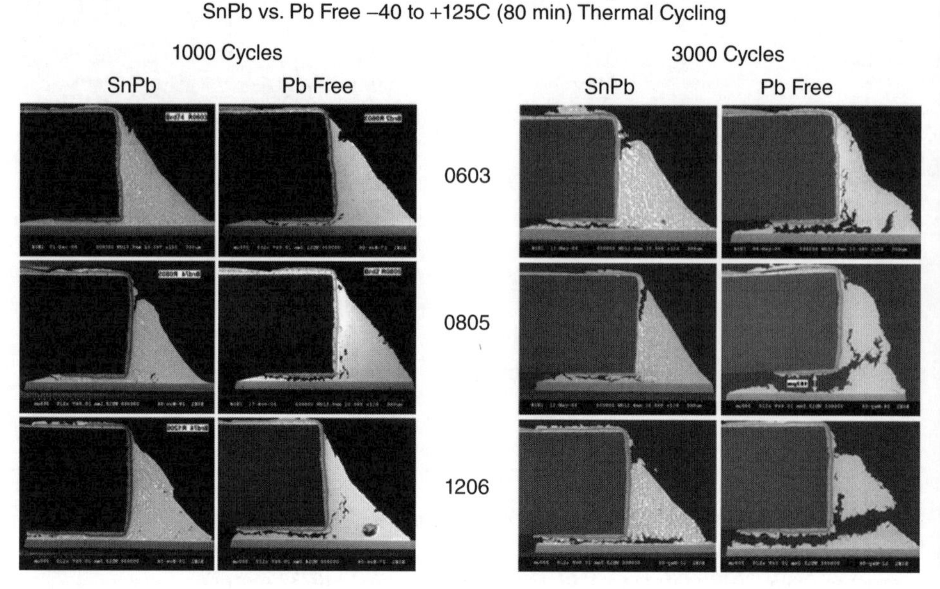

Figure 10.3 Comparison of the thermal cycling cross-sectional results for SnPb and Sn3Ag0.5Cu soldered joints on SMT chip components for 0603 (1608 metric), 0805 (2012 metric), and 1206 (3216 metric) chip sizes

(CTE) between the component and the substrate it is soldered to. The CTE of the ceramic base of the surface mount resistors is dramatically lower in this case than in that of the printed circuit board (~6 ppm/K vs. 17 ppm/K). Other types of components would need to be evaluated for this situation as well, including SAW (surface acoustic wave) filters, and all ceramic-based devices and crystals. In some cases such as plastic BGA components, particularly at lower thermal cycling ranges (delta T) (4), the CTE mismatch may be less or even reversed. Each component type will thus need to be evaluated.

10.6 IMPACT TO LEAD-FREE COMPONENT PROCUREMENT AND MANAGEMENT

In addition to purely technical concerns, the transition to lead-free automotive electronics will have a number of impacts on the procurement and management of components.

The ideal situation is to convert all components to lead-free versions ahead of lead-free product design cycles. Ideally old programs will continue to use "leaded" components until the end of their life cycles. However, sometimes there will be no choice but to use lead-free components prior to conversion of an assembly to lead-free.

Many component suppliers provide identical components to both consumer and high-reliability industries and are driving automotive OEMs among others to convert to lead-free components before they are ready. If automotive OEMs attempt to procure SnPb plated or SnPb solder balled components, availability cannot be guaranteed. This situation has been particularly problematic for legacy (existing) product programs that may never convert to lead-free processing.

Stress in the circuit board due to warpage and bending can result in unexpected failure modes in lead-free designs, and must be avoided.

Most automotive OEMs prefer to do the lead-free conversion on new car platform rollouts. For fully lead-free assemblies, the component solder process thermal capability has to be checked. The ability to withstand temperatures of 260°C for reflow soldering and 270°C for components exposed to wave solder would be required. Furthermore, for components with plated terminations, tin whisker mitigation and testing to the JEDEC JESD201 standard [5] (or equivalent) would be required. Compliance to these requirements would, of course, need to be managed and tracked.

Overall, tracking the inventory of converted parts would be required to assure compliance with lead-free regulations. This would include tracking of:

- Component finish for components with plated terminations
- Temperature capability of components, including MSL (moisture sensitivity level)
- Conformance to tin whisker testing standard (e.g., JEDEC JESD201 [5]) or other customer-specific requirements
- Legacy parts that are not to be converted need to be coded as “do not use in new designs”

To achieve such tracking and to validate that assemblies are compliant to regulations, generating new part numbers would be strongly preferred when switching components to lead-free finishes.Whereas the component supplier could create a new part number or the user could do this, an overlap in inventory is a risk that would need to be managed, and this could add cost. To make matters worse, many components transitioned to lead-free in different stages by first changing their termination finish to pure tin (the years 2004–2006). This was followed by changes to component thermal heat durability in the lead-free transition for some components with some packaged components changing moisture sensitivity level (MSL). The OEM component database would need to track all of these factors and changes while still having an orderly flow to the assembly operations. Finally error proofing the entire flow will be important.

10.7 CHANGE VERSUS RISKS

The automotive component manufacturer will need to accommodate the automotive customer’s expectations when the transition to lead-free is made. Many of the design-related changes that will be made, either in the assembly process or the product itself, will need to be transparent to the automotive customer.

For example, making the transition to lead-free solders may necessitate design changes to better support solder joints subjected to shock or vibration in the final product and also during the assembly and test processes of manufacturing. From the design perspective, the transition may force changes to a smaller size or to a different type of component that puts less stress on the solder joints. In manufacturing, changes may be necessary to circuit board handling during process transfers and test fixturing for automated equipment. Some lead-free solder alloys will result in higher dissolution of the circuit board pads by the solder wave or selectively soldered processes. This could be compensated for in the design of the circuit board or with the selection of a different lead-free solder alloy such as those containing a higher copper content and/or trace levels of nickel.

There are many other design parameters and properties that must be taken into account when changing to a lead-free compatible design. The following would be a partial list of potential impacts

- CTE impacts of components
- MSL (moisture sensitivity level) of packaged integrated circuits
- Vibration and shock performance
- Improved heat resistance of plastics for connectors and structural supports
- Manufacturing costs
- Component costs
- Warranty and service life re-certification

Automotive manufacturers are risk averse, so they have a difficult time accepting all the changes that a lead-free design will require. Since there has been a lack of lead-free product field data in harsh environments, the automotive product manufacturer will have to generate their own data. These data would be used to justify the risks that the final customer would have to absorb in the newly designed systems. The extra testing, and in some cases re-testing, will add to the product cost. This cost has to be absorbed or transferred up the value chain and will most be noticed during the lead-free transition years. As new products are designed for future programs, the cost impact of lead-free solder will not be noticed and will just be part of the assembly costs for the product.

All these considerations point to the need for a slower adoption of lead-free electronics in the automotive environment. Figure 10.4 reflects such a type of product introduction, starting with the products that are of minimal safety concern or inconvenience risk for the end consumer. This would allow some lead-free products to be phased into the market under true field conditions, thus building a database of experience that can lower the risks going forward. As experience is gained both in the manufacturing process and in the field, more complex and mission critical products will be assessed and transitioned to lead-free designs.

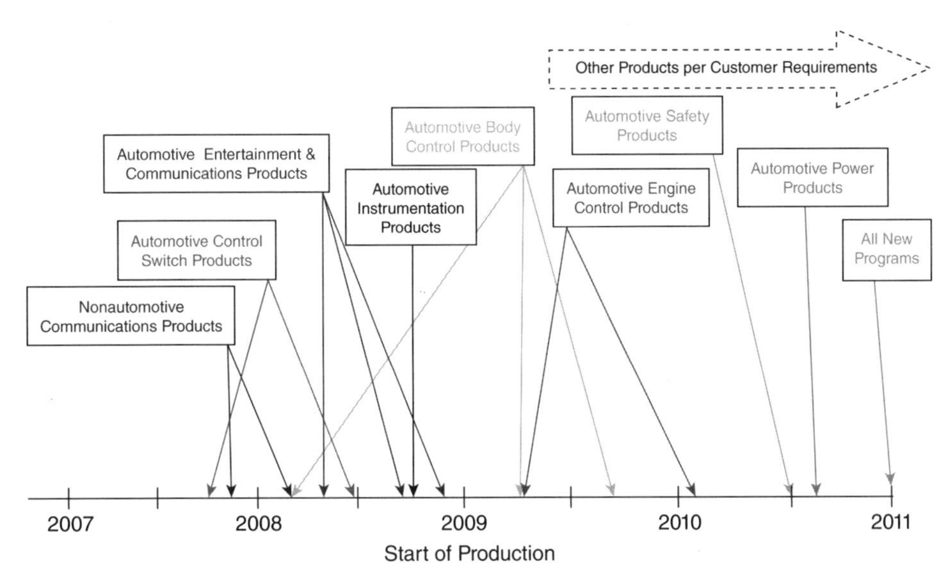

Figure 10.4 Example of a lead-free product implementation plan by an automotive electronics supplier

10.8 SUMMARY AND CONCLUSIONS

The world is changing focus to lead-free electronics. The consumer electronics industry has successfully made the transition and embraced the change, capitalizing on the notion that their products are "greener" and better for the environment. The automotive industry has intentionally lagged behind this segment of the electronics industry. The perceived risks to safety and reliability in the automotive environment have made it necessary to be cautious.

In this chapter we discussed risk management as the overriding factor in the transitioning of automotive electronics to a lead-free product. Automotive manufacturers are now driven by legislation that has set the timing for transition to lead-free. It is now up to the electronics manufacturers to design and develop lead-free automotive systems that can stand up to the rigors of the vehicle service. This is more easily done with new programs but because of legislative changes, will have to be done for some existing or carryover programs. This situation will impact the choice of components for any particular design, and may result in increased material and processing costs. It is expected that the transition of automotive electronics to lead-free will continue through 2015 and beyond.

Managing the transition effectively will require discipline in the procurement portion of the business, with component database management, tracking, and error proofing. The design of new electronics will need to be based on new data for lead-free components and circuit board layout requirements. Converting all components needed in a design ahead of the actual need will reduce the possibility for errors on the manufacturing line.

Assessing and testing all aspects of an electronic assembly with new lead-free components will ensure a successful transition. Even minor considerations, like the mounting location of screws on a circuit board, need to be re-evaluated in a lead-free design.

REFERENCES

1. R. S. Pandher, B. G. Lewis, R. Vangaveti, and B. Singh, "Drop Shock Reliability of Lead-Free Alloys — Effect of Micro-Additives", *ECTC*, IEEC Cat No. 07CH37875C pp. 669, 2007.
2. Prismark Partners LLC 2009–2010 Short Term Outlook, Feb. 2009.
3. ELV Annex II European Parliament. http://eur-lex.europa.eu/LexUriServ/LexUriServ.do?uri=OJ:L:2008:225:0010:0013:EN:PDF.
4. J.-P. Clech, "Lead-Free and Mixed Assembly Solder Joint Reliability Trends," IPC Printed Circuits Expo SMEMA Council APEX Designer Summit, 2004.
5. JEDEC Standard JESD201, *Environmental Acceptance Requirements for Tin Whisker Susceptibility of Tin and Tin Alloy Surface Finishes*, 2006.

INDEX

$(Cu,Ni)_6Sn_5$, 56
01005, 21, 22
0201, 18, 30, 35
0402, 21, 22, 30
0603, 64
0805, 64
AATC, 126, 127, 130, 132, 134, 135, 139, 140, 141, 144
accelerated thermal cycling, 208, 209, 210, 211, 212, 239
acceleration factors, 111, 112, 118, 120, 139, 212
activators, 16, 17, 25, 26, 27, 39, 49
aerospace, 211, 212, 215, 205–209, 235–239
Ag_3Sn, 104, 110
air, 15, 29–31, 40, 48, 50, 58, 59, 60, 61
air atmosphere, 29, 30
Alcohol based, 48, 49, 50, 51, 63, 65, 66, 68
alloy 42, 126, 127, 135, 137, 141, 144, 210, 215, 234
alloy acceptability, 97, 113
alloy choice, 96, 97, 98, 99, 112, 116, 118, 119, 120
alloy testing, 95, 102, 112, 113, 118
alternate alloys, 97, 102, 106, 118
aluminum, 215, 238
annealing, 237
ANSI-GEIA-STD-0005-1, 207, 208, 211
ANSI-GEIA-STD-0005-2, 207, 211
ANSI-GEIA-STD-0005-3, 208
antimony, 9, 238
antioxidant, 139, 141
AOI, 37, 38, 40
aperture wall, 19, 20
assembly yield, 108, 120
Astatine, 16
ATC, 110, 111, 112, 116
Atmosphere, 59, 60, 61
$AuSn_4$, 56
automotive, 243, 245, 246, 247, 248, 249, 250, 251, 252, 253
Automotive electronics, 246, 247, 248, 250, 253

backward compatibility, 108, 109
baking, 28–29
ball alloy, 96, 99, 100, 101, 108, 111, 112, 113, 114, 115, 119, 120
ball alloy composition, 33
ball coplanarity, 31, 34
ball shear/pull, 181, 182, 198, 200, 201
Barrel erosion, 88–89, 90, 92
barrel fill, 38
batteries. EPR (Extended Producer Responsibility), 7, 11, 12
BBP, 4, 5, 6
Bend flexure, 151
bend resistance, 168
bending, 106, 251
Beryllium, 9

Lead-Free Solder Process Development, Edited by Gregory Henshall, Jasbir Bath, and Carol A. Handwerker

BGA, 20, 28, 31–34, 38, 40, 41, 76–85, 91, 101, 108, 109, 173, 191, 210, 211, 220, 221, 222, 228, 231, 232, 234, 235, 236, 245, 246, 250
BGA ball alloys, 96, 101, 108, 109, 114
BGA rework, 32, 76, 85, 91
BGA socket rework, 84, 91
BGA solder joint reliability, 106
BGA solder joint strength, 102, 106
BGAs, 101, 108, 109, 210, 211, 221, 228, 231, 234, 235, 236
bismuth, 101, 109, 111, 114, 116, 238
board finish, 210, 211
Board flexure, 168, 171
board pad design, 18–20
Board re-test, 154, 155
Board strain, 152
boundary scan, 39, 41
bridging, 19, 20, 22, 24, 25, 33, 47, 48, 49, 53, 54, 55, 57, 58, 61, 62, 68
bright tin, 127
brittle fracture, 99, 102
brittle fracture strength, 174, 177, 180, 184, 185, 187, 191, 195, 196, 198, 199, 200, 201
brittle joint fracture, 173, 174, 177, 178, 180, 184, 185, 186, 191, 194, 195, 196, 198, 200, 201
brominated flame retardants, 3, 6, 8
bromine, 9, 16

cadmium, 2, 3, 4, 8, 12, 125
CALCE, 208
California Department of Toxic Substances Control, 8
California Green Chemistry Initiative, 8
CE mark, 4
Cerium, 109, 236
CFC, 1
chemical etch stencil, 20
China-RoHS, 6–7
1206 chip, 212
2512 chip, 234
chip component, 20–21, 24, 30
chip resistor, 36, 37, 249
Chip Wave, 52, 53, 54, 66
chisel-type probe, 38
Chlorine, 9, 16
Chromium, 101, 102
CLCC, 210, 211, 214, 215, 221, 223
CMR, 5, 6
Coating thickness, 58
Cobalt, 101
Coble creep, 128
coefficient of thermal expansion (CTE), 34, 127, 144, 249
combined vibration/thermal cycling, 209, 211
component finish, 251
component leadframe, 125, 142
component warpage, 246
compressive stresses, 128, 137
conformal coating, 208, 237, 238
Contact Length, 45, 46, 52
contact repeatability, 151, 152, 154, 158, 159, 161, 162, 163, 164, 166, 167, 170, 171
Contact time, 52–53, 55, 58, 65, 67
contamination, 31, 34, 151, 166, 167, 171, 231
control plan, 207, 211
Conventional probe, 152, 153, 155, 171
Conveyor Speed, 45, 47, 52, 53, 62, 66
cooling, 25–27, 32, 33, 40, 46, 55
cooling gradient, 27
copper, 95, 96, 101, 104, 119, 127, 128, 136, 137, 138, 141, 142, 148, 196, 211, 236, 238
copper dissolution, 95, 211
copper leadframe, 127, 137, 138, 141, 142, 148
corner bond, 200–201
corrosion, 133, 137, 138, 148
covalently bonded halide, 16–17
Crane/SAIC Repair Project, 211
crown-type probe, 38, 39
CSP, 20, 21, 28, 31–32, 38, 40, 41, 95, 99, 101, 113, 115, 116, 117, 119, 120, 211, 234
Cu_3Sn, 56, 104, 196
Cu_6Sn_5, 56, 104, 105
Cu-OSP, 105, 106, 161, 196
Current density, 135, 139–140
cyclic bend test, 185, 186, 200

DBP, 4, 5, 6
Deca BDE, 3
DEHP, 4, 5, 6
delta Temperature (delta *T*), 32, 51, 58, 63, 80, 85, 250

design, 1, 2, 7, 8, 9, 13, 45–46, 52, 54–55, 56, 61, 62, 63, 65, 66, 68, 245, 246, 248, 249, 250, 251–252, 253, 254
Design for the Environment, 2, 9, 13
design guidelines, 19
diarsenic trioxide, 5, 6
die size, 34
Dissolution, 88–89, 90, 92
DoD, 206, 207, 209, 210, 211
dopants, 99, 105, 109, 117
drop, 175, 177, 178, 179, 180, 181, 182, 183, 187, 189, 190, 191, 193, 194, 196, 197, 200, 201
drop impact, 175, 177, 189, 194, 201
drop reliability, 105
drop shock, 235, 236
Dross, 52, 53, 54
dwell time, 52, 56, 58, 67, 72, 135, 212
dynamic PWB strain, 189

EBSD, 135
ECHA, 5
ECO, 75, 91
ECO wiring, 75, 91
edge bond, 200–201
EIA, 8
Elastic Modulus, 248
Electrical contact, 151, 152, 153, 155
electroform stencil, 20
electroless nickel/immersion gold, 96
electroless Ni-P, 181, 194
electrolyte chemistry, 139
electrolytic nickel-gold, 195, 157, 158, 161
electromigration, 209
Electronic Information Products (EIP), 6
electroplating, 135
electropolishing, 20
ELF IPT, 206–207
EMSF, 115
EN 14582, 17
End-of-Life Vehicles (ELV), 246, 247
Energy Efficiency, 1, 9–10
Energy Star, 10
ENIG, 96, 194, 195, 199, 201
Environmental Compliance, 4, 7, 9, 13
Environmental Protection Use Period, 7
EPA, 2, 10
EPA SW-846 5050/9056, 17
epoxy, 200–201
escape rate, 37
EU ELV, 7
EuP, 9
EU-RoHS, 6–8
exemptions, 3–4, 7, 205

failure mode, 102, 119, 175, 178, 181, 182, 184, 186, 192, 236
failure rate, 175, 177
failure strain, 104
false call, 30, 37
fatigue life, 110, 111, 112, 113, 120
FEA, 191, 192, 193
Fe-Ni substrate, 137
Fillet, 48, 62
film microstructure, 130
finite element analysis, 132
flex, 106, 118
flexure, 151, 167, 168, 169, 170
floor life, 28
Fluorine, 16
flux, 15–18, 21–22, 23–24, 25–27, 30, 31–34, 38–40, 45–56, 58, 59, 61, 62, 63, 65, 66, 68, 73, 76, 81, 82, 83, 85, 89, 90, 152, 153, 157, 161, 164, 165, 166, 167, 171
Flux activity, 46, 47, 48, 49, 53, 54, 58
flux layer, 32
Flux penetration, 49, 50
flux residue, 15–17, 27, 30, 38–39, 40, 152–155, 161, 165, 166–167, 171
flux residue accumulation, 38, 39
Foam Fluxer, 47
focused ion beam, 195
focused ion beam milling, 130
four-point bend, 106
four-point monotonic bend, 168, 169, 170
fracture, 173, 174, 177, 178, 179, 180, 181, 182, 184, 185, 186, 187, 189, 191, 192, 193, 194, 195, 196, 198, 199, 200, 201
fracture prediction, 174, 201
fracture strength, 174, 177, 178, 179, 180, 182, 184, 185, 187, 191, 194, 195, 196, 198, 199, 200, 201
fracture surface, 173, 181, 194, 195, 198, 199, 201
Functional test, 151, 154
functional testing, 32, 38

gallium, 238
GEIA-HB-0005-1, 207
GEIA-HB-0005-2, 208
GEIA-HB-0005-3, 208
GEIA-HB-0005-4, 208
GEIA standards, 207–208, 211
germanium, 88, 101, 238
gold, 165, 238
Grain boundary diffusion, 130
Ground planes, 56, 63

H1, 17
halide content, 16, 17
halogen, 9, 16
halogen-containing, 16
halogen-free, 9, 15, 16–17, 39, 40, 49, 167
Hand solder rework, 86
hardness, 187
harsh environments, 205, 215, 221, 228, 245, 252
HASL, 165, 195, 210
Hazardous Substances, 1–9
HBCDD, 4, 5, 6
HDPUG, 116
head-in-pillow, 15, 31–34, 40–41
Hexagonal close packed (hcp), 104
hexavalent chromium, 2, 3, 8, 12
high Ag alloys, 96, 100, 101, 102, 110, 111, 113, 115, 119, 120
high frequency, 17
high reliability, 208–209, 236, 239, 245, 246, 248
High residue, 48
HIP, 31–34, 40
hole fill, 21, 30, 38, 40, 48, 49, 51, 53, 58, 62, 63, 64, 65, 67, 68, 95
homogeneous, 3, 6
Hot air rework, 71, 73, 75
hot pin pull, 187
hot tearing, 95
humidity, 209
humidity indicator cards, 28

Icicles, 47
ICT probes, 151
IEC, 16
IMC, 102, 104, 105, 120, 136, 137, 141, 142, 181, 195, 198
IMC layer, 102, 104, 120
immersion silver, 58, 157, 158, 159, 161, 210, 211
immersion tin, 58, 157, 158, 161, 195
in circuit test (ICT), 30, 38, 39, 40, 41, 49, 106, 151, 188
incubation time, 126
indium, 101, 238
iNEMI, 34, 39, 106, 110, 111, 112, 114, 115, 116, 117, 118, 121, 168, 169, 210
inspection, 15, 22–23, 32, 35, 37, 38, 39–40
insufficient, 19, 22–23, 32, 34
interfacial, 177, 178, 191, 195
Intermetallic, 56
intermetallic compound, 26, 128, 136
intermetallic growth, 134, 136
Iodine, 16
Ion chromatography, 16
ionic halides, 16
IPC, 9, 16, 17, 28–29, 34, 35–36, 46, 67, 68, 108, 114, 115, 116, 118
IPC 610, 35, 46
IPC Class 2, 67
IPC Class 3, 212, 215, 235
IPC standards, 212, 215, 235
IPC/JEDEC 9702, 168, 177, 178, 201
IPC/JEDEC 9704, 171
IPC/JEDEC 9707, 169, 171
IPC/JEDEC J-STD-020, 28, 34
IPC/JEDEC J-STD-033, 28
IPC/JEDEC J-STD-075, 28
IPC-1601, 29
IPC-A-610DC, 67
IPC-TM-650 2.3.28, 16
IPC-TM-650 2.3.28.1, 16
IPC-TM-650 2.3.35, 16
iron, 238

JCAA/JG-PP, 209, 210–211, 212–225, 227, 229, 230, 232, 233, 235
JEDEC, 8, 28, 33, 34, 67, 108, 114, 115
JEDEC JC-14, 115
JEDEC JEP-95, 33
JEDEC JESD201, 251
JEDEC JESD22-B106, 89, 90
JEITA, 33, 34
JESD22-A111, 67
JESD22-B106-D, 67
JGPSSI, 8
JIG, 8, 9

joint test protocol, 209
JP002, 139
JPCA ES-01-2003, 17
J-STD-001, 46, 48
J-STD-004, 16, 17
J-STD-006, 116
J-STD-075, 67
J-STD-609, 114, 115

Korea-RoHS, 7

L0, 16, 17
L1, 17
laminate damage, 168
laser cut stencil, 20
LCA, 9
lead, 2, 3, 4, 6, 7–8, 12, 125, 126, 127, 129, 135, 136, 137, 138, 139, 141, 142, 143, 144, 147, 148, 245, 246, 247, 249, 250
leadframe, 126, 127, 141, 144, 215
lead-free, 15, 16, 17, 18–20, 21–22, 23–24, 25–28, 29–30, 32–34, 35–36, 37, 38–41, 95, 96–100, 106, 107–109, 110, 111, 113, 114, 115, 116, 117, 118, 119, 120, 151, 152, 153, 154, 157, 158, 161, 164, 165, 166, 167, 168, 171, 186, 187, 191, 195, 196, 197, 198, 201, 205, 206, 207, 208, 209, 210, 211, 212, 215, 219, 228, 231, 234, 235, 236, 239
lead-free control plan, 207, 211
lead-free electronics, 246, 248, 252, 253
lead-free HASL, 165
Lead-free probe, 152–153, 161, 165, 171
Lead-free rework, 71, 73, 81
Lead-free rework profiling, 71–79
lead-free solders, 73, 76, 82, 86, 88, 89, 90, 91, 92, 95, 97, 106, 107, 108, 119, 187, 195, 196, 197, 198, 201, 205–206, 207–208, 209–212, 215, 219, 228, 230, 234, 235, 236, 239, 245–246, 248, 249, 252
lead-free solder reliability, 187. 195, 198
lead-free surface finishes, 126
lead free test, 151, 152, 157, 158, 161, 164, 165, 166, 167, 168, 171
lead-free transition, 245, 246, 251, 252
legislation, 246–247, 253
Level 1, 28
Level 3, 28
LF35, 101, 105, 106, 116
Liquid flux, 73, 83
liquidus, 107, 108
low Ag, 96, 97, 99, 100, 101, 102, 104, 105, 106, 107, 108, 109, 110, 111, 113, 115, 116, 119, 120
low Ag SAC, 99, 100, 101, 102, 119
low profile, 17
Low residue, 48
low silver alloys, 96, 99, 101, 102, 104, 105, 106, 108, 110, 111, 115, 119, 120, 208, 210, 211, 215, 217, 218, 219, 221, 225, 231, 234, 235, 236, 238, 239
LQFP, 143

M1, 17
machine wipe, 22
Main Wave, 54
Manganese, 109, 236
Manufacturing test, 165, 171
Mask, 46, 47, 49, 66
Matte, 66
matte tin, 126, 128, 130, 134
maximum allowable strain, 175
maximum concentration values, 3–4, 7–8, 12
Mechanical damage, 152, 168, 170
mechanical loading, 173, 174, 178, 191, 201
mechanical reliability, 99, 101, 105, 174, 198
mechanical shock, 96, 97, 99, 100, 101–102, 104, 105–106, 113, 116, 119–120, 175, 178, 179, 180, 181, 189, 192, 198, 200, 208, 209, 235, 236, 239
melting point, 95, 96, 107, 109, 115, 120
mercury, 2, 3, 4, 8, 12
micro alloying, 101, 104, 116, 119
micro-alloy additions, 99, 111, 116, 119
microstructure, 107, 109, 110, 112
microvoid, 194, 195
military, 205–206, 211, 212, 215, 235, 236, 237, 239
Mini-pot, 86, 87, 88, 89, 90, 92
Mini-stencil, 74, 75, 83, 85, 91
mitigation, 125, 126, 134, 137, 142, 146, 147, 148, 207, 208, 209, 236, 237, 240
mixing, 206, 208, 210, 211, 236
modeling, 191–194, 201
models, 208, 212
modulus, 102, 104, 106
moisture, 17, 18, 28–29, 34

moisture barrier bags, 28
moisture sensitivity, 28, 34
Moisture Sensitivity Level (MSL), 34, 251, 252
mold compound thickness, 34
monotonic bend test, 177, 178, 180, 201

NASA/DoD Project, 210–211
NCMS, 210
near-eutectic SAC, 95, 96, 97, 98, 99, 106, 107, 113, 119
Ni additions, 96, 99, 101, 105, 106
Ni layer, 137, 142
Ni underplate, 141
Ni/Au, 96, 99, 105, 106, 157, 160–163,
Ni/Au finish, 96, 99, 105, 106
Ni/Pd, 29
Ni_3Sn_4, 56
nickel, 88, 96, 101, 102, 104, 105, 109, 111, 116, 117, 137, 141, 196, 237, 238
nickel barrier, 237
nickel-gold, 157, 158, 159, 161
Ni-P, 181, 194
NiPdAu, 211, 221, 224, 225, 230, 231, 233
nitrogen, 15, 29–31, 38, 40, 45, 48, 54, 58, 59, 60, 61, 68
nitrogen atmosphere, 29, 30, 40
no-clean, 1, 15–17, 48, 49, 62, 63, 65, 66, 165
non-wetting, 31, 34

O_2, 29, 30–31
OctaBDE, 8
Opens, 66, 20
Organic acid, 48
OSP, 16, 19, 20, 30, 38–39, 40–41, 58, 59, 60, 61, 63, 64, 65, 68, 105, 106, 157–159, 161, 166, 171, 195
oxidation, 17, 25, 27, 31, 32, 33, 34, 40
oxidation resistance, 99
oxide, 16, 25, 26, 31, 33, 46, 47, 48, 58, 59
Oxygen, 58, 59, 60
Oxygen bomb combustion, 16

Package body temperature, 80, 85
packaged system drop, 190, 191
packaging, 12
Pad, 156, 157, 158
pad cratering, 108, 167–168, 178, 186, 236
pad finish, 105
Pad repair, 76, 91
pad rupture strength, 187
Passive Component, 72, 91
paste clogging, 22
paste flux bleeding, 21
paste inspection, 22–23, 32, 38, 40
paste printing, 15, 19, 21–23, 38
paste release, 18, 19–20
paste slumping, 17, 18
paste smearing, 21, 22
paste transfer efficiency stencil fabrication, 19, 20
Paste-in-hole, 156, 157, 161
pasty range, 107, 109, 120
PBB, 3
PBDE, 3, 8
Pb-free alloys, 114
PbSn, 247
PBT, 5, 6
PCB, 20, 21, 26, 32, 33, 104, 105, 106, 108, 114, 119
PCB laminate, 104, 108
PDIP, 210, 211, 219, 221, 224, 225, 226, 228, 231, 233
peak reflow temperature, 145, 146, 147
peak temperature, 26–28, 30, 34
PentaBDE, 8
performance testing, 206, 207–208, 210, 211, 212, 215, 228, 231, 235, 236, 239
PERM, 206–207
phosphorus, 238
Pin wetted length, 67
pin-to-hole area ratio, 21, 63, 64, 65, 68
pitch, 18–21, 33–34
plating, 125, 126, 135, 137, 139, 141, 142
plating chemistry, 126, 133, 135, 139
PLCC, 210, 228
post-plating, 139, 142
Pot temperature, 51, 53, 54, 55, 57, 58, 63, 65, 67, 88–89, 90
Pot times, 90
Potential Alternatives Report (PAR), 209
Power Spectral Density (PSD), 217, 220, 221, 222, 223, 224, 225, 227, 229, 230, 232, 233
ppm, 16, 17, 29, 30–31, 49, 59, 60
preheat, 25–27, 31, 40, 50, 51, 59, 62, 63, 65, 81, 89, 91

press-fit, 38
principal strain, 188
print pressure, 22
printability, 17, 19, 21–22
Printed Circuit Assembly (PCA), 151
Printed Wiring Assembly, 210, 235
printing speed, 21
printing window, 19
probe, 151–161, 164, 165, 166, 167, 170, 171
Probe contamination, 166
Probe force, 152, 158, 167, 170
probeability, 30, 38, 152, 156, 164
process temperature, 211
product labeling, 7
pull, 181, 182, 183, 184, 187, 199
PVC, 4, 6
PWB, 173, 174, 175, 176, 177, 178, 179, 180, 186, 187, 188, 189, 190, 194, 195, 196, 198, 200, 201, 236, 237
PWB laminate, 173, 186, 187, 200, 201
PWB strain, 174, 175, 176, 177, 178, 179, 180, 186, 187, 188, 189, 190, 194, 200, 201

QFN, 38, 211
QFP, 20, 35, 72, 73, 74, 91

ramp rate, 80, 85, 212
ramp-up/heating rate, 27
REACH, 4–6
Recycling, 1, 7, 9, 10–12
reflow, 15–17, 19–21, 22, 23–24, 25–34, 37, 38–41, 143, 144, 145, 146, 147, 156, 157, 158, 166, 171, 210
reflow process, 96, 113
reflow profile, 15, 25–28, 32–34, 40–41
regulatory, 1, 2, 8
Reject rate, 37
reliability, 17, 18–19, 26, 40, 41, 205, 206, 208, 209, 210, 211, 212, 215, 231, 234, 235, 236, 239, 245–250, 253
Removal, 72, 73, 75, 76, 81, 85, 90
repair, 75, 76, 80, 88, 89, 91, 206, 208
Replacement, 72, 73, 76, 82, 83, 85, 89, 90
residual stress, 133
Resist, 46, 47, 48, 49, 54, 64, 66
Re-test, 152, 154, 155
Rework times, 73
rheological, 17, 22
risk mitigation, 125, 126, 139
RoHS, 2–9, 95, 108, 205, 246–247
rosins, 25, 48
Rosin emulsion, 48

SAC, 25, 144, 213, 214
SAC101, 99, 119
SAC105, 99, 100, 101, 102, 109, 116, 117, 119, 236
SAC125, 99, 105, 119
SAC205, 116
SAC305, 88, 90, 95, 99, 100, 101, 105, 106, 107, 109, 116, 117, 118, 144, 235, 236
SAC387, 95, 110, 235, 236
SAC396, 210, 213, 215, 220, 221, 222, 223, 224, 227, 229, 233, 235
SAC405, 95, 96, 99, 101, 102, 107, 109, 117, 220, 221, 222, 231, 232
SAC solders, 102, 213
SACX, 101, 102, 116
salt fog, 209
SCCP, 5, 6
screen printing, 21, 40
Selective wave, 154, 156, 157, 158, 164, 165
separation speed, 21–22, 40
shadow moire, 32, 34
shear, 177, 180, 181, 182, 183, 184, 187, 191, 192
shear thinning, 18
shelf life, 18
Shiny, 66
shock, 205, 208, 209, 211, 235, 236, 239, 240, 246, 252
short circuiting, 142
shorting, 125
silver, 95, 96, 97, 99, 100, 101, 102, 104, 105, 106, 107, 108, 109, 110, 111, 113, 115, 116, 119, 120, 148, 210, 211, 236, 238
SIR, 209
Skip, 66
slumping, 17–18, 22
SMT, 15, 17, 22, 28, 30–31, 33, 39–40, 58, 59, 63, 67
SMT reflow, 79, 187, 198
Sn0.3Ag0.7Cu, 58, 101, 116
Sn0.3Ag0.7CuNi, 58
Sn0.5Cu, 238
Sn0.7Cu, 101, 110, 88, 90, 231

Sn0.7Cu0.05Ni, 210, 211, 219, 221, 224, 225, 231, 233, 234
Sn0.7CuNi, 58
Sn0.7CuNiGe, 88, 90
Sn1.2Ag0.5Cu, 99, 101, 116, 119
Sn100C, 96, 99, 101, 116, 119, 210, 221, 225
Sn1Ag0.1Cu, 99, 119
Sn1Ag0.5Cu, 41, 96, 99, 101, 119, 191, 211, 245, 246
Sn1Ag0.7Cu, 58
Sn2.1Ag0.9Cu, 110
Sn2.3Ag0.5Cu0.2Bi, 110
Sn2.5Ag0.9Cu, 110
Sn3.4Ag1Cu3.3Bi, 210
Sn3.5Ag, 29, 100, 116, 238
Sn3.8Ag0.7Cu, 110
Sn3.9Ag0.6Cu, 210
Sn3-4Ag0.5Cu, 26, 57
Sn3-4AgCu, 245
Sn37Pb, 26–27, 58, 116, 166, 194, 195, 196, 198, 201, 210, 220, 221, 222, 223, 224, 225, 227, 229, 230, 231, 232, 233, 234
Sn3Ag, 57
Sn3Ag0.5Cu, 41, 57, 60, 88, 90, 98, 166, 144, 145, 146, 147, 148, 211, 219, 234, 249, 250
Sn4Ag0.5Cu, 98, 182, 211, 231
Sn9Zn, 238
SnAgCu, 25, 27, 28, 29–30, 35, 41, 53, 56, 57, 58, 60, 61, 63, 64, 65, 67, 68, 72, 73, 74, 75, 76, 80, 82, 83, 85, 95, 96, 97, 98, 99, 101, 102, 104, 106, 107, 109, 110, 113, 114, 119, 120, 144, 196, 166, 167, 169, 215, 219, 221, 226, 228, 231, 234, 235, 236, 239, 245, 248, 249
SnBi, 97, 211
SnCu, 56, 57, 58, 63, 65, 68, 227, 231
SnO_2, 137, 138
SnPb, 26–27, 28, 58, 63, 64, 65, 67, 95, 106, 108, 109, 111, 113, 117, 118, 120, 143, 145, 167, 170, 208, 210, 211, 213, 214, 215, 219, 221, 225, 226, 228, 231, 234, 235, 236, 239, 245, 246, 248, 249, 250, 251
Sn-rich, 95
SnZn, 97
soak, 25–27, 31, 39, 40
Soak time, 80, 85
SOIC, 29, 72
solder alloys, 6, 88, 92, 97, 113, 116, 117, 119, 173, 175, 177, 182, 196, 198, 200, 201, 210, 215, 221, 225, 231, 235, 239
solder ball alloys, 78, 82, 83, 100, 220, 222, 236
solder ball shear test, 177, 180, 184, 191
solder balling, 17, 20, 21
solder balls, 20, 25, 27, 48, 49, 66, 68
Solder fountain, 87, 88, 90, 92
Solder fountain rework, 87–88, 92
solder joint, 173, 174, 175, 177, 178, 179, 180, 181, 182, 184, 185, 186, 187, 189, 191, 192, 193, 194, 195, 196, 198, 199, 200, 201
Solder Joint Peak Temperature, 80, 85
solder joint reliability, 173, 175, 178, 185, 194, 195, 245, 246, 249
Solder mask, 76, 81–82
Solder mini-pot rework, 86, 87
solder paste, 15–27, 29–34, 38–41, 151, 152, 156, 157, 158, 159, 161, 164, 165, 166, 171
solder paste handling, 15, 17, 40
Solder wick, 72, 73, 81
solder/flux splatter, 17
solderability, 16, 27, 29–30
Soldering iron tip, 72–73, 75, 76, 87
Soldering iron tip temperature, 72, 73, 75, 76
solidification, 26
solidification temperature, 109
solids, 25, 38
solids content, 38, 48, 49, 65
solvent, 22, 24, 25
solvent evaporation rate, 24
SOT, 72
sphere size, 31–33
spherical bend, 168, 169, 170
spherical bend test, 186
Spray Fluxer, 47, 65, 68
Spray Fluxes, 46, 47, 50
SPVC, 116, 118
squeegee blade, 21
squeegee down stop, 21
standards, 97, 113, 114, 118, 119, 120
stencil, 15, 18–22, 32, 33, 34, 38, 40, 74–75, 82–83, 85, 91
stencil aperture, 18–20, 21–22, 32–34
stencil aperture shape, 20

stencil area ratio, 19, 20
stencil aspect ratio, 19, 20
stencil design, 15, 18–19, 21, 22, 40
stencil idle time, 22
stencil life, 22
step-up/step-down stencil, 21
stiffness, 96, 102, 104, 166
strain energy density, 132
strain gage, 177, 187, 188, 189, 201
strain limits, 174, 177, 200
strain rates, 175, 177, 178, 181, 201
strength, 96, 102, 104, 106
supply chain, 205, 208, 211
Surface diffusion, 130
surface finishes, 95, 99, 105, 106, 119, 157, 158–161, 165, 166, 171, 173, 175, 182, 194, 195, 196, 198, 199, 200, 201
surface finish effects, 206, 207, 208, 209, 210, 211, 217, 219, 221, 223, 224, 225, 226, 227, 228, 229, 230, 231, 233, 236, 237, 240
surface tension, 25
SVHC, 5–6

TAC, 3
tack force, 23–24
tack time, 17, 23
tackiness, 23–24
Tacky flux, 73, 83
take back, 1, 11–12
taper, 20
Telecordia/Bellcore GR-78, 67
TEM, 104
temperature cycling, 111, 112, 116, 117, 118
temperature gradient, 27, 32, 33
Tented vias, 81–82
test, 15, 16–17, 18–19, 29–30, 38–41
Test pads, 152, 156–162, 164, 165, 166, 171
test probes, 151, 152, 166
Test vias, 155, 156, 157, 159–162, 164–165, 166
Testability, 151, 161, 165, 167, 171
Tg, 210
TGA, 27, 187
thermal aging, 173, 181, 198
thermal cycle, 110
thermal cycling, 245, 249, 250
thermal expansion, 127
thermal fatigue, 95, 96, 100, 110, 111, 112, 113, 116, 117, 118, 120, 200
thermal fatigue reliability, 95, 112
thermal fatigue resistance, 212, 215
thermal shock, 209, 211
thermocouple, 27–28, 78–79, 80, 85, 89
thermo-gravimetric analysis, 27
thixotropic agent, 22, 25
thixotropy, 21
through-hole, 21, 38
Through-hole rework, 87, 90, 92
Time above 217°C, 80, 85
time above liquidous, 26, 27, 32, 33
time above liquidus, 84, 108
tin finish, 126, 127, 128, 129, 132–133, 134, 135–137, 139–142, 144, 147, 148
tin pest, 238, 239
tin whisker alert group, 208
tin whiskers, 134, 137, 142, 148, 149, 207–209, 211, 236–238, 239, 245, 251
tin-lead, 17, 18–20, 21–22, 23–24, 25–28, 29–30, 34, 35–36, 37, 38–40, 144, 145, 146, 147, 152, 206, 208, 210, 219, 236
tin-plated, 208, 236, 238, 240
Tip blunting, 166
Titration, 16
TMA, 187
tombstoning, 20, 24, 30
total heating time, 27
Touch-up, 72, 91
toxicity, 2, 5, 8
TQFP, 210, 211, 214, 228, 229, 230, 231, 234
Trace cracking, 167
TSOP, 72, 210, 211, 214, 215, 225, 226, 227, 228, 231, 234
Type 3, 22
Type 4, 21, 22

underfill, 200–201
unpackaged drop, 189–190, 200, 201
Unpackaged system drop, 190
Untented vias, 81–82

Via, 156, 157, 158
vibration, 205, 208, 209, 210–211, 215–216, 218–225, 227–230, 232–235, 239, 240, 245, 246, 248, 252
vibration fatigue resistance, 215

viscosity, 21, 22
VOC, 48, 62, 64
VOC-free, 1, 48, 62, 64
voiding, 17, 30
voluntary, 2, 8, 9, 10
vPvB, 5, 6

warpage, 31, 32, 33, 34, 40, 41, 108
water based, 48, 49, 50, 51, 62, 65, 66
water wash, 48
water-clean, 1
water-soluble, 15, 17, 30, 48, 49
wave, 45–54, 56–68, 156, 157, 165, 171, 210, 211, 219, 221, 225, 231, 237
Wave Contact Time, 55, 65, 67
Wave Height, 51, 52, 54, 66, 67
Wear, 151, 166, 171
WEEE, 3, 7, 11, 12
Weibull plot, 212, 249
wetting, 16, 17, 25, 29, 30, 31, 33, 34, 35, 38, 46, 48, 54, 56, 57, 62, 66, 143, 146, 147
wetting time, 29
whisker density, 127, 135, 139, 140
whisker growth, 125, 126, 127, 130, 132, 133–134, 135–136, 137–138, 139, 140, 141, 142, 144, 145–149
whisker length, 127, 134, 135, 144
whisker nucleation, 132, 137, 149

X-ray, 38, 40, 75, 84, 86, 92
x-ray diffraction, 135

yield, 19, 23, 31, 151, 154, 155, 160, 162, 166, 171
Yield strength, 166, 171

zinc, 125, 238